AF704216

WITHDRAWN

PARASITIC WEEDS OF THE WORLD
Biology and Control

Parasitic Weeds of the World
Biology and Control

C. Parker and C.R. Riches

CAB INTERNATIONAL

SB
611
.P36
1993

CAB INTERNATIONAL
Wallingford
Oxon OX10 8DE
UK

Tel: Wallingford (0491) 832111
Telex: 847964 (COMAGG G)
Telecom Gold/Dialcom: 84: CAU001
Fax: (0491) 833508

© CAB INTERNATIONAL 1993. All rights reserved. No part of this publication may be reproduced in any form or by any means, electronically, mechanically, by photocopying, recording or otherwise, without the prior permission of the copyright owners.

A catalogue entry for this book is available from the British Library.

ISBN 0 85198 873 3

Typeset by Castlefield Press Limited, Kettering, Northants
Printed and bound in the UK

Contents

List of Boxes

(Note: Boxes relate to all parasitic groups, not just to those in the chapter where the box is placed.)

About the Authors

Chris Parker is now an independent weed science consultant. While based at the Weed Research Organization at Oxford, UK, and later at Long Ashton Research Station, Bristol, his research on weed problems of developing countries covered many different topics but became increasingly concentrated on the parasitic weed *Striga*. His work for the UK Overseas Development Administration (ODA) and other agencies has taken him to more than 50 countries in an advisory capacity and has also included three years research and training on *Striga* in Ethiopia. He has been concerned with other parasitic plants through specialized research groups and has been instrumental in initiating a series of international symposia on parasitic weeds.

Charlie Riches is on the staff of the UK Natural Resources Institute. He was involved in on-farm adaptive research for animal draught tillage systems for eight years while working on ODA assignments in Botswana. With EEC funding he also conducted research on resistance to, and control of, *Alectra vogelii* and *Striga asiatica*. This work resulted in the award of a PhD by the University of Reading in 1989, for a thesis entitled 'The Biology and Control of *Alectra vogelii* in Botswana'. Based at Long Ashton Research Station, University of Bristol, since 1987 he is now involved with research projects in a number of developing countries, maintaining a particular interest in weed control in smallholder cropping systems. Work on the control of *Striga* continues through ODA-funded projects in Africa and at Long Ashton Research Station.

Preface

Parasitic plants have been the subject of intensive research in recent decades, not only as objects of intrinsic botanical and physiological interest, but also because of increased awareness of their importance as pests of agriculture, horticulture and forestry. Parasites are quite distinct from other weeds in forming intimate attachments to their host crops, whether to roots as in *Striga* and *Orobanche* species, or to shoots as in *Cuscuta* and mistletoe species. Through these attachments they draw water and minerals and varying proportions of their carbohydrate requirements. As a result of this parasitic feeding, the host crops are often weakened or suffer more readily from water stress. Furthermore, a number of these parasites, especially the *Striga* species, profoundly alter their hosts' physiology, resulting in severe debilitation or even total crop failure.

Interest in the academic and practical aspects of parasitic plants has resulted in the publication of several thousand technical papers in journals and in the proceedings of both general and specialized conferences and symposia. A number of books and guides have been devoted to different aspects of parasitic plants as a whole, or to individual families or genera; the most important are listed in the following Introduction. It may then be questioned why this book is needed. Previous publications have all served a valuable purpose and many will continue to do so. Most, however, are no longer fully up-to-date and the more practical of them have limited background on biology and physiology. The need, we believe, is for a comprehensive and up-to date review of the biology and control of all the major parasitic weed groups, in a form which will allow those readers concerned with only one group to extract what is of direct practical relevance to them. At the same time they will have the opportunity to learn what may be relevant from the other groups. We anticipate that the enterprising researcher will find useful ideas and leads from the chapters on the 'other' groups, all within this single volume.

The aim of this book is thus to ensure that all those involved in helping farmers, directly or indirectly, have a means of recognizing and understanding the biology of these weeds, the precautions that need to be taken to prevent their further spread, and the combination of control measures that might be applied in order to limit their damage. Each of the major parasitic groups is dealt with separately and discussed in detail. Identification keys cover the main species that are likely to occur. The biology and ecology of each group or individual species is then considered with emphasis on how this knowledge may be used in devising and understanding the best possible control measures. More than 850 references are given to the publications which might need to be consulted for further detail.

In spite of all the research, there are few simple solutions to these problems and they are of increasing importance in many parts of the world. In much of sub-Saharan Africa in particular, parasitic weeds pose a major threat to the food security of rural communities. In this and many other areas the parasites we describe are weeds of smallholder farming systems. The challenge offered by these situations, as pointed out by Doggett in 1984, is not so much how to control a parasite, as how the small farmer can control a parasite *and* find it worthwhile to do so. This challenge may be met by the discovery of resistant cultivars, as has now been largely achieved for cowpea, with resistance to both *Striga* and *Alectra* species. For other crops, progress has been less dramatic and development of parasite-resistant or herbicide-resistant cultivars may require sophisticated genetic engineering techniques. We hope that the necessary funds will be found to support such high-tech approaches, but simpler techniques must not be neglected and priority should be given to multi-disciplinary work involving crop physiologists, agronomists and economists working together to develop integrated systems, of relevance to local conditions. Such an approach has rarely been applied to any of the major parasitic problems, even *Striga*, and we hope that this book will provide the basis for development of such research activities and the alleviation, however gradually, of the severe problems faced by so many farmers throughout the world.

Acknowledgements

We acknowledge with thanks the important financial contribution of HM Overseas Development Administration (ODA) which has enabled us to include colour plates of the main parasite species. Furthermore, both authors owe most of their experience of parasitic weeds to their involvement in ODA-financed research projects in the UK and overseas. These projects have contributed significantly to progress in the understanding and control of *Striga*, *Alectra* and *Orobanche* species.

For the use of original photographs or slides we thank Dr Athene Lane (Plate 1A); Ms Tess Moore (Plate 4A); ICRISAT, India (Plate 6C); Mr Bob Harvey (Plate 11B, Figs 1.3 and 1.8); Dr Steve Moss (Plate 11C); Mr T.G. Denham (Fig. 6.1); Dr Bob Scharpf (Plate 15B); and Dr J.S. Beatty (Plate 16A).

For the use of line drawings from other publications we acknowledge the following publishers: G. Bell and Sons, London (Fig. 4.1); Cairo University, Egypt (Fig. 4.4); Her Majesty's Stationery Office and Royal Botanic Gardens, Kew (Fig. 4.5); Torrey Botanical Club, New York (Figs 5.1 to 5.14) and *La Hulotte*, 08240 Boult-aux-Bois, France (Fig. 6.2).

Access to the excellent library of weed science literature at the Long Ashton Research Station, Bristol, has been an invaluable help throughout the preparation of the book and we thank the Head of Station for allowing this use and Mr Steve Smith and his library staff for their assistance.

We thank many other colleagues and friends at Long Ashton and elsewhere throughout the world who have been kind enough to provide help and information. In particular we thank Mr John Terry for his support and encouragement at all stages, for assistance with line drawings and for critical reading of parts of the text; Mr Bob Harvey for preparing the black and white prints; Ms Karen John for the painstaking task of measuring seed weights; Ms Tess Moore for her comments on part of the text and Mrs Lynn Roome for typing tables. We thank Helen Parker for welcome help with the English language.

Finally we acknowledge and thank CAB International, Wallingford, for the availability of *Weed Abstracts*, through which such a large proportion of the world literature could be traced, and Tim Hardwick and his colleagues in the Book Publishing Department for their support and guidance towards the completion of this volume.

CHRIS PARKER

CHARLIE RICHES

A. B.

C.

Plate 1

A. *Striga hermonthica* on sorghum (courtesy Dr A. Lane). **B.** *Striga aspera* on rice. **C.** Comparison of flowers and bracts of (L.) *S. hermonthica* and (R.) *S. aspera.*

A.

B.

C.

Plate 2

A. *Striga asiatica* scarlet form with yellow variant in Malawi. **B.** *S. densiflora* (L.) with (R.) white form of *S. asiatica* in India. **C.** *S. angustifolia* in India.

A. B. C.

Plate 3

A. *Striga forbesii.* **B.** *S. latericea.* **C.** *S. latericea* root system with adventitious shoots.

Plate 4

A. Photomicrograph of *Striga hermonthica* on a sorghum root showing xylem connections (courtesy Ms T. Moore). **B.** Young plant of *S. hermonthica* on a maize root. **C.** Typical chlorotic symptoms from *S. hermonthica* in the leaf of sorghum. **D.** Typical wilting symptoms from *S. asiatica* on sorghum.

A.

B.

Plate 5

A. *Striga hermonthica* being hand-pulled by a farmer in Ethiopia. **B.** *S. hermonthica* causing the complete failure of early sown sorghum in Eritrea while later sown crops in the background are less affected.

A. B. C.

Plate 6

A. *Striga hermonthica* reduced by intercropping with cowpea in the row (L.) compared with (R.) no intercropping . **B.** Galls caused by a *Smicronyx* species on *S. hermonthica.* **C.** *S. asiatica* on two sorghum varieties in India; the susceptible CSH-1 (L.) and the resistant SAR-1 (R.) (courtesy ICRISAT).

A. B. C.

Plate 7

A. *Striga gesnerioides* in the typical form attacking cowpea. **B.** Other forms of *S. gesnerioides* from wild hosts. **C.** Uprooted plants showing the large haustorium of *S. gesnerioides* on the root of cowpea.

A. B. C. D. E.

Plate 8

A. *Alectra vogelii* on cowpea. **B.** *A. vogelii* flower showing glabrous filaments. **C.** *A. picta* flower showing bearded filaments. **D.** *A. orobanchoides.* **E.** Cowpea crop killed by *A. vogelii.*

A. B. C. D.

Plate 9

A. *Orobanche ramosa* on tomato. **B.** *O. aegyptiaca* (L.) and *O. crenata* (R.) on chickpea. **C.** *O. minor.* **D.** *A. cernua* on tobacco.

A.

B.

C.

Plate 10

A. *Orobanche crenata* killing faba bean. **B.** *O. crenata* close-up of inflorescence. **C.** Control of *O. crenata* by glyphosate (L.) in faba bean and (R.) untreated.

Plate 11

A. *Buchnera hispida*. **B.** *Ramphicarpha fistulosa* on rice. **C.** *Odontites verna* in wheat (courtesy Dr S. Moss). **D**. *Rhinanthus angustifolius*.

A. B. C. D.

Plate 12

A. Flowers of *Cuscuta reflexa*. **B.** *C. epilinum* on linseed. **C.** *C. campestris* on lucerne. **D.** *C. campestris* flowering on *Artemisia* sp.

A.

B.

C.

Plate 13

A. Crop of niger seed seriously damaged by *Cuscuta campestris* in Ethiopia. **B.** *Cassytha filiformis* with flowers and fruit. **C.** Tree covered by *C. filifomis* in Mombasa, Kenya.

A. B.

C. D.

Plate 14

A. *Scurrula elata.* **B.** Seedlings of *Tapinanthus globiferus.* **C.** *T. globiferus.* **D**. Ornamental tree *(Azadirachta indica)* killed by *T. globiferus.*

A.

B.

C.

D.

Plate 15

A. *Taxillus kaempferi.* **B.** Plant of *Phoradendron pauciflorum* on white fir in the USA (courtesy Dr R.F. Scharpf). **C.** *Viscum album* close-up. **D**. Plant of *V. album* seriously damaging the branch of an apple tree.

A. B. C. D.

Plate 16

A. *Arceuthobium douglasii* on douglas fir in the USA (courtesy Dr J.S. Beatty). **B.** *A. minutissimum* fruiting on blue pine in Bhutan. **C.** *A. pini* on spruce in Bhutan. **D**. Blue pine killed by *A. minutissimum.*

Introduction

The Main Parasitic Plant Families

Parasitism occurs in at least 17 different plant families. These are described in full in the excellent monograph on parasitic plants by Kuijt (1969). But the families which include species of importance as weeds number not more than eight. These are:

- **Scrophulariaceae**, containing *Striga* (witchweeds), *Alectra* and a number of other less important genera of green hemi-parasites, attacking the roots of host plants including a number of widely grown tropical cereal and legume crops.
- **Orobanchaceae**, containing *Orobanche* (broomrapes), parasitic on legumes and vegetables, especially in the Mediterranean region; also *Aeginetia* and *Christisonia*. All are holo-parasites, also attaching to roots.
- **Santalaceae** (sandalwoods), containing a number of woody, green hemi-parasites which affect forest trees, but also *Thesium*, herbaceous root parasites on some agricultural crops.
- **Balanophoraceae**, including *Thonningia sanguinea*, a relatively massive perennial root parasite occasionally affecting tree crops in the humid tropics.
- **Convolvulaceae**, containing *Cuscuta* (dodders), mainly holoparasitic stem parasites with twining yellow stems (sometimes put in their own family, the Cuscutaceae).
- **Lauraceae**, also stem parasites containing *Cassytha*, superficially resembling *Cuscuta*.
- **Viscaceae** and **Loranthaceae**, two families of green hemi-parasites (mistletoes) attacking the aerial parts of trees and shrubs. One or two

other families of mistletoes exist but are not known to include species of economic importance.

Some Definitions of Terms Used in the Description of Parasitic Plants

- **Parasite** A plant depending on another plant for part or all of its nutrition.
- **Hemi-parasite or semi-parasite** A plant which is only partially parasitic, possessing its own chlorophyll (green colour) and photosynthetic ability (may be facultative or obligate – see below).
- **Holo-parasite** A plant which is totally parasitic, lacking chlorophyll and thus unable to synthesize organic carbon.
- **Obligate parasite** A plant which cannot establish and develop independently.
- **Facultative parasite** A plant which can establish and grow independently but which normally behaves as a parasite to obtain some of its nutrition.
- **Haustorium** The bridge of tissue connecting the host and parasite, usually a swollen mass consisting of both host and parasite tissue. This acts as a conduit for the flow of water and nutrient from host to parasite.

The Main Items of Literature on Parasitic Weeds

Major books of the past 25 years have included *The Biology of Parasitic Flowering Plants* by Kuijt (1969). This remains the outstanding authority on many aspects of parasitic plants (not just weeds) and is unrivalled in its descriptive coverage of taxonomy, evolution and morphology of all groups; *South African Parasitic Flowering Plants* by the late Dr Johann Visser (1981) is not only a beautifully illustrated account of all the main parasitic plants of southern Africa but includes many original observations on their biology and ecology; *Parasitic Weeds in Agriculture. Volume I. Striga* edited by Musselman (1987) has 19 chapters on different aspects of the biology and control of *Striga* species; *Witchweed Research and Control in the United States*, edited by Sand *et al.* (1990), a monograph published by Weed Science Society of America, includes 17 chapters covering all aspects of the very valuable research conducted on *Striga asiatica* in the USA; *The Biology of Mistletoes*, edited by Calder and Bernhardt (1983), has 17 chapters on the biology of both Viscaceae and Loranthaceae; *Mistel, Arzneipflanze, Brauchtum, Kunstmotiv in Jugenstil* by Becker and Schmoll (1986), and *Die Mistel. Botanik, Lektine, Medizinische Anwendung* by Luther and Becker (1987), provide a wide range of information on many aspects of *Viscum album*, including its mythology and potential for medical use.

Small guides to the biology and control of the major groups include *Striga Identification and Control Handbook* by Ramaiah *et al.* (1983), published by ICRISAT; and *Orobanche Field Guide* by Linke *et al.* (1989), prepared in conjunction with ICARDA.

Symposium proceedings include the five general symposia on parasitic weeds, arranged under the aegis of the International Parasitic Seed Plant Research Group, the first in *Malta, 1973* (European Weed Research Council (EWRC), 1973 – 32 papers); the second in *Raleigh, 1979* (Musselman *et al.*, 1979 – 46 papers); the third in *Aleppo, 1984* (Parker *et al.*, 1984 – 33 papers); the fourth in *Marburg, 1987* (Weber and Forstreuter, 1987 – 87 papers); and the fifth in *Nairobi, 1991* (Ransom *et al.*, 1991 – 97 papers).

Proceedings of more specialized symposia have included those of the *Second International Workshop on Striga*, Oagadougou, 1981 (Ramaiah and Vasudeva Rao, 1983 – 17 papers) (there is no published record from the first workshop, at Khartoum, 1978); *Workshop on Biology and Control of Striga*, Dakar, 1983 (Ayensu *et al.*, 1984 – 12 papers plus recommendations); *OAU/FAO Workshop on Striga*, Yaounde, 1985 (FAO, 1985 – 11 papers plus reports); *FAO/OAU All-Africa Government Consultation on Striga Control*, Maroua, 1986 (Robson and Broad, 1989 – 12 papers plus reports); *International Workshop, Combating Striga in Africa*, Ibadan, 1988 (Kim, 1991a – 17 papers); *Biology and Control of Orobanche*, proceedings of a workshop, Wageningen, 1986 (ter Borg, 1986a – 27 papers); *Progress in Orobanche Research*, proceedings of the international workshop, Obermarchtal, 1989 (Wegmann and Musselman, 1991 – 43 papers); *Proceedings of the Symposium on Dwarf Mistletoe Control through Forest Management*, Berkeley, 1978 (Scharpf and Parmeter, 1978 – 30 papers); *Biology of Dwarf Mistletoes*, proceedings of a symposium, Fort Collins, 1984 (Hawksworth and Scharpf, 1984 – 16 papers).

A valuable summary of the early literature on *Striga*, up to 1957, is *Witchweed (Striga asiatica) – A New Parasitic Plant in the United States* (McGrath *et al.*, 1957) in which several hundred papers are very informatively abstracted.

Review papers used in the preparation of this volume include those on *root parasites* by Musselman (1980); on *Striga* by Pieterse and Pesch (1983); on *Alectra* by Visser (1978); on *Orobanche* by Kasasian (1971), Pieterse (1979), Puzzilli (1983), and Foy *et al.* (1989); on *Cuscuta* by Gaertner (1950),

Ashton and Santana (1976), Dawson *et al.* (1984) and Cooke and Black (1987); and on *mistletoes* by Gill and Hawksworth (1961) and Hawksworth and Wiens (1972).

Newsletters on parasitic weeds include *Haustorium* (on all parasite groups – contact Prof L.J. Musselman, ODU, Norfolk, VA 23529-0266, USA); *Golden Bough* (on mistletoes – contact Dr R.M. Polhill, Herbarium, Kew, Surrey TW9 3AE, UK); and the *Striga Newsletter of the Pan-African Striga Control Network*, distributed by FAO (contact FAO, P.O. Box 1628, Accra, Ghana).

1

STRIGA, THE WITCHWEEDS, ON CEREAL CROPS

There are a number of possible origins of the name *Striga*. Raynal-Roques and Sallé (1987) point out that it could have been derived from Latin words meaning variously 'striate', 'harshly hairy', 'lean' or 'witch'. All of these are appropriate, suggesting perhaps they have a common root, but the term 'witch' is especially expressive and reflects the almost magical way in which the parasite often damages the host crop even before it is itself visible above ground. This must certainly be the origin of the common name 'witchweed' in English. The exceptional degree to which the parasite damages its host is just one of several characteristics which make *Striga* species in cereals the most serious of all parasitic weed problems. Other characteristics include the vast numbers of seed produced, several hundred thousand per plant; the plant's longevity, reputed to be up to 20 years in the field; and the preference of most species for erratic rainfall and soils of low fertility. These and other factors help to make it a severe problem for millions of smallholders through semi-arid areas of Africa and parts of Asia, while increased population pressure and more continuous cropping result in the problem continuing to spread and intensify. Concerted efforts are urgently needed to prevent still further threats to the farming systems of these areas.

TAXONOMY AND IDENTIFICATION

The family Scrophulariaceae includes both free-living (autotrophic) and parasitic (heterotrophic) species. The non-parasites include the garden snapdragon (*Antirrhinum*) and the weedy speedwells (*Veronica* species). The parasitic members are mostly included in the exclusively hemiparasitic tribe the Rhinantheae. This tribe includes many familiar genera of wild flowers such as *Euphrasia* (eyebright) in Europe, *Castilleja* (Indian paintbrush) in North America and *Pedicularis* (louseworts) in both. These parasites are not covered in this book as they rarely if ever attack any crop plants. The most notable genus of parasitic plants in the family is *Striga*, but a number of others will be covered later in Chapters 2 and 3, including the important *Alectra*.

The genus *Striga* is currently thought to include 35 species, of which at least 11 are known to attack crops (Raynal-Roques, 1991). The following key covers only the *Striga* species most likely to be found parasitizing crops in Africa, the USA or Asia (not Australia). For species occurring on wild hosts this key may be quite unreliable, and a local flora or taxonomist will need to be consulted. Identification of the main *Striga* species is not usually difficult but there can be some problem separating *S. hermonthica* and *S. aspera* in Africa, while the three white-flowered species of *Striga* commonly attacking cereals in India are confusingly similar at first sight. Useful diagnostic features are indicated in the key and in the following text. The number of ribs on the calyx is an especially important character, though somewhat variable within species and even within a single plant, partly due to variation in the number of calyx teeth (Fig. 1.1). The minimum is one rib to the tip of each calyx tooth, while there may be additionally one or two extending up to the sinus between the teeth (sometimes continuing from the sinus up to the tip). Unless otherwise indicated in the key, all species are annual and have normal green foliage.

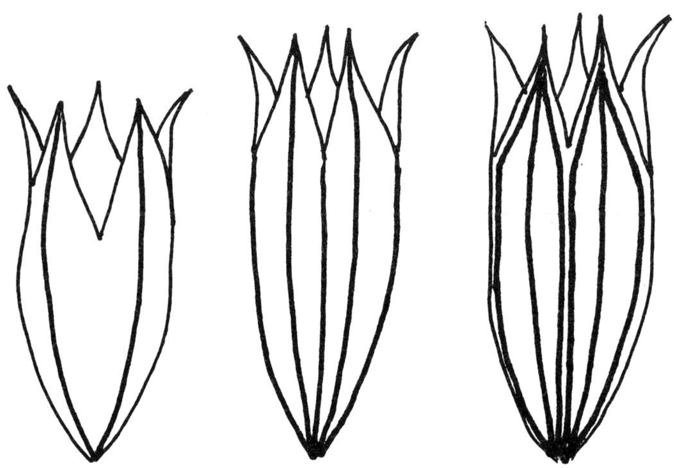

Fig. 1.1 Calyces of *Striga* with (l. to r.) 5, 10 and 15 ribs.

Three further species occurring in Australia and Southeast Asia are noted below.

Key to the Main *Striga* Species

A. Calyx ribs usually 5 or 6 (1 per calyx lobe)

Plant about 50 cm high (sometimes up to 1 m), spike of many *pink* flowers, 1–2 cm across, corolla tube 1.5–2 cm long, **bent at or just above halfway (two-thirds up in East Africa)**, bracts below each flower 2–3 mm wide, **fringed with hairs**. Parasitic on cereal crops in Africa and Arabia . . . . *S. hermonthica*

Plant up to 50 cm high, spike of few to many *pink* flowers about 1 cm across, corolla tube about 1–1.5 cm long, **bent at least two-thirds up** from the base, bracts 1–2 mm wide only, mostly **without a fringe of hairs**. Parasitic on cereals in Africa only . . . . *S. aspera*

Plant 15–30 cm high, flowers **white** in relatively dense spike, 2–6 open at a time, 0.5–1 cm across, **drying bluish**. Parasitic on cereal crops in Asia only. . . . . *S. densiflora*

Plant 15–30 cm high, flowers **creamy white to yellow**, occasionally pale pink, **drying black**, bracts **much longer than calyx**, occasionally parasitic on cereal crops in Africa only . . . . *S. passargei*

Plant 15–30 cm high, **leaves reduced to scales** and **whole plant often pale green or purplish with little chlorophyll**. Flowers usually **mauve/purple but sometimes white**, in spike with only 2–6 open at a time, 0.5–1 cm across. Parasitic on cowpea, tobacco, sweet potato in Africa only . . . . *S. gesnerioides*

B. Calyx ribs usually 10–14 (mainly 2 per calyx lobe)

Plant 15–30 cm high, flowers **red, white or yellow**, in spike with up to 6 open at a time, 0.5–1 cm across. Parasitic on cereals in Africa, Asia and USA . . . . *S. asiatica*

C. Calyx ribs usually 15 (3 per calyx lobe)

Plant 15–30 cm high, leaves narrow without conspicuous teeth, flowers **white**, 2–6 open at a time, 0.5–1 cm across. Parasitic on cereals in Asia only . . . . *S. angustifolia*

Plant about 50 cm high, **annual**, leaves up to 1.5 cm wide, coarsely toothed, flowers 1–2 cm across, **pale salmon-pink**, occasionally white;

corolla tube 2 cm long bent just below the corolla lobes, sparsely, **coarsely hairy** on the calyx ribs. Parasitic on some cereals in Africa only . ***S. forbesii***

Shoots up to 1 m high, arising from a perennial rootstock, leaves up to 1 cm wide with some obscure teeth, flowers **pink/brick red** 2 cm across, corolla tube 2 cm long, bent just below the corolla lobes, calyx **covered in short fine hairs**. Parasitic on sugarcane in Africa only . . . . ***S. latericea***

The Main *Striga* Species in Cereals

Striga hermonthica Benth. (= *S. senegalensis* Benth.)

Striga hermonthica (see Plate 1A) is the largest of the species occurring commonly as weeds, usually at least 30–40 cm high and sometimes 100 cm or more. Each flower is 1–2 cm across, and the inflorescence of 6–10 open flowers can be very striking. Occasionally the spike has many more than 10 open flowers but this is usually an indication of impaired pollination as the corolla normally drops within a few days after fertilization. The flowers are similar in size to those of *Striga forbesii* and *Striga latericea* but the colour of these is normally paler and more salmon-pink and they can be distinguished by the broader leaves and greater number of ribs on the calyx. Confusion with *Striga aspera* is a somewhat greater problem, and although *S. aspera* is generally a less robust plant with flowers rarely exceeding 1 cm across, the bract and corolla tube characters need to be looked at closely (see Plate 1C). Although the typical West African, Sudanese and Ethiopian *S. hermonthica* have the bend in the corolla tube almost exactly half-way, those in Kenya, Uganda and Tanzania have the bend distinctly above half-way, making it closer to the *S. aspera* pattern (Parker, 1991a). The length of the corolla tube visible beyond the calyx is not a reliable character but the actual length of the corolla tube from the bend to the opening of the lobes generally exceeds 5 mm in *S. hermonthica* while it is 5 mm or less in *S. aspera*. Chromosome number (2n) = 64.

Distribution *S. hermonthica* is mainly distributed through semi-arid areas of northern tropical Africa, but extending just into southwest Arabia and southern tropical Africa, including Angola and Namibia; also Madagascar. Occurrence is less widespread than the related *S. asiatica* in southern tropical Africa for reasons not fully understood. Musselman and Hepper (1986) believe the species originates from Sudan/northeast Africa on the basis of its commoner occurrence there on wild grass hosts. It is then a little surprising that its development and spread in Ethiopia has mainly occurred in the past 100 years, but this could be due to the relatively recent intensification of agriculture in the country. *S. hermonthica* occurs on the wild host *Setaria incrassata* in western Ethiopia, but whether this is the origin of the infestations on cereal crops in that area, or the result of them is not certain.

Host range The range of crops attacked by *S. hermonthica* is indicated in Table 1.1 and includes all the major tropical cereals and sugarcane.[1] Occurrence is also reported on some more temperate crops, including teff (*Eragrostis tef*) and barley in Ethiopia. Parasitization of broad-leaved crops has been reported, but Andrews (1946) obtained only very weak development on legume crop species, including groundnut, cowpea, lablab and soyabean in pots, the *Striga* shoots never exceeding 1 cm high when emergence occurred at all. Similar limited development was observed on groundnut in early experiments at the Weed Research Organization, UK (WRO). There have been a number of recent reports of apparently normal development on sesamum and groundnut in the field in West Africa. The fragility of the original attachment to host roots makes any confirmation or disproof of these reports almost impossible but, as normal development on such hosts cannot be confirmed under ideal pot experiment conditions, we believe that most if not all such reports arise from attachment of *S. hermonthica* to grass weeds or the roots of cereal crops already harvested. Matthies (1989) has shown how *S. hermonthica* can continue to develop and flower for many weeks after the death of its maize host.

Occurrence on wild hosts is not especially common, but apart from the *S. incrassata* mentioned above, *Eleusine indica*, *Cynodon* spp., and

Table 1.1. The main cereal crops attacked by *Striga* species.

Striga species	Maize	Sorghum	Pearl millet	Finger millet	Fonio*	Rice	Sugarcane
S. hermonthica	xxx	xxx	xx	xxx	—	xx	xx
S. aspera	xx	x	—	x	xx	xx	x
S. forbesii	xx	xx	—	—	—	—	xx
S. latericea	—	—	—	—	—	—	x
S. asiatica	xxx	xxx	xx	xx	—	xx	x
S. densiflora	—	xx	x	—	—	x	—
S. angustifolia	—	xx	—	—	—	—	xx
S. curviflora	—	—	—	—	—	—	x
S. multiflora	—	—	—	—	—	—	x
S. parviflora	—	—	—	—	—	—	x

* Fonio = *Digitaria exilis*. xxx widely and seriously affected; xx seriously affected locally only; x locally affected to a moderate degree; — susceptibility not known.

[1] Crops are generally referred to in the text by their English names. Appendix I provides a table of their scientific names.

Rottboellia cochinchinensis are almost certain wild hosts in Ethiopia. Musselman and Hepper (1986) recorded *Cymbopogon* and *Ophiuros* as wild hosts in eastern Sudan. In West Africa, *Brachiaria* spp., *Dactyloctenium aegyptium*, *Eleusine indica*, *Panicum walense*, *Cymbopogon* spp., *Rottboellia cochinchinensis*, *Paspalum scrobiculatum* and *Pennisetum violaceum* are sporadically attacked, while in all localities, wild sorghum species, especially *S. arundinaceum*, very commonly occur as alternate hosts. In general, persistence on wild hosts during a fallow period is of minor importance, but it may be locally significant.

Variation *S. hermonthica* is a highly variable species, even within localities. This is to be expected from its obligate out-crossing requirement (Musselman *et al.*, 1981). Variability is demonstrated in a wide range of shape and colour of the corolla, and less obvious variation in leaf size and shape. Physiological variation is seen in differences of host range. Table 1.1 represents the total range of crops attacked by *S. hermonthica* as a species but not all these crops are necessarily attacked in any particular locality, i.e. by any individual population. There is evidence for distinct host preference in many areas. For instance, there are localities in West Africa where only pearl millet (*P. americanum*) is attacked and sorghum is quite immune. Elsewhere, sorghum is attacked and millet very little or not at all. Isoenzyme studies on the variation within *S. hermonthica* have confirmed that the genetic divergence between populations in different countries is wider than the divergence between populations with different host-preference within one country (Bharathalakshmi *et al.*, 1990; Musselman *et al.*, 1991). This suggests that adaptation to a particular host crop can occur on a local basis and this has indeed been observed. Where sorghum has been attacked seriously for many years, maize may be introduced as a new crop and be very little attacked initially, but after some years the *Striga* population apparently adapts to attack maize also.

The mechanism for host preference has been studied in populations attacking either millet or sorghum only (Parker and Reid, 1979). In this study it was deduced that the preference arose from different response of the two populations to different germination stimulants in the host root exudates (see Fig. 1.10, p. 44). In most populations, even those showing strong host preference, it is suspected that individuals occur which have the ability to respond to some stimulants other than those from the preferred host crop. When exposed to a change of crop these individuals are able to germinate and lead to a gradual build-up of a subpopulation with an ability to attack the new crop. Ability to attack the original crop can apparently be retained, but there is as yet inadequate understanding of either the complex of stimulant substances involved, or the genetic basis for their control, to predict the processes of adaptation/de-adaptation. One apparent generalization is that there is rarely any long delay in adaptation from sorghum to maize while the adaptation from sorghum to millet or vice-versa can be very long delayed, suggesting the almost total loss of the respective genes in the population. In eastern Sudan it is only quite recently that millet is being attacked in the main sorghum-growing area, while in parts of Niger, sorghum continues to be unaffected even where it has been grown as a mixed crop for some decades in heavily infested millet fields.

Importance *S. hermonthica* must surely be the most important parasitic weed species on a world scale. The area of crops affected is over 5 million ha in six countries of West Africa alone (Sauerborn, 1991a) and may be over 10 million ha in Africa as a whole.

The problems of measuring crop losses due to *Striga* and other parasitic weeds is discussed in Box 1.1. and in spite of demands from research administrators for quantification of losses due to *Striga*, the estimates can never be fully comprehensive or reliable.

Using hand-pulling after emergence, Carson (1988) recorded losses varying from 0–3% per emerged *Striga* per m^2 in millet and 1–10% per *Striga* per m^2 in sorghum. Bebawi and Farah (1981), also hand-pulling, recorded a 56% loss from a density of 11 emerged parasites per m^2 (5% per *Striga* per m^2). These losses must inevitably be underestimates, as damage will have occurred before the hand-pulling.

Doggett (1965) created different levels of infestation by hand-pulling *S. hermonthica* in sorghum over a period of years and arrived at an estimate of a 1 kg loss per 2200 plants per ha or about 5 kg loss per *Striga* plant per m^2. As yields were little more than 100 kg ha^{-1}, this represented about 5% yield loss per *Striga* per m^2.

Carson (1989a), comparing adjacent micro-plots with and without infestation, deduced that losses due to *S. hermonthica* in sorghum were 20–35% from infestations of 3–5 emerged parasites per m^2(6–7% per *Striga* per m^2). In the case of sorghum there is always the danger of confounding due to the tendency for *Striga* intensity to be higher under low soil fertility; but using very small plots as Carson did should have minimized the risk of this effect.

The recurring figure of 5% loss per *S. hermonthica* plant per crop plant or per m^2 may be of some value as a yardstick, but it should mean that crop plants with 20 emerged parasites have zero yield. This is often the case, but there are also many times when such dense infestations fail to prevent the crop from developing and yielding moderately well. Such instances could result from usefully tolerant crop varieties, but are also noted with varieties known to have normal sensitivity. Presumably there are local conditions which have helped to minimize the damaging effect. Soil fertility may be important in this context if only by delaying attachment and development of the parasite. More studies are also

Box 1.1. The problem of estimating crop losses due to parasitic weeds.

Estimating losses due to parasitic weeds is problematic for a number of reasons, including the difficulty of creating plots with and without the parasite, and the serious non-uniformity of natural infestations. Removal of weeds to create 'weed-free' plots is not a fully satisfactory approach in the case of root parasites, because of the damage known to be done before parasites emerge from the soil. Nor is it altogether satisfactory with the shoot parasites, *Cuscuta* or mistletoes, as their removal almost inevitably entails some damage to the host.

Creating weed-free plots by chemical control is possible for *Cuscuta* in some crops but is difficult in the case of root parasites and mistletoes for lack of suitably selective herbicides. Methyl bromide fumigation has been used successfully for control of both *Striga* and *Orobanche* species but estimates obtained this way are almost certainly overestimates if the fumigant controlled other pathogens in the soil. The use of ethylene to eliminate *Striga* by suicidal germination would have fewer side-effects of this sort but there have been no reports of its use for this purpose.

Different levels of infestation can be created over a period of 2–3 years by preventing seeding on some plots and allowing it on others. Doggett (1965) used this method to estimate losses due to *Striga hermonthica* in East Africa.

A further technique is to select individual plants, hills or microplots with infestation and compare them with adjacent or nearby plants/plots with less or no infestation. Carson (1989a) used this method to measure losses due to *S. hermonthica* in sorghum. But in the case of *Striga* there is always the danger of confounding with soil fertility; low yields on highly infested plots may be at least partly due to low fertility which is favouring high infestation. Hence there is again a danger of overestimation of loss by this method. But using very small adjacent plots as Carson did should have minimized the risk of this effect. A similar reservation must apply to this method of measurement of losses due to mistletoe, where a high infestation of parasite may be partly the result, as well as the cause, of weak host growth. In the case of *Orobanche* there is less concern on this account.

An interesting technique has been used by Zaitoun *et al.* (1991a) with *Orobanche*. On a range of relatively comparable sites, the percentage yield loss of faba bean was very closely correlated with the percentage of plants infected by *O. crenata*, based on a sample of 100 uprooted crop plants. They propose an equation such that % yield loss = 0.987 × % infection −5.37. This technique depends on easy identification of infected plants by examination of the roots and would not be suitable for *Striga* on cereals, but could be for *S. gesnerioides* or *Alectra* on cowpea.

needed on the importance of the time of attachment on the damaging effect. Conversely, severe damage and virtual crop failure can occur where there are many fewer than 20 emerged *Striga* per m^2, or even none at all. This arises from exceptionally heavy infestation of the roots on poor soils when intra-specific competition between the *Striga* plants combines with very weak host growth to prevent any emergence at all before the crop is killed.

Scaling up from loss estimates per plant or per hectare to those for a region or country require detailed surveys of infestation levels which have not generally proved practicable for *Striga*, given the extremely variable infestation both within and between fields. Sauerborn (1991a) quotes a number of estimates for countries in West Africa, ranging from 21% loss of sorghum in northern Ghana, through 10% for all cereals in Nigeria, to 8% for the Gambia and 6% in Benin. The reliability of these estimates and of most other published figures are frankly suspect, but there is no doubting the immense importance of the losses suffered over large areas of semi-arid Africa.

Striga aspera (Willd.) Benth.

As noted above, the flowers of this species (see Plate 1B) are almost identical in colour to those of *S. hermonthica*, and there is often confusion between the two species where they overlap in West Africa. Apart from its generally more delicate appearance and somewhat smaller flowers, the characters to look for in *S. aspera* (see Plate 1C) are the narrower bracts without a distinct fringe of hairs (most easily seen on the inflorescence above the open flowers) and the position of the bend at least two thirds up the corolla tube. This gives a relatively long 'leg' below the bend and a short length of tube, only 4–5 mm, above the bend (versus at least 6 mm in *S. hermonthica*). Musselman *et al.* (1991) also note that the outside of the corolla tube is more glandular in *S. aspera*, and that the lower lip of the corolla is straight in this species, reflexed in *S. hermonthica*. *S. aspera* resembles *S. hermonthica* in most aspects of its biology and physiology and in being self-incompatible, thus requiring cross-pollination. It is not known to hybridize with *S. hermonthica* in the field: Musselman *et al.* (1991) claim to have made successful crosses but in view of the different chromosome numbers of the two species, $2n = 54$ in *S. aspera* and $2n = 64$ in *S. hermonthica*, it seems unlikely that progeny will be viable.

Distribution *S. aspera* has a more restricted distribution than *S. hermonthica*, occurring mainly in western Africa (including Cameroon) though present sporadically also in Sudan, Ethiopia and Tanzania. Within West Africa it commonly overlaps with *S. hermonthica* especially in the drier sub-Sahelian zone, but also occurs in wetter situations than *S. hermonthica*, sometimes in irrigated crops. It also replaces *S. hermonthica* as the commoner species at higher altitude (over 1000 m) on the Jos Plateau in Nigeria. In northern Ivory Coast it is the dominant species in the west of the savanna but is not seen 50 km further east.

Host range Although *S. aspera* has been recorded on most of the major cereal crops and sugarcane (see Table 1.1), it appears to have a somewhat different pattern to that of *S. hermonthica*, occurring most commonly on maize and rarely on sorghum (probably more rarely than is apparent from published accounts, in which there has been some known mis-identification). The isolated infestations in Ethiopia occur only on maize and not on sorghum at all. Both upland and irrigated rice are affected in parts of West Africa, especially in Senegal and Ivory Coast. Occurrence on sugarcane, and on the minor crop findo or fonio (*Digitaria exilis*), is also more common than that of *S. hermonthica*. In Ivory Coast it has been reported to attack groundnut but further work is needed to confirm this.

Occurrence on wild hosts is the norm in West Africa where it is generally seen on grasses, such as *Dactyloctenium aegyptium*, *Schizachyrium exile*, *Cenchrus biflorus*, *Brachiaria ramosa*, much more frequently than it is seen in crops. Other probable wild hosts include *Aristida* species, *Andropogon pseudapricus*, *Loudetia togoensis*, *Setaria pumila*, *Paspalum scrobiculatum* and *Elyonurus elegans*. In Ethiopia, the wild hosts could not be positively determined, but were thought to include *Brachiaria* and *Hyparrhenia* species.

As an out-crossing species, *S. aspera* resembles *S. hermonthica* in its variability in flower form and colour. Whether it has the same tendency to physiological variation is not yet certain. So far there has not been published evidence of populations varying in their host-preference, and seeds collected from a wild host (*Schizachyrium exile*) have been shown to attack maize as readily as a population from maize (LARS, 1988).

Importance *S. aspera* is a minor problem compared to *S. hermonthica*, but locally there are quite severe infestations of maize, as in Nigeria, Cameroon, Ivory Coast and Ethiopia, and of rice, as in Ivory Coast and Senegal. In each case it can be a significant problem, causing serious symptoms and presumably yield loss, comparable to those caused by *S. hermonthica*.

Striga densiflora (Benth.) Benth.

S. densiflora (see Plate 2B) normally grows to 15–30 cm, somewhat less robust than *S. thermonthica* but at least as large as *S. asiatica*. Foliage consists of normal green entire leaves, but these are somewhat coarser and more scabrid than in *S. asiatica*. The flowers are white, similar to those of *S. asiatica* in India, but drying a characteristic bluish colour, instead of black or brown; also the anthers are bluish-black versus brownish yellow as in *S. asiatica*. The other diagnostic difference from *S. asiatica* is the number of calyx ribs, rarely more than five in *S. densiflora* and never less than ten in *S. asiatica*. Like *S. asiatica* it is known to be autogamous (Musselman *et al*., 1991). Chromosome number (2n) = 40.

Distribution *S. densiflora* occurs through much of Asia from Pakistan to Indonesia, but is commonest in India. An isolated infestation has been recorded in Oman (Arabia), presumably introduced from India, with sugarcane setts.

Host range The main host crops are sorghum, maize, sugar cane and millets, including finger millet and Italian millet (*Setaria italica*), as well as pearl millet. There is only one report on rice, from Bangladesh.

Wild hosts include a wide range of grasses including *Paspalum scrobiculatum*, *Brachiaria eruciformis*, *Tragus racemosus*, *Eragrostis wildenoviana* and possibly also some broadleaved species but the latter records are not well documented.

Importance Damage to the host has not been measured but in India it is regarded as no less a menace than *S. asiatica*.

Striga passargei Engl.

This is a little-known species (see Fig. 1.2), relatively small, up to 30 cm, somewhat resembling *S. asiatica*, with normal green, entire leaves. Flowers are usually cream but are reported to vary in colour from white through cream and yellow to pink. The main distinguishing feature of *S. passargei* is the elongated bracts, often up to 2 cm long. Very little is known of its biology but it overlaps with *S. asiatica* through most of its range and is assumed to have similar ecological requirements. Like *S. asiatica*, it is known to be autogamous (Musselman *et al.*, 1991).

Distribution The distribution of this species is not well-known, but it occurs sporadically across northern tropical Africa, from Senegal east to Sudan and south to Tanzania; also locally in Arabia.

Host range *S. passargei* is usually restricted to wild grasses including *Schizachyrium exile*. The only crop host to be reported is maize (Parkinson, 1985). No work has been done to determine whether the infestation on maize is due to the development of a newly virulent population, but this possibility cannot be ruled out.

Importance Parkinson (1985) reported that maize was occasionally seriously damaged in northern Nigeria, but exact symptoms and amount of damage have not been recorded. The species is included and described here in case it may prove to be a significant new problem for African farmers.

Fig. 1.2. *Striga passargei*.

Striga gesnerioides

This species is covered later, in Chapter 2 on root parasites of cowpea.

Striga asiatica (L.) Kuntze (= *S. lutea* Lour.; *S. hirsuta* Benth.)

S. asiatica (see Plates 2A and 2B) is a much smaller plant than *S. hermonthica*, normally growing only up to 15 to 30 cm high. On suitable crop hosts the plant may be vigorous and much branched, whereas on wild hosts it is typically much smaller and often unbranched. Leaves are green and entire, but somewhat smaller than those of *S. hermonthica*. Flowers are smaller than those of *S. hermonthica*, and usually fewer (up to six) are open at any one time. They are less showy, perhaps because, unlike *S. hermonthica*, this species is autogamous and has no need to attract pollinators. Flower colour varies greatly between regions and from one population to another. The commonest colours in Africa are scarlet, orange or yellow, while in Asia they are mainly white, yellow, pink or purple. White-flowered populations are readily confused with *S. densiflora* or with *S. angustifolia*, but the number of calyx ribs provides a reliable means of distinguishing them (see Fig. 1.1); the 'basic' number in *S. asiatica* is 10, 2 per calyx lobe, one to each tip and one to each sinus, but in most plants there are at least 11 ribs, resulting from a pair of parallel ribs to one sinus. It is suspected that this duplication of ribs develops as the calyx ages, and is thus commoner in the lower half of the inflorescence. The pairing can occur more frequently in individual flowers, and the calyx may develop one or more additional lobes resulting in up to 15 ribs (Ralston *et al.*, 1987), but this is never a consistent character throughout the inflorescence, as it is in *S. angustifolia*. Chromosome number (2n) = 38 or 40.

Distribution *S. asiatica* has a much wider distribution than *S. hermonthica*, not only occurring throughout most semi-arid regions of Africa, including Botswana, Swaziland and South Africa, well outside the tropics (down to 30 °S), but also extensively in Asia. It occurs in Arabia and eastwards through much of southern and eastern Asia, including India, islands of the Indian Ocean, Pakistan, Bangladesh, Thailand, Vietnam, Indonesia, Australia and China. There is also the well known, well researched infestation in North and South Carolina, USA, also well outside the tropics (about 35 °N). This infestation was first recognized in the 1950s when it had already spread to cover about 175,000 ha.

Host range Host crops attacked (see Table 1.1) are virtually the same as those affected by *S. hermonthica*, sorghum and maize being the most widely damaged, especially in India and Southern Africa, including

Botswana, Swaziland and Zimbabwe, but increasingly also in West Africa, including Togo, Nigeria and Benin (Parkinson, 1985), and also in Cameroon and Zaire in Central Africa and in Tanzania and Kenya in East Africa. Upland rice is seriously affected locally in Ivory Coast, also in South Africa, Mozambique and the Indian Ocean islands, especially Madagascar. Pearl millet is less consistently attacked, mainly in northwestern India and relatively rarely in Southern Africa, though infestations are now reported from Zimbabwe. Sugarcane is affected in India, Pakistan and Bangladesh, and minor millets including finger millet, Italian millet and *Panicum* species are also liable to be attacked in India and in parts of Africa.

S. asiatica differs from *S. hermonthica* in occurring much more widely on wild hosts, often without attacking any crop. A great many of these wild hosts have been listed for the USA (Nelson, 1958a), where *Digitaria sanguinalis* is perhaps the most common. In Africa, a similarly wide range is affected. Kumar and Solomon (1941) list a number of broadleaved hosts for *S. asiatica*, apparently on the basis that haustoria were found attached to the roots of these species, following germination by a grass species. This raises the question of whether the attachment of such secondary haustoria should constitute evidence for a species being a host in the full sense of the term (see Box 5.1, p. 198).

Variation This is even greater within this species than in *S. hermonthica* but is usually less pronounced within populations. In the US population of *S. asiatica* it is remarkably small, presumably because the whole infestation was derived from a small number of introduced individuals (Werth *et al.*, 1984). Conversely, variation is shown most strikingly between populations, which can form distinct ecotypes, differing in their morphology, especially vigour, leaf size and flower colour. This variation is in addition to that resulting from different growing conditions and host species, as recorded by Ralston *et al.*, (1987). Physiological variation also occurs as expressed in host range. Work in India suggests that there are populations which attack mainly sorghum, pearl millet or *Paspalum scrobiculatum* respectively and that this preference is based upon different germination stimulant requirements (Bharathalakshmi and Jayachandra, 1984; Bharathalakshmi *et al.*, 1990). The autogamous nature of *S. asiatica* may largely explain why this species apparently adapts less readily from wild host to crop or from crop to crop; also the common overlap of different morphotypes in many areas.

In West Africa, white-, yellow- and red-flowered forms all occur. It is usually a red-flowered form which is responsible for attack on maize in Togo, and yellow in Ivory Coast, but white and yellow forms attack crops in Benin and yellow and red forms in Nigeria and Cameroon. There is clearly more than one 'race' within each colour form, as in these and other areas red- or yellow-flowered populations are found on wild grasses without any attack occurring on crops. There is, for instance, a distinct yellow-flowered population with very reduced leaves which occurs only on wild grasses in Mali, Sierra Leone, Zimbabwe and elsewhere. In eastern and southern Africa where crops are more commonly attacked it is almost always red-flowered, with a small proportion of the population yellow-flowered but otherwise morphologically similar (see Plate 2A). In Ethiopia the population on maize and sorghum is mixed, with typical red, and also duller reddish-orange, flowers. The infestation in the USA is red-flowered and is presumed to have originated from Africa. In Arabia, the usual colour of the form attacking crops is white, though red and yellow forms also occur. In India, Pakistan, Bangladesh and Burma the populations attacking crops are almost always white-flowered, but those on wild hosts are sometimes yellow or even deep purple. In Thailand and other parts of southeast Asia and China including Sumatra in Indonesia, crops are mostly attacked by yellow-flowered forms, though other very small forms with pink or purple flowers also occur on wild grasses in that region.

Importance The importance of *S. asiatica* as a crop weed is greatest in India, the Indian Ocean islands, and in East and southern Africa. In West Africa, it overlaps with *S. hermonthica* in many countries, and is now increasingly reported as a significant parasite on crops, especially on maize in Togo, perhaps because of the increasing intensity of maize cropping. In eastern Asia it is sporadically troublesome in maize in Thailand and on rice in Indonesia, but its full extent and importance is not at all well documented in that region.

Crop losses due to *S. asiatica* can be as serious as those from *S. hermonthica*, but there has been the same problem of quantifying them. Vasudeva Rao *et al.* (1989a) used regression analysis to arrive at a wide range of figures varying from less than 0.1 up to 1% crop loss per *Striga* plant per m^2 in different localities in India. Agbobli and Huguenin (1987) also arrived at an estimate of about 1% sorghum yield loss per *S. asiatica* plant per m^2 in Togo. As the individual plants of *S. asiatica* are a great deal smaller than those of *S. hermonthica* it is to be expected that losses per *Striga* per m^2 would be lower. In a series of pot experiments at the Weed Research Organization, UK (unpublished), it was concluded that damage to sorghum was closely related to the weight of emerged parasite, regardless of the species involved. On the basis that the average weight of *S. asiatica* is about one fifth that of *S. hermonthica*, a rough rule of thumb might be that each *S. asiatica* per m^2 might cause a 1% yield reduction. The conclusion from the survey by Vasudeva Rao *et al.* was that the average loss of yield of sorghum in infested fields in southern India was about 21%. While it is immensely difficult to estimate losses over a region, it can, nevertheless, be said with confidence that yields may be severely reduced, even to zero, if the problem is allowed to get out of hand. Land may also be abandoned, or the preferred cereal not grown on the most infested land.

Striga angustifolia (Don) Saldanha (= *S. euphrasioides* (Vahl) Benth.)

This species (see Plate 2C) is sometimes more robust than *S. asiatica*, but superficially quite comparable, with white flowers and green entire leaves. The only simple means to distinguish it from white-flowered forms of *S. asiatica* is the presence of *consistently* 15 ribs per calyx, the paired parallel ribs running up to the sinus and continuing up the sides of the calyx lobes to meet at the tip (see Fig 1.1, p. 2). In details of its biology and physiology, this species differs quite significantly from *S. asiatica*. The seeds have a characteristic surface pattern, and are larger, about 0.5 mm long. These larger seeds allow it to establish without a host and the germination biology differs in the lack of requirement for a host-exuded stimulant, though light is required. Ecologically, it appears to be better adapted to clay soils and wet conditions than *S. asiatica*. Chromosome number (2n) = 40.

Distribution *S. angustifolia* is reported sporadically from East and southern Africa and from southern Asia through to Indonesia, but occurs mainly in India, Burma and Sri Lanka.

Host range Host crops affected include sugarcane, sorghum, maize and rice but apparently not any of the millets. It is not reported to attack crops in Africa, and most incidence on crops appears to be in India where it commonly overlaps with both *S. asiatica* and *S. densiflora*. A range of 23 wild grass and dicot hosts are recorded by Rao *et al.* (1986), of which *Aristida hysterix* appeared to be the most important.

Importance Damage can be severe, especially on sugarcane.

Striga forbesii Benth.

S. forbesii (see Plate 3A) is intermediate in size between *S. asiatica* and *S. hermonthica*, often 30–40 cm high, and differing from both those species in the shape of the leaves which are broader, with a few distinct small teeth. Flowers are large, comparable with those of *S. hermonthica* but fewer, only 2–6 open at any time. The corolla tube is about 2 cm long and bent only just below the expanded lobes. Flower colour is a delicate salmon pink and occasionally white. As in *S. asiatica*, no cross-pollination is required and flowers are self-pollinated before opening (Musselman *et al.*, 1991). Biology and physiology are generally similar to those of *S. hermonthica* but it differs ecologically in being adapted mainly to relatively wet soil conditions. Seed size is significantly larger than in most other *Striga* species but there is no evidence that this affects its germination biology or obligate parasitism. Chromosome number (2n) = 44.

Distribution *S. forbesii* is widespread in Africa and islands of the Indian Ocean including Madagascar.

Host range Crop hosts attacked by *S. forbesii* are indicated in Table 1.1 and include many of the tropical cereals and sugarcane, also the grass *Setaria sphacelata* where it is grown as a forage crop as in Kenya, but attack on crops is relatively rare. This species, like *S. aspera*, is much more commonly seen on wild hosts than on crops. Host grasses favoured include the *Setaria* species, especially *S. sphacelata*; also *Echinochloa pyramidalis* in Ethiopia.

Variation The degree of variation in this species has not been closely studied, and so far there have been no claims for populations differing in host preference. In the Kwekwe district of Zimbabwe it attacks sorghum and maize equally, while a population from *E. pyramidalis* in Ethiopia is able to attack maize in pots, even though it was not seen attacking this crop locally.

Importance The importance of this species is very localized but moderately acute in parts of Zimbabwe and Tanzania on maize and sorghum, in Somalia on sugarcane and possibly on rice in Ivory Coast. Symptoms of damage to the host are similar to those of *S. hermonthica*, and yield reductions are thought to be equally serious.

Striga latericea Vatke (including *S. somaliensis* Skan)

S. latericea (see Plate 3B) is quite similar to *S. forbesii* and may be confused with it. It is robust, up to 1 m high, and the flowers are large, up to 2 cm across. Both species are favoured by wet conditions and are both known from the same sugarcane estate in Somalia. Specimens with slightly smaller flowers have been named *S. somaliensis*, but it now seems most probable that these also belong to *S. latericea*. The main distinguishing features from *S. forbesii* are that flower colour is a deep, rich brick red, the hairs on the 15-ribbed calyx are velvety, uniform and short, rather than bristly on the ribs only. The whole plant is densely though shortly hairy, and the leaves, although similarly wide (about 1 cm), have few, if any, distinct teeth. The most unusual character, distinguishing it from all the other main *Striga* species, is that it is perennial, the shoots emerging from adventitious buds on a spreading, horizontal root system (see Plate 3C). This can result in shoots being arranged in straight lines, 10–15 cm apart, and infestations are typically in patches several metres across, presumably established from a single parasite plant. Little is known of its biology and physiology, but seeds respond to root exudates from sorghum and presumably have generally comparable behaviour. It is also subject to similar predation by insects, and heavy attack by *Stenoptilodes taprobanes* was noted in Ethiopia. The very limited seed production at this site could have been due to this attack, or to poor pollination, assuming it requires cross-pollination. The perennial habit, however, may make seeds less important and it has been noted that patches of *S. latericea* quickly reappear in sugarcane fields after ratoons have been ploughed out and replanted, presumably re-establishing from surviving roots.

Distribution *S. latericea* has a limited distribution in eastern Africa, being recorded only from Ethiopia, Kenya, Tanzania and Somalia.

Host range The only crop reported to be attacked by *S. latericea* is irrigated sugarcane, in Ethiopia and Somalia. Wild hosts are not well documented but include *Cynodon nlemfuensis* in Ethiopia, both in non-crop situations and around cane fields, so it is assumed a single population is able to attack both hosts.

Importance *S. latericea* is sufficiently damaging to be of concern on individual sugarcane estates in Ethiopia and Somalia, and severe symptoms and even death of the crop can occur.

Additional Species Occurring in Australia and Southeast Asia

***Striga multiflora* Benth.** This species has a 5-ribbed calyx, orange-yellow flowers up to 1.5 cm long, drying to blue-grey. The upper lip of the corolla is longer than the lower lip. It is described in one report as taller, more slender than *S. asiatica*, with longer leaves and more branching, but this may be by comparison with a local depauperate form of *S. asiatica*.

***Striga parviflora* Benth.** This species also has a 5-ribbed calyx, very small reddish/violet/purple flowers less than 1 cm long, again drying blue. The upper lip is longer than the lower.

***Striga curviflora* Benth.** This also has a 5-ribbed calyx, but flowers are larger, about 2 cm long, white, drying blue, with the upper lip shorter than the lower.

Information on these three species is extremely limited but all have been reported occasionally attacking sugarcane in Queensland, Australia and are known collectively as 'cane-killing weed'. All three species occur in New Guinea, and *S. multiflora* also in the Philippines.

BIOLOGY AND PHYSIOLOGY

In this section, the discussion is based mainly on *Striga hermonthica*, with comments added on other species where information is available and appropriate.

The Seeds

The seeds of *S. hermonthica*, in common with most other obligate root parasites, are extremely small, about 0.3 mm × 0.2 mm, weighing about 7 μg (see Fig. 1.3). As noted in Box 1.2, this very small size is associated with, but possibly not altogether responsible for, the obligate nature of their parasitism. Table 1.2 compares the sizes of a range of parasite species in Scrophulariaceae, showing that most of the *Striga* species other

Box 1.2. Seed size in relation to obligate and facultative parasitism.

Kuijt (1969) has expressed doubts about the usefulness of distinguishing between obligate and facultative parasites, considering that in nature all facultative parasites are likely to be attached to hosts and behaving as parasites even if they strictly speaking do not need to. It is interesting, however, to consider the characteristics which are critical in making the difference. Lack of photosynthetic capacity is one such characteristic and this adequately explains the obligate parasitism of *Cuscuta* and *Orobanche* species. Although mistletoes generally have an adequate photosynthetic apparatus, most species have apparently developed such a modified root system that they cannot establish and survive directly by rooting in soil. All the parasitic Scrophulariaceae have functional photosynthesis and moderately well developed roots, and the critical factor separating facultative from obligate parasites appears, in this group, to be seed size. The seeds of the facultative parasites *Odontites verna* and *Rhinanthus minor* are relatively massive, 3 mm long, while those of most *Striga* species are less than 0.4 mm long. Those of *S. angustifolia* are only slightly, but apparently significantly, larger than most other *Striga* species, 0.45 mm × 0.3 mm. They develop minute cotyledons, lacking in other species, and are just able to establish without a host (Sharma *et al.*, 1954; Reddy and Rao, 1980). The related *Parentucellia* species also have seeds of about this size. They are assumed to be facultative parasites but this is not certain. Those of *S. forbesii* are also of a similar size but there is no evidence this is a facultative parasite. Does this suggest that *S. forbesii* is evolutionarily more primitive than other *Striga* species? The seeds of many orchid species are even smaller than those of *Striga* but have the assistance of a symbiotic mycorrhyzal association in place of a higher plant host.

Fig. 1.3. Seeds of *Striga hermonthica* compared with those of maize and sorghum.

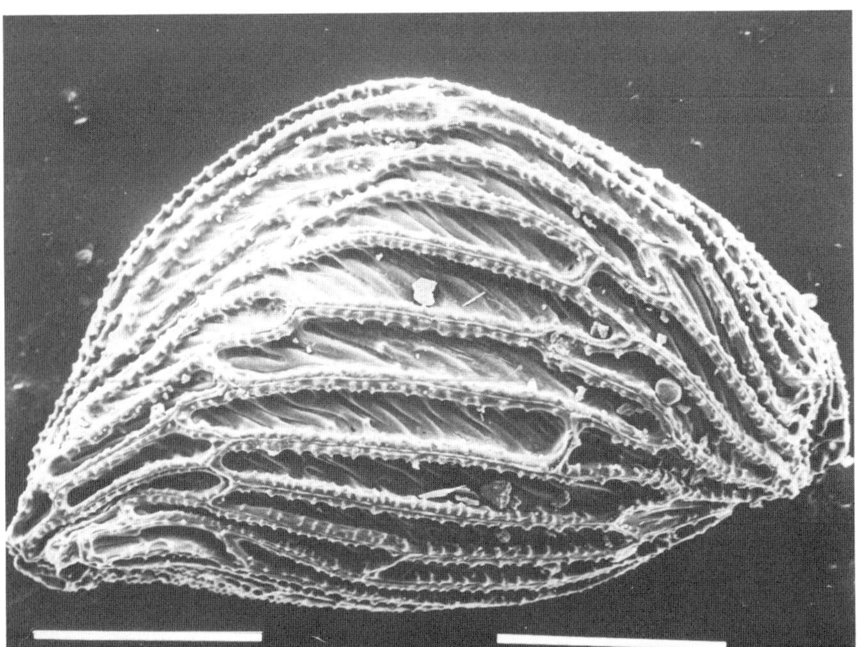

Fig. 1.4. SEM photograph of *Striga asiatica* (scale intervals are 0.1 mm).

than *S. angustifolia* are comparable to *S. hermonthica* except for *S. forbesii* which is distinctly larger, even though it is still an obligate parasite.

The seeds of *Striga* species have a characteristic surface pattern of ridges (Musselman and Parker, 1981a). Figure 1.4 shows the seed of *S. asiatica*; those of other species are generally similar. The pattern may be distinctive and uniform within a single plant of *S. hermonthica* (Jones and Safa, 1982) but varies considerably from plant to plant. The seeds of *S. aspera* have a comparable pattern, and the range of variation in each has so far made it impossible to distinguish these two species by their seeds. Similarly, those of *S. asiatica*, *S. gesnerioides* and *S. densiflora* are somewhat distinct from *S. hermonthica* but not from each other. Consequently it has not been possible to identify individual *Striga* species by these patterns, other than the larger, more ornamented seeds of *S. angustifolia*.

Table 1.2. Comparison of seed size in some parasitic Scrophulariaceae.

Obligate parasites	Seed length (mm)	Seed weight (μg)	Facultative parasites	Seed length (mm)
Alectra vogelii	0.20*	0.3	*Buchnera hispida*	0.55
Ramphicarpa fistulosa	0.55		*Odontites verna*	1.50
Striga asiatica	0.33	3.7	*Parentucellia viscosa*	0.44
S. aspera	0.30	5.5	*Rhinanthus minor*	3.50*
S. densiflora	0.30		*Striga angustifolia*	0.45
S. forbesii	0.50	12.4		
S. gesnerioides	0.33	5.8		
S. hermonthica	0.38	7.1		
S. multiflora	0.35			
S. parviflora	0.25			
S. passargei	0.30			

* excluding expanded seed coat.

The number of seeds per capsule varies from about 700 in *S. hermonthica* (Andrews, 1945) to 800 in *S. asiatica* and 1800 in *S. forbesii* (Obilana *et al.*, 1987; Knepper *et al.*, 1991). Number of capsules per plant may be only about 14 in *S. forbesii*, but average 60–70 in *S. hermonthica* and *S. asiatica*. Hence numbers of seeds per plant can be massive, averaging 25,000 for *S. forbesii* and 58,000 in *S. asiatica*, and numbers over 200,000 can almost certainly occur in well-grown *S. hermonthica*.

Longevity

It is often quoted that seeds of *Striga* can survive 20 years in the field. There does not seem to be good evidence for this figure but Saunders (1933) reports longevity up to 14 years for *S. asiatica* in South Africa and this figure is supported by direct experimentation on the same species in the USA (Bebawi *et al.*, 1984). No comparable data are available for *S. hermonthica* but observations by a number of authors suggest survival of this species can be at least as long.

Germination

The germination of *Striga* and other obligate root parasites is highly complex and significant to the ecology of the species, and to the understanding of control methods. *S. hermonthica* and most other *Striga* species are obligate parasites, capable of only limited seedling growth before the seed resources are exhausted. Without a host the seedling will die after only a few days; thus it is essential that seeds germinate within a few millimetres of a suitable host root. *Striga* species have consequently evolved a germination requirement that is only satisfied by a stimulant substance exuded by the roots of a potential host.

Natural germination stimulants

The chemical nature of the stimulants in the main host species is only now being determined after efforts spanning several decades from the 1940s onwards. The first natural stimulant to have been successfully separated and identified is known as 'strigol' (Cook *et al.*, 1972) (Fig. 1.5a). This is active on *S. asiatica* down to 10^{-16}M but came not from a normal host plant but from cotton, one of the many non-host (or 'false-host') species which curiously exude *Striga*-stimulating substances from their roots but are not themselves parasitized. Such species are of potential value in control systems and are discussed later under 'trap-crops'. It has not yet been explained how evolution has allowed *Striga* to respond to substances from non-host species or how the aberration has persisted even when such crops have been grown regularly in *Striga*-infested regions. The fact that most *Striga* individuals or populations able to respond to these false signals have not been eliminated is perhaps an indication of the complexities of germination response and the resultant limited effectiveness of trap-crops which will be discussed further in the relevant section.

Many years after the discovery of strigol a further class of natural stimulant substances was identified from the roots of the true host

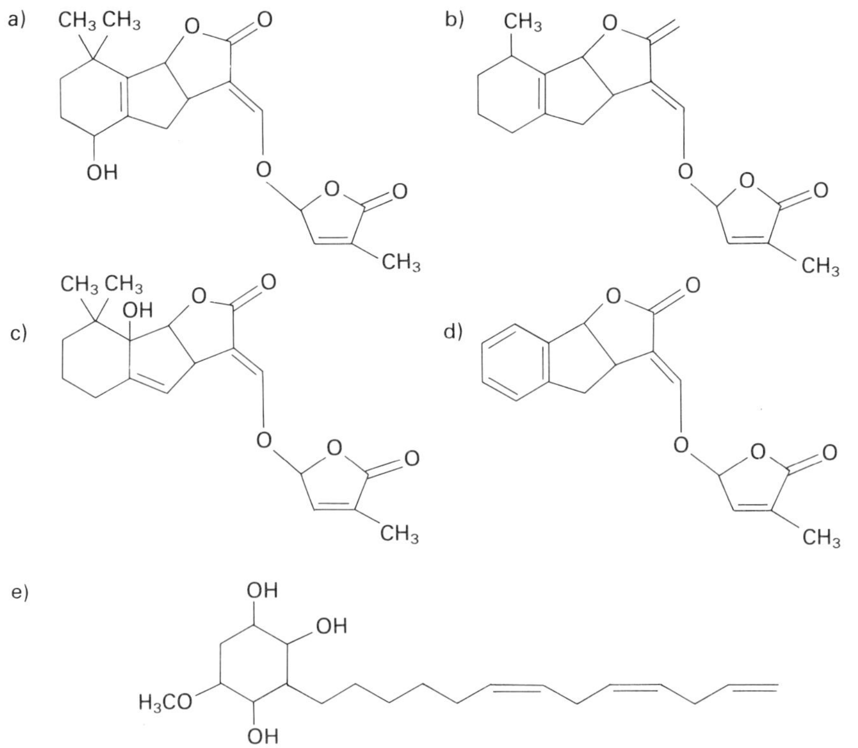

Fig. 1.5. Structural diagrams of germination stimulants: (a) strigol; (b) sorgolactone; (c) alectrol; (d) GR 24; and (e) sorgoleone.

sorghum (Chang *et al.*, 1986; Netzly *et al.*, 1988). This group of substances, known as sorgoleones, have a benzoquinone structure (Fig. 1.5e). They have extremely high germination stimulant activity on *S. asiatica* in the reduced dihydroquinone form (half-maximum germination caused at 10–100 nM solution) but none in the oxidized form. It has been claimed (Chang *et al.*, 1986; Fate *et al.*, 1990) that these substances can be detected in agar media around sorghum roots and are largely if not completely responsible for the normal stimulation by sorghum. These authors believe that the oxidation and hence inactivation of the dihydroquinone is influenced by oxidation inhibitors, also exuded from the root, thus exaggerating the natural concentration gradient around the root and ensuring that parasite seeds germinate only within a few millimetres of the root. However, Hess *et al.* (1991) have found that under wet conditions exudation of sorgoleone is suppressed; they also find no correlation between the sorgoleone production of different varieties, and their susceptibility to *Striga*. Thus there is some continuing uncertainty about the importance of sorgoleones in nature.

Most recently a further natural stimulant substance has been identified from sorghum by Hauck *et al.* (1992). This is, chemically, very closely related to strigol and has been given the name sorgolactone (Fig. 1.5b). It has been shown to have activity not only on *S. asiatica* and *S. hermonthica*, but also on *Orobanche aegyptiaca* and *Alectra vogelii*. The relative importance of this and of sorgeolone remains to be resolved.

Sorgoleone does not occur in maize roots, and the isolation and identification of any other stimulant substances from maize exudates is still proving extremely difficult, partly because of their very low concentration in roots or root exudates, but especially because of their instability. It has now been suggested that this very instability is an essential characteristic, at least in the activity of strigol and its analogues (Mangnus and Zwanenburg, 1991, 1992; Mangnus, 1992). Evidence so far suggests that a considerable number of different substances are involved, several in each species, some of these perhaps common to more than one host (e.g. Visser and Botha, 1974). It is fairly certain that different populations of *S. hermonthica* vary in the range of substances to which they will respond. In particular, there are populations which will respond to stimulants from sorghum but not from millet, and others conversely which are only able to respond to those from millet (see Fig. 1.10, p. 44). The main stimulant from pearl millet is apparently distinct from those from sorghum and maize. It is presumed but not certain that each of the *Striga* species occurring on cereals is responding to the same components of the roots exudate.

Strigol analogues

It is at least known that most of the important species will respond to strigol and to the 'strigol-analogues', related compounds which have been synthesized and tested as a possible means of control (Johnson *et al.*, 1976, 1981) (see Fig. 1.5d for structure of one of these analogues, GR-24). Species stimulated by these analogues include *S. hermonthica*, *S. asiatica*, *S. aspera* and *S. forbesii*; also *Alectra vogelii*, but **not** *S. densiflora*.

Ethylene

The possible role of ethylene in the natural germination process has recently been demonstrated. Exogenous ethylene is known to stimulate germination of both *S. hermonthica* and *S. asiatica* (Egley and Dale, 1970; Chancellor *et al.*, 1971) and is being used as a soil treatment in control systems in the USA. It is now apparent (Jackson and Parker, 1991) that the stimulatory effect of the strigol analogue GR-24 on *S. hermonthica* and perhaps also *S. asiatica* and *S. aspera*, but not *S. forbesii*, depends very largely on its stimulation of ethylene synthesis within the seeds. Such synthesis has been measured and when ethylene action is inhibited by 2,5-norbornadiene, germination in the presence of GR-24 is reduced or prevented. It has also been confirmed that the natural stimulants in host root exudates act in the same way via ethylene synthesis within the seed of *S. hermonthica* (Logan and Stewart, 1991). Information on other *Striga* species suggests that most respond to ethylene, including *S. densiflora* and *S. gesnerioides*, but that *S. forbesii* does not.

Other compounds

Worsham and Egley (1990) summarized available information on substances causing germination of *S. asiatica*, including a number of cytokinins, coumarin derivatives, scopoletin, thiourea and allythiourea, sulphuric acid, sodium hypochlorite and L-methionine. Most of these cause abnormal germination and none is thought to have any relationship with natural stimulants, other than those such as cytokinins and L-methionine which may act by release of ethylene. It is likewise assumed that the activity of the growth regulator thidiazuron is also via ethylene release (Babiker *et al.*, 1992).

Egley (1972) reports that conditioned seeds will germinate in the absence of a stimulant if they are punctured through the aleurone layer, though not if punctured at the side. Scarification by sulphuric acid could have the same effect.

Temperature

The optimum temperature for germination of *S. hermonthica* is about 35°C but it is capable of germinating over a wide range of temperatures from as low as 20° up to 40°C, above which damage occurs. *S. asiatica* has similar requirements. Other *Striga* species have not been studied in detail but are not known to differ significantly.

Conditioning and secondary dormancy

A further complication in the germination of *S. hermonthica* and other *Striga* (and *Orobanche*) species studied is their inability to germinate, even in the presence of a suitable stimulant, until they have been imbibed ('conditioned' or 'pre-conditioned') for at least a few days, and ideally for 1–2 weeks, at a suitable temperature. Optimum temperature of conditioning for the highest eventual germination is between 25 and 30°C for *S. hermonthica* but somewhat higher, about 35°C, for *S. asiatica* and *S. densiflora*. Higher temperatures will result in more rapid conditioning of *S. hermonthica* but percentage germination will not be as high even after several weeks (Vallance, 1950; Reid and Parker, 1979) (see Fig. 1.6). There is evidence that prolonged conditioning of *S. hermonthica* beyond a few weeks, especially at higher temperatures, will result in secondary or wet dormancy, which may or may not be overcome by drying and reconditioning (Vallance, 1950; Reid and Parker, 1979; Pieterse *et al.*, 1984). This phenomenon may be of importance in the natural germination pattern of *Striga* species in the field. Evidence of wet dormancy from prolonged conditioning is less convincing for other *Striga* species, but may also occur at least with *S. asiatica* (Reid and Parker, 1979).

The depth and persistence of this secondary dormancy in *S. hermonthica* may be further enhanced by conditioning at depth in the soil (Babiker *et al.*, 1987), presumably because of particular gas exchange conditions.

The importance of carbon dioxide levels during conditioning and germination of *S. hermonthica* has not been well studied but preliminary unpublished work at Long Ashton Research Station suggested that concentrations up to 8% had only slight effect on the germination of seeds following normal aerobic conditioning. However, much lower concentrations of 0.5 to 2% carbon dioxide during conditioning may have caused some reduction in later germination. On *S. asiatica*, Worsham and Egley (1990) report that germination is inhibited by levels of 10% carbon dioxide but give no information on the effect of carbon dioxide during conditioning. Lack of oxygen during germination is inhibitory after aerobic conditioning. The converse, anaerobic conditioning, followed by aerobic conditions for germination, caused less serious reduction in germination and seedling vigour; while, very surprisingly, anaerobic

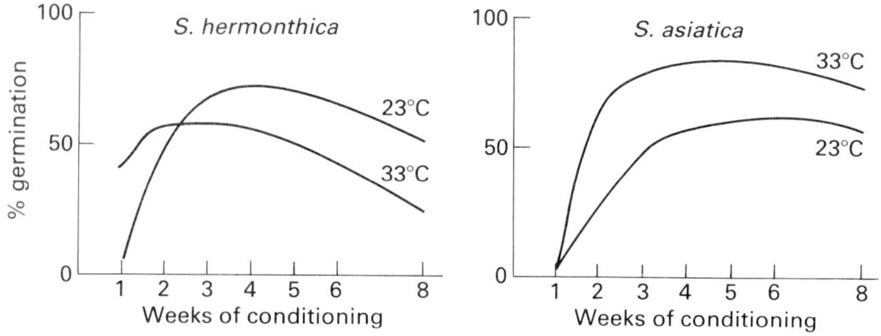

Fig. 1.6. Germination of *Striga* species following different times and temperatures of conditioning (from Reid and Parker, 1979).

conditions during both conditioning and germination phases resulted in almost normal germination and seedling growth.

A further complexity in the behaviour of both *S. hermonthica* and *S. asiatica* is that the application of germination stimulants, either natural or synthetic (including ethylene), before or during conditioning, results in reduced germination even when stimulant is reapplied at the normal time (Hsiao *et al.*, 1981; Babiker and Hamdoun, 1982, 1983).

The biochemical processes involved in the conditioning and germination of *Striga* species have not yet been fully resolved. There have been suggestions for the involvement of inhibitors that have to be leached out or degraded during conditioning; also the possibility that some stimulant co-factor has to be synthesized during conditioning. Vallance (1951a,b) suggests that this co-factor acts as a substrate for respiration and is thus degraded by the enhanced respiration which is stimulated by the premature addition of the natural germination stimulant. This would help explain the detrimental effect of the presence of germination stimulant during the conditioning phase. Only when the co-factor has been fully formed is the germination stimulant able to have its positive effect.

Light

Light can reduce the germination of *S. asiatica* seeds when they are exposed within one or two days before, or up to 4 hours after exposure to the germination stimulant. In each case red light is inhibitory and its effect can be reversed by far-red light, indicating the involvement of phytochrome. The inhibition is not usually complete, however, unless the stimulant concentration is very low (Egley, 1972; Worsham and Egley, 1990). Most other species are believed to respond in the same way, except for *S. angustifolia*. Germination of this species is apparently enhanced to some extent by sorghum root exudate (Rao *et al.*, 1985a) but does not fully depend on such stimulation, and it **does** need light (Kumar and Solomon, 1940). This requirement is understandable as their very limited seed resources would not be adequate for them to reach the soil surface from more than a few millimetres depth. Again, far-red light reverses the effect of red, implicating phytochrome.

Once conditioned and then exposed to stimulant under suitable temperature and moisture conditions, the seeds of most *Striga* species will germinate within 24 hours, but *S. forbesii* may require a little longer, and *S. gesnerioides* takes 2–3 days.

Further useful reviews on germination of *Striga*, not already quoted, include Okonkwo (1991) and Logan and Stewart (1992).

Contact, Attachment and Penetration of the Host

Once the seeds of *S. hermonthica* have germinated, up to several millimetres from a host root, they have still to make contact with that root in order to parasitize it. Given purely random directional growth after germination, it has been estimated (Dixon and Parker, 1984) that the chances of contact with a single root of small diameter may be less than 10%. It is to be expected that in the course of evolution these chances might have been improved by the development of some chemotropic or other direction-seeking assistance. Indeed, Williams (1961) presents evidence for chemotropism in *S. hermonthica*, while Saunders (1933) also observed it in *S. asiatica*. The effect is far from complete (e.g. seedlings initially growing away from the root may be unable to respond) and some other workers have failed to confirm what is nevertheless a presumably important phenomenon in nature. Williams (1961) has suggested that effective chemotropism could result from a gradient in pH around the root.

On contact with the host root the elongation of the radicle stops and the development of the organ usually known as the haustorium immediately begins (see Box 1.3). This process has been shown to depend on haustorium-initiating substances. That responsible for initiating haustorial development in *S. asiatica* has been identified as 2,6-dimethoxy-*p*-benzoquinone (2,6-DMBQ) (Chang and Lynn, 1986; Smith *et al.*, 1990) (Fig. 1.7).

The latest studies (Lynn and Chang, 1990; Riopel *et al.*, 1990) reveal the interesting observation that 2,6-DMBQ cannot normally be detected

---

Box 1.3. The haustorium and endophyte.

There have been various definitions of the term 'haustorium' but perhaps the simplest, and at the same time most authoritative, is that of Kuijt (1977): 'A specialized organ of absorption of a parasitic plant'. It has sometimes been argued that it should only apply to the parasite tissue within the host. In that case the swelling at the tip of the radicle before it penetrates the host is referred to as the 'pre-haustorium', 'upper haustorium' or 'holdfast' (especially in mistletoes), but we have not felt it necessary to make this distinction. A primary haustorium is that formed from the tip of the radicle in the obligate parasites of *Striga* and *Orobanche* and mistletoes. Secondary haustoria are those formed on any organ other than the radicle.

The 'endophyte' is defined by Kuijt as 'that portion of a haustorium surrounded by host tissues'. It can also be termed the 'intrusive organ'. Kuijt defines a great many other terms used by plant anatomists in describing the structure of haustoria and those interested should refer to his review.

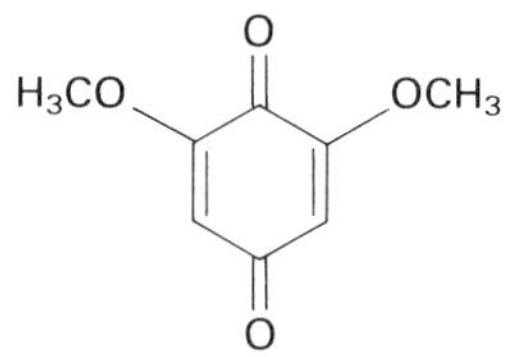

Fig. 1.7. Structural diagram of the haustorial-initiating substance 2,6-dimethoxy-*p*-benzoquinone.

in the exudate from sorghum roots although it is present in extracts of the sorghum root. They propose that *Striga* itself triggers the synthesis and release of this substance at the root surface by exuding extracellular enzymes (including laccases) which act on phenylpropanoid precursors in the host root surface, to yield the substituted quinone, 2,6-DMBQ. Many other phenolic and flavanoid substances can also initiate haustoria in both *S. asiatica* and *S. hermonthica* (Macqueen, 1984; Chang and Lynn, 1986), presumably acting as the substrate for the production of 2,6-DMBQ via this enzyme system. Cytokinins are well known to be initiators of haustorial development, as also is thidiazuron which has some cytokinin type activity (Babiker *et al.*, 1992).

In *S. asiatica*, the radicles are most responsive and best able to form haustoria within 4 days of germination, and at 30°C may begin to lose this ability even 2 days after germination.

The process of haustorial development and penetration of the host is closely similar in *S. hermonthica* (Okonkwo, 1966a; Olivier *et al.*, 1991a), and *S. asiatica* (Saunders, 1933; Rogers and Nelson 1962; Musselman, 1973; Ramaiah *et al.*, 1991). Sticky hairs develop on the young haustorium which help it to adhere to any surface. After attachment by these hairs, intrusive cells develop at the root tip which penetrate the cortex of the host root, apparently producing enzymic secretions so that host cells are separated rather than intra-cellularly penetrated. The haustorium sometimes fails to complete its penetration of the cortex (Ramaiah *et al.*, 1991) and may also fail on reaching the endodermis, which sometimes provides a barrier (Saunders, 1933). Okonkwo observes that the breaching of the endodermis may involve the breakdown of cell walls and intra-cellular penetration, but most observations suggest an enzymic separation of the cells. Whether this separation of the endodermal ring is hindered by the degree of suberization and development of the Casparian strip has not been established, but this could be a factor limiting the penetration of parasites on older roots. A novel idea was proposed by Smith and Stewart (1987) for the penetration of cowpea roots by *S. gesnerioides*, that the penetration of the haustorium towards the stele triggered root initial development in the pericycle, within the stele. The root initial was then responsible for breaching the endodermis from within, thus making easy access for the haustorium. However, this theory has not been clearly proven.

Once the haustorium is inside the host stele there is rapid development of direct links between parasite and host xylem systems, but no corresponding development of phloem connection (see Plate 4A). Thus the transfer of nutrients apparently depends on the xylem bridge and some limited diffusion through parenchyma tissues. In *S. asiatica* the time from first penetration of the epidermis to a connection established with the host stele is 60 hours (Ramaiah *et al.*, 1991). Plate 4B shows a young plant of *S. hermonthica* developing as a total parasite prior to emergence from the soil.

In most *Striga* species the primary haustorium remains small, never larger than 1–2 mm in diameter, but as the seedling grows, adventitious roots develop from the base of the elongating stem. Where these roots make contact with other host roots, secondary haustoria are formed, providing the parasite with further points of contact with the host (see Fig. 1.8). Okonkwo (1966a) has studied the structure of these secondary haustoria and although they have a different origin, developing as lateral rather than terminal organs on the root, he finds they have very similar internal structure and organization. The importance of the secondary haustoria in the success, vigour and host range of *S. hermonthica* and *S. asiatica* has not yet been fully studied but there may be many hundred to a single *Striga* plant. This must surely make them very important and deserving of more study. Does it matter how many develop? Also, can they develop on hosts other than those which can act as primary hosts and, if so, can such 'secondary' hosts be damaged to the same extent as the primary host? For discussion of 'primary' and 'secondary' hosts see Box 5.1, p. 198.

Dependence on the Host

What does the parasite need from the host? Working with *S. hermonthica* in sterile culture, Okonkwo (1966b) showed that the parasite needed only sugars and inorganic minerals. Yoshikawa *et al.* (1978) suggested that *S. asiatica* in the USA required cytokinin for normal development, but Okonkwo (1991) has recently contradicted this finding and Chidley and Drennan (1987a) confirmed that an Indian (white-flowered) population of *S. asiatica* could, like *S. hermonthica*, be grown to maturity without added cytokinin, using only sucrose and inorganic nitrogen.

In nature, the source of nitrogen for the parasite is not fully understood. Stewart *et al.* (1984) considered that inorganic nitrate was unlikely to be an important source, because very little nitrate reductase activity could be detected in either *S. hermonthica* or *S. asiatica*. It is possible, however, that such activity might have been overlooked because it is concentrated in the haustoria. Ba (1979) and Ba and Kahlem (1979) draw attention to the presence of highly staining inter-tracheidal

Fig. 1.8. Root systems of millet and *Striga hermonthica* showing secondary haustoria (small black spots) linking the roots.

parenchyma in the heart of the haustorium which is assumed to be responsible for biochemical transformations of material from the host, emphasizing that the xylem bridge is not a simple tubular connection between host and parasite. Stewart *et al.* concluded also that ammonium ions were an unlikely source and suggest that much of the nitrogen (and carbon) acquired by the parasite from the host xylem was likely to be in the form of amino acids and amides, especially glutamine and glutamate which are commonly detected in *Striga* xylem sap. Although *S.*

hermonthica and *S. asiatica* absorb such elaborated nitrogen compounds from the host, they do have their own ability to synthesize amino acids (Press *et al.*, 1986).

While the source of sugars in culture and in the host root is normally a combination of glucose, fructose and sucrose, the sugar in xylem exudates of *S. hermonthica*, apart from being five times more concentrated than that in the host xylem, is found to be largely mannitol (which is barely present in the host), again emphasizing the transformations that must take place in the haustorium (Press *et al.*, 1989). Nour *et al.* (1984) propose that the high concentrations of mannitol found in *S. hermonthica* and *S. asiatica* as well as other parasitic species have a role in osmoregulation and/or protection against heat stress.

Although it is obvious that the parasite depends totally on the host for all its organic nutrition up to the time of emergence, it was long supposed that once emerged, the typical green-leaved species such as *S. hermonthica* and *S. asiatica* were able to photosynthesize most of the sugars they needed. It is now quite clear that this is not so. A series of studies on several species have shown that chlorophyll content is lower in the parasite and its efficiency only 20–30% that of a comparable non-parasitic plant (de la Harpe *et al.*, 1979, 1981; Tuohy *et al.*, 1986; Stewart and Press, 1990). It has further been shown that, as a C_3 plant, *S. hermonthica* has a high rate of photorespiration and may respire so much that there is little or no net gain from its own photosynthesis. Without the alternative source from the host it could make little or no growth at all. Sophisticated studies using the ^{13}C isotope of carbon have shown that about 35% of the organic carbon in the parasite is derived from the sorghum host (Press *et al.*, 1987) and even more from millet (Graves *et al.*, 1990).

One possible reason for the parasite's loss of photosynthetic efficiency is hinted at by Press *et al.* (1990) when they point out that the amount of amide and amino acid that the parasite needs, to provide its nitrogen requirement, carries with it a highly significant quantity of elaborated carbon. Hence the parasite does not need to elaborate much carbon so long as it is getting its requirement of nitrogen. This suggests that nitrogen is the limiting factor for the growth rate of the parasite making it difficult to explain why a nitrogen-rich host supports weaker rather than stronger parasites. The detrimental influence of high nitrogen conditions on most species of *Striga* will be discussed under cultural control methods. It is a phenomenon repeatedly demonstrated but still not fully explained.

The ability of the parasite to draw the necessary amounts of elaborated carbon and nitrogen from the host almost certainly depends on the maximum flux of sap across the xylem bridge. This may be achieved to some degree by the differential osmotic pressure (OP) which is generally much higher in the parasite than in the host. Gworgwor and

Weber (1991) have recently shown that the optimal N level for reducing *Striga* attack is one which also results in a balancing of OP between host and parasite. But perhaps more important are the ever-open stomata which ensure a high rate of transpiration in all but highly humid conditions. Studies by Shah *et al.* (1987) and Smith (1991) show that the stomata of *S. hermonthica* differ from normal stomata in not closing in response to drought stress, darkness, or abscisic acid. The importance of high humidity in reducing the flow of nutrition and hence the parasite vigour has only been guessed at in *S. hermonthica*, but in *S. asiatica* it has been clearly demonstrated by Egley (1971) that high humidity can reduce the growth rate by more than 70%, a likely, at least partial, explanation for the effect of mixed cropping in reducing *Striga* survival (see later).

Where there are favourable conditions for parasite growth, *S. hermonthica* will normally germinate and attach to the root system within 2–3 weeks after host germination, emerge after 4–5 weeks and flower within 7–8 weeks. Viable seeds are probably produced within 2 weeks of the flower opening and being pollinated, but are not fully matured and shed until about 2 weeks later. The corresponding timings for *S. asiatica* are about the same up to emergence, but development after emergence is somewhat more rapid and seed may be shed rather earlier. The exact sequence and timing of seed maturation will be discussed later, under 'Hand-pulling'.

Effects on the Host

Any parasite will have some detrimental effect on its host by reason of the materials that it removes. In this way a heavy population of *S. hermonthica* must inevitably deplete the elaborated nitrogen and carbon compounds available to the host shoot and result in reduced growth. Quite simple measurements, however, can show that the reduction of host growth can be enormously greater than is explained by simple removal of resources. When the *S. hermonthica* on a sorghum plant were only a few millimetres long, it was estimated that the reduction in host weight was at least 30 times the weight of the parasite (Parker, 1984). Some apparently 'toxic' effect is therefore indicated, but the precise nature of this toxin or effect is still not fully understood. The flow of materials between host and parasite is apparently very largely, if not completely, one-way, from host to parasite, and any transfer of toxin from parasite to host must be in very small quantities. One striking effect of infection by *S. hermonthica* on cereals, difficult to detect in the field but readily observed in pot experiments, is a marked change in root : shoot balance, the root systems of infected plants being greatly stimulated, at least for a period after infection, while the shoot systems are stunted and weaker. Where the root : shoot ratio in healthy sorghum plants was 0.25

at 7 weeks after planting, the ratio in *S. hermonthica*-infected plants at the same stage was 0.61. Figure 1.9 shows how the total root mass is increased as a result of attack by *S. hermonthica* in pots. Although some of the additional roots are those of the parasite, the great majority are those of the host, sorghum. A similar effect occurs with *S. asiatica* on maize, Patterson (1990) reporting root : shoot ratios changing from 0.35 in uninfested crop to 0.90 in infested crop at 10 weeks after sowing. This induced change can be seen to be an important way in which the parasite ensures that the resources of the host are diverted downwards to its benefit.

The change in root : shoot balance and the resultant reduction in host shoot growth can occur at a very early stage of parasite growth, even before it emerges from the soil. It is this very early effect, even before the weed is visible, which has no doubt given rise to the common and Latin names for the weed, involving 'witch'. The crop appears indeed to be bewitched.

Fig. 1.9. The soil/root mass from pots in which sorghum had grown (l.) with and (r.) without *Striga hermonthica*, showing stimulation of host root development by the parasite.

Some relevant studies have shown that in sorghum plants infested by *S. hermonthica*, the proportions of different growth regulating substances reaching the shoot via the xylem sap were drastically altered, cytokinins and gibberellins being greatly reduced, while inhibitors, abscisic acid and farnesol were increased (Drennan and El-Hiweris, 1979). The change in root : shoot ratio can probably be largely attributed to this change in growth regulator signals from the root, though other mechanisms might also be involved. An important finding in the same study was that the effects of infection by *S. hermonthica* closely mimicked those of drought stress. It appears possible therefore that the profound physiological effects of *Striga* may be mediated through an existing mechanism in the host for detecting drought. *S. asiatica* causes much the same degree of damage to its host, given approximately the same mass of parasite per host (Parker and Wilson, unpublished), and there is virtually the same tendency for change in root : shoot balance. Other symptoms in the host shoot, however, are quite different for these two species. In hosts affected by *S. hermonthica* there are almost always some chlorotic symptoms in the form of small elongated yellowish blotches (Plate 4C), while those affected by *S. asiatica* show no localized chlorosis, but a generalized wilting, even when soils are still quite wet (Plate 4D). Most other *Striga* species, including *S. aspera*, *S. densiflora* and *S. forbesii*, cause symptoms similar to those of *S. hermonthica*. Severe infestations by any of these species can lead eventually to generalized chlorosis, scorch and collapse of leaf tissue, resulting in numerous local names which translate as 'fire-weed' suggesting that the host has been burnt.

Apart from the damage to the host shoot system caused by diversion of food to the host roots and parasite (whether by the influence of growth regulators from the root, or by more direct hormone-directed transport or physiological sink effects), there is now thought to be a reduction of photosynthetic efficiency in the host due to both *S. hermonthica* and *S. asiatica*, such that photosynthetic capacities may be reduced to less than half those in healthy cereals (Press and Graves, 1991). It has been further estimated that 80–85% of the growth reduction recorded in sorghum and millet is due to reduced photosynthesis in the host and only about 20% of the damage due to actual removal of carbon by the parasite (Graves *et al.*, 1989, 1990). It is not quite clear by what mechanism the photosynthesis is reduced, but one contributory mechanism may be the interference with water balance in the host, resulting in closure of stomata. Another may be the loss of nitrogen. It is puzzling that the parasite causes this reduction in host function when it must surely be in the best interests of the *Striga* for the host to continue producing the food that the parasite needs.

One further form of damage to the host, recorded for *S. hermonthica* on pearl millet, is a marked change in the amino acid content of the grain from infested crops, almost certainly reducing the nutritional quality of the produce (Wylde, 1991).

ECOLOGY

Temperature

Relatively high temperatures, around 30°C, are known to be optimal for conditioning and germination of *S. hermonthica* and other *Striga* species and for their rapid growth. Patterson (1990), however, has shown how *S. asiatica* can germinate, develop and mature on sorghum at a mean daily temperature of only 22°C (alternating 20/26°C), while dormant seed over winter could survive freezing at −7°C or −15°C for 49 days. Such detailed information is not available for *S. hermonthica* but perhaps this species may have somewhat less latitude in its tolerance of low temperatures, as it rarely occurs outside the tropics, while *S. asiatica* flourishes outside the tropics in both North America and in southern Africa.

Light/Daylength

S. asiatica can mature normally in a range of photoperiods up to 16 hours (Kust, 1964), and there is no evidence, for any species, of any dependence on specific photoperiods for either development or flowering. Light intensity should not be critical, as it has been shown that completely albino plants of *S. hermonthica* can grow to maturity, and *S. asiatica* has matured in complete darkness (Rogers and Nelson, 1962). As photosynthetically active plants, however, they do show maximum growth in full sunlight. It is a common observation that *S. hermonthica* is conspicuous as a fringe around the edges of fields where light is highest, though it is also possible that lower humidity is contributing to their greater vigour in this situation.

Moisture

In general, *S. hermonthica* and *S. asiatica* both thrive best under conditions of erratic or limited rainfall, and may be suppressed by irrigation (see e.g. Andrews, 1945) or by heavy continuous rain (Ogborn, 1972a). Ogborn pointed out that the pattern of the rainfall season in northern Nigeria could have a profound influence, early continuous rains from the beginning of the season delaying emergence of *S. hermonthica* for about two months, while erratic, fitful rains at the beginning of the season encouraged a serious early attack. *S. hermonthica* does not occur as a problem in flooded or irrigated rice, though it can be serious under the drier conditions of upland rice. Any generalization, however, is subject to many qualifications. Andrews showed no influence of even heavy irrigation on *S. hermonthica* in a November planting in Sudan, and in pot culture continuous watering, even by sub-irrigation, does not always prevent vigorous germination and development. In the field the normal, often very heavy rains, causing short-term wetness and even waterlogging, may have no detrimental influence. Wet dormancy has been the commonest explanation for a suppressive effect of wet conditions, but there is no experimental proof under field conditions and other factors are almost certainly involved. Possibilities that have been suggested include the leaching and/or dilution of the stimulants; reduced soil temperatures affecting the exudation of stimulants and/or the germination response of the *Striga*; and increased susceptibility of the unemerged *Striga* to soil-borne pathogens. High humidity could also be of critical importance by reducing transpiration and hence the vigour of the parasite, leading in turn to increased susceptibility to pathogens after emergence.

At the other extreme, very dry conditions can also result in reduced occurrence of *S. hermonthica* and in the most arid zones, heavy infestations are associated with the wetter seasons (see e.g. Delft and Stroud, 1992).

Among other species of *Striga*, *S. passargei* and *S. densiflora* are also associated with relatively dry conditions, and rarely if ever with irrigation, while several other species are far less strictly associated with dryness. Although *S. aspera* is common in the dry sub-Sahelian zone on wild grasses, it also occurs under higher rainfall conditions, further south than *S. hermonthica*, and in some irrigated crops, including sugarcane and rice. *S. forbesii* and *S. latericea* are quite different from *S. hermonthica* in occurring mainly in wet, even waterlogged conditions, on wild grasses in swamps and in irrigated sugarcane. *S. forbesii* does, however, occur also under drier upland conditions, often in mixture with *S. asiatica* on sorghum and maize as in parts of Tanzania and Zimbabwe. *S. angustifolia* is also associated with irrigated sugarcane as well as with upland cereal crops in India.

Soil Type

S. hermonthica occurs on a wide range of soil types from heavy cracking clays, to very light sands. Some individual strains or populations of the species may have more restricted tolerance, but this could relate more to the most suitable soils for the host rather than any direct influence on the parasite. For instance, *S. hermonthica* on millet in West Africa is associated with light sandy soils (typical of millet-growing areas) while that on sorghum is more often on heavier soils. *S. asiatica* has been shown to develop on a wide range of soil types (Nelson, 1958b; Robinson, 1960) but on heavier soils it required higher temperatures than on light soils and there was no establishment in heavy clays. In Africa, especially in Tanzania, it has been observed that *S. asiatica* occurs mainly on the lighter soils, while *S. hermonthica* is commonest on the heavier soils (Doggett, 1965).

Fertility

One of the safer generalizations that can be made for both *S. hermonthica* and *S. asiatica* is that they are favoured by low soil fertility. In some of the earliest work on *S. asiatica*, in Burma (Sawyer, 1925) and in South Africa (Pearson, 1913), the association of the problem with poor soils and the corresponding benefits of fertilization were already recognized. Later, the same was shown for *S. hermonthica* in Kenya (Watt, 1936) and in Sudan (Andrews, 1945), and it has been confirmed by many other workers since, including quite recently in Ghana and Togo (Vogt *et al.*, 1991), where intensity of *S. hermonthica* was positively correlated with population density and intensity of land use, and with stone and gravel content, while there was negative correlation with organic matter content. Some of the correlation with intensity of land use can be attributed to the greater opportunity for build-up of the population, but it has also been repeatedly, though not consistently, demonstrated that *Striga* may be suppressed to some degree by increased application of fertilizer. The most important element has almost invariably been shown to be nitrogen. Phosphorus has rarely shown any marked effect, while potassium has occasionally been shown to cause increased infestation of *S. asiatica* (Farina *et al.*, 1985). Sherif and Parker (1988) found no influence of organic matter in the absence of associated nitrogen.

The influence of soil fertility on other *Striga* species and other genera in Scrophulariaceae is not well documented, but it has been hypothesized that parasitism may have developed in this and other families as a means of overcoming limited nitrogen availability in the environment (Press *et al.*, 1990). It is probably true, therefore, for most if not all parasitic members of the genus, that they are at least associated with low nitrogen soils if not positively favoured by them. The practical importance and exploitation of the nitrogen effect are discussed in more detail later.

Insects

Insect pollination Most species of *Striga* are autogamous, having systems for self-pollination even before the flower opens, but *S. hermonthica* and *S. aspera* are out-crossers, requiring cross-pollination for fertilization (Musselman *et al.*, 1981; Musselman *et al.*, 1991). In *S. hermonthica*, self-incompatibility of the pollen has been confirmed (Safa *et al.*, 1984). Many insects have been recorded as visitors to *S. hermonthica* and *S. aspera*. Which of these are the most important pollinators is not easily determined but they are thought to include butterflies such as

Vanessa cardui, Colotis antevippe and *Borbo* spp., also the moth *Cephanodes* spp. and perhaps bee-flies (*Litorhina* and *Systoechus* spp.)(Musselman *et al.*, 1983; Musselman and Hepper, 1986).

Insect predation *Striga* species are damaged by a range of insects, as documented by e.g. Sankaran and Rao (1966), Greathead and Milner (1971), Greathead (1984), Markham (1985), Bashir (1987). Two commonly occurring genera are the *Junonia* (= *Precis*) butterflies and the plume moths, *Stenoptilodes* spp. *J. orythia* and other species are widespread in India, Africa, southeast Asia and the USA. The larvae can cause serious defoliation of *Striga* but are not specific to it. The eggs of *S. taprobanes* are laid on young shoot tissue and sometimes cause noticeable damage to the youngest leaves. More often the damage is to the anthers and style, frequently resulting in sterility of the flower. The presence of the larva is not obvious, but may be detected by the presence of frass in the corolla tube, often in the place of pollen. Most, if not all, of the *Striga* species (and *Alectra*) are damaged by this non-specific predator, but the exact extent of its effects are not well recorded.

The most important insect predators are the gall-weevils, *Smicronyx*, of which several species occur in both Africa and Asia. The commonest species, widespread in West Africa, is thought to be *S. umbrinus*. The adults lay their eggs in the unopened flower bud, causing the base of the corolla and ovary to form a swollen gall up to 5 mm in diameter, completely preventing any seed formation in that flower. The larva develops in the gall to about 5 or 6 mm long, before emerging through an exit hole and falling to the ground, where it remains in diapause (as a larva) in the soil over the dry season. In the following season, rains trigger pupation and emergence of the adult. There appears to be only one generation per season, but the biology has yet to be adequately studied. There are certainly several other species, including *S. quadritubulatus*, recorded from *S. hermonthica* in Burkina Faso (Greathead, 1984) and at least one so far unnamed species occurring in East Africa and Ethiopia whose larvae feed on the fruits, sometimes eating the developing seeds in the capsule, causing only mild deformation of the capsule without always causing galling. In West Africa, larvae are likewise sometimes found in normal capsules of *S. hermonthica*. Galling on *S. aspera* has been observed occasionally. A distinct species, *S. albo-variegatus*, occurs on *S. asiatica*, *S. densiflora* and *S. angustifolia* in India, causing galls in stems and roots as well as capsules.

The overall effect of insect predation, combined perhaps with incomplete pollination under some conditions, is that the proportion of good capsules on *S. hermonthica* may be greatly reduced. In West Africa and Sudan, commonly recorded figures are about 35%. In Ethiopia, however, where *Smicronyx umbrinus* is generally absent, the

S. hermonthica is relatively very healthy and over 90% of healthy capsules are commonly observed.

Some further insects and the potential for biological control are discussed later.

Pathogens

Many fungi have been recorded on *Striga* (Greathead, 1984), the most frequently reported being a powdery mildew, *Sphaerotheca fuliginea*, in both Africa and India. Fungal disease of *Striga* is not often conspicuous but it is possible that pathogens, including *Fusarium* species, may be important in suppressing *Striga* both before and after emergence under marginal conditions, especially when soil moisture and humidity are excessive. The range of species reported and their potential for biocontrol will be discussed later.

CONTROL OF *STRIGA* IN CEREALS

As in the previous section, *Striga hermonthica* is the main species being considered. Most of the methods described will be equally relevant to other species but some exceptions will be pointed out.

Controlling *Striga* and other root parasites presents special difficulty due to the fact that this type of weed can do much of its damage to the host crop before emerging above ground. In addition to the practical difficulty of preventing attachment of the parasite before it does damage, there is the additional problem that controlling the emerged parasite, when it does become possible, gives limited immediate benefit. It emerges after the main weedings and it is difficult to convince the farmer to spend the extra effort needed to hand-pull and so prevent seeding, even though this is one of the most important ways of preventing further intensification of the problem for the future.

In the following sections all possible means of controlling *S. hermonthica* and related problem species are considered, but few if any of these techniques can provide a complete answer. It is usually necessary to use a combination of those methods most relevant to the farming system and infestation level. The integration of control methods is considered at the end of the chapter.

Cultural and Mechanical Control Methods

Hand-pulling

This is the one option open to virtually all farmers and its use should be the first priority of extension effort, though its exact function and limitations should be clearly understood. Firstly, hand-pulling has two possible objectives: (i) to reduce damage to the current crop; and (ii) to prevent seeding of the weed and so reduce future infestation.

Although much of the damage from *S. hermonthica* is done before the weed emerges, it is often possible to show a benefit from hand-pulling emerged plants (e.g. Carson, 1988), and even from destruction at the time of *Striga* flowering (Press *et al.*, 1989). Removal of *Striga* at the early vegetative stage is likely to be more beneficial in terms of crop yield than later removal (Ogborn, 1984), but the general experience is that there is rapid regrowth, presumably from the broken shoot bases below ground, and farmers find it very discouraging to have to go through their crops repeatedly many times. The extent to which hoeing or hand-pulling can reduce damage to the current crop is far from consistent, however, and this possible benefit should **not** be regarded as the main objective of hand-pulling.

Ogborn (1984) quotes alarming figures for labour costs and the numbers of *S. hermonthica* resulting from repeated pulling of young plants, but if we aim only to prevent seeding, then pulling need not begin until 2–3 weeks after the *S. hermonthica* begins to flower (e.g. in September). The stems are then woody at the base and there is little problem of regrowth. Trials in Ethiopia have shown that the labour needed for such later pulling is less than half that required for early pulling (Parker, 1990). Hoeing is hardly suitable at this later stage, partly because of greater likelihood of regrowth, but also because the plants need to be collected for removal from the field. Provided all flowering plants are pulled at not longer than 3 week intervals, there should be little or no seed shed in the field, but the green capsules will contain mature seeds and the plants must be carried out of the field and eventually burnt. If all plants **in bud** are also removed, perhaps the maximum interval might be a little more than 3 weeks but this requires experimental confirmation. To avoid the risk of seed being blown or washed back onto the field, it is suggested they be placed in a shallow pit until they are dry enough to burn. Many farmers practice such pulling and carrying out from their fields, as in Ethiopia (see Plate 5A) and in parts of West Africa, but where they often fail is in not continuing to do it through harvest time, when seeds mature most rapidly and prolifically. Pulling therefore needs to be continued through and beyond harvest.

In the short term, economic analysis of hand-pulling is likely to show no benefit, the extra labour being offset by little or no yield increase. It is vital, therefore, that any economic assessment of hand-pulling (and other *Striga*-control practices) be made on a longer-term basis which not only takes account of the benefits of reduced infestation on future cereal crops but also the detrimental effects of increased infestation in untreated crops.

Although Ramaiah (1985a) has reported a striking benefit from hand-pulling *S. hermonthica* in millet in a single season, it will usually need to be continued for 3–4 years before a big benefit can be seen. Farmers should be aware of this so that they are not disappointed and give up the practice.

Hand-pulling is very often dismissed as impractical and far too demanding in labour. This is certainly true for dense infestations and farmers should **not** be pressed to use this method on severely infested land. Hand-pulling is most economical and worthwhile on the **least-infested** fields, and these should be the first to be covered. Row-planting allows much more rapid and systematic pulling and should be encouraged where not already practised. Furthermore, we suggest below how the use of other techniques can greatly reduce the numbers of emerged *Striga* to be pulled and so make hand-pulling more feasible. Without hand-pulling (or herbicide) to prevent the seeding of residual populations after other control measures have been applied, it is unlikely that any long-term reduction of the problem can be achieved.

Other *Striga* species such as *S. forbesii* and *S. aspera* can and should be treated in the same way as *S. hermonthica*. Some, however, including *S. asiatica* may prove more difficult and expensive to control by hand-pulling. Plants are individually smaller and less conspicuous and may be much more numerous. Seeds become viable within 10 days from flowering (Sand, 1990) and maturation may be more rapid, requiring more frequent cycles of removal, resulting in correspondingly greater time and effort for complete prevention of seeding in the field. Hoeing, at shorter intervals of about 2 weeks, might prove less time-consuming than attempting removal by hand every 3 weeks. Herbicide use will be a correspondingly more attractive alternative where this is at all feasible, though, as with hand-pulling, it may be some years before the cost is recovered.

Rotation, trap-crops and catch-crops

Rotation of infested land into non-susceptible crops or into fallow is theoretically the simplest of all solutions, but also one that is now very rarely at all simple or acceptable. It was once the traditional farming system of most of the affected areas, land being allowed to return to bush fallow for 10 to 20 years, by which time, incidentally, the *Striga* seed bank in the soil was very largely depleted and cereal crops could be grown again with negligible infestation. Unfortunately, with increased population pressures this option is now limited to very few areas, and the length of the fallow can rarely be sufficient to allow adequate decline in infestation, given the longevity of *Striga* seeds in the soil of at least 5–10 years.

Rotation into crops which are not attacked by *Striga* has two functions: (i) to interrupt the production of further *Striga* seed; and (ii) to allow and/or encourage the seed population in the soil to decline.

Any resistant crop, or a fallow, will achieve (i), but crops may vary in their effectiveness in reducing the seed population in the soil. Some so-called trap-crops are known to produce *Striga*-germinating stimulants but are not susceptible to attack (see Box 3.1, p. 136). These may therefore be preferable to others. A great many such crops have been reported, whose roots can cause stimulation of different populations of *Striga*. Parker and Reid (1979) showed, for example, that roots of cotton, cowpea, jute, soyabean, pigeon pea, chickpea, kenaf, and groundnut all stimulated germination of a Sudanese population of *S. hermonthica* at least as well as the natural host sorghum. Of the crops tested, only rice, tomato and sesamum were much less effective. However, using a sample from Nigeria, only cotton and jute caused germination comparable to that from sorghum while cowpea, pigeon pea and groundnut all proved relatively ineffective. There was the further important finding that a sample of *S. hermonthica* from millet responded totally differently and only germinated in response to cowpea and pigeon pea and hardly at all to cotton or groundnut (or sorghum – see Fig. 1.10). This helps to confirm that populations of *Striga* specific to millet are responding to different germination stimulants and are likely to require a quite different range of trap-crops. Additional crops reported effective on *S. hermonthica* from sorghum in Sudan (Andrews, 1945) included sunflower, lab-lab, lupin, lucerne, *Phaseolus* bean, pea and faba bean. There are relatively few reports, however, of their successful use in practice for control of *S. hermonthica*. In Uganda, Doggett (1965) showed little benefit in terms of *Striga* numbers or crop yield from a range of 5-year rotational treatments, involving cotton, velvet bean, millet, sunflower and mung bean. The one outstanding treatment after this period was the regular growing (**and hand-pulling**) of the susceptible sorghum though even this treatment, because of the very high initial infestation, still showed quite high *Striga* counts. Groundnut and cowpea were not included in this work because of unsuitability of the soil type. One of the few more recently published reports, from Burkina Faso (Ramaiah, 1981), showed little or no benefit from two years of cropping with cotton or groundnut on subsequent infestation of sorghum by *S. hermonthica*.

Results on *S. asiatica* have sometimes been more encouraging, in particular those in the USA published by Robinson and Dowler (1966),

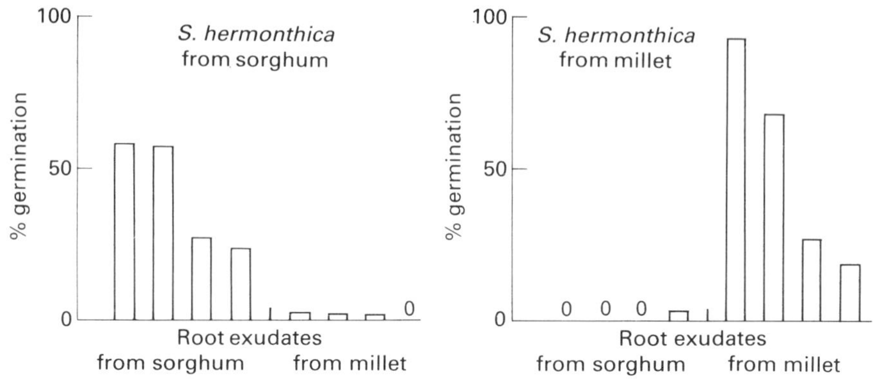

Fig. 1.10. Histograms showing the host specific germination response of *Striga hermonthica* seed from sorghum and millet hosts respectively to root exudates from four varieties each of sorghum and millet (from Parker and Reid, 1979).

who showed that after three years of rotation into pearl millet (not attacked in the USA), there were very large benefits in terms of both *Striga* numbers and maize yields compared with continuous cropping with maize. Soyabean and pea also gave good benefits but required four years to achieve an effect equivalent to that of millet after three years. Eplee and Norris (1990) have compared the performance of many potential trap-crops and shown cotton to be capable of providing up to 90% reduction of *S. asiatica* in a single season, and many others over 50%, but they were using artificial infestations.

The fact that at least three years of rotation are likely to be needed emphasizes the practical limitations of the technique. Few farmers will be prepared to give up growing their preferred cereal for a three year period, and in most infested areas, the choice of alternative crops will be extremely limited. Where a farmer is prepared to rotate then it should not be assumed that only the recognized trap-crops would be worth growing. **Any** non-susceptible crop which prevents new seed production will be preferable to the growing of a susceptible crop species in which new seeding is allowed. Choice of rotational crop should be based first on its suitability to the local conditions and only secondarily on its potential as a trap-crop.

Wherever rotation is used as a *Striga*-control technique, the land should be rested for at least two seasons if at all feasible as it is unlikely to show clear benefit in less. Any *Striga* seen growing on weeds or volunteer crop plants must of course be pulled. Even after several years there is still likely to be *Striga* persisting and farmers must be aware of this; but it should at least be less intense and feasible to manage by a combination of approaches as noted below.

Catch-cropping is quite distinct from trap-cropping in that it uses susceptible crops which are then destroyed or harvested before the parasite can reproduce (see Box 3.1, p. 136). In a monomodal rainfall system, catch-crops replace the normal cereal and are, in general, economically undesirable as they involve more cost than a fallow and may give no immediate return. In Rhodesia (now Zimbabwe) in the 1940s, however, it was shown that repeated catch-cropping with Sudan grass (*Sorghum sudanese*) twice per season was no more costly than green-manuring with sunn-hemp (*Crotalaria* spp.) and brought a severe *S. asiatica* problem under control in 4–5 years. Each thickly-sown crop was grown for about 6 weeks before being ploughed in. In some years a third catch-crop or a crop of sunn-hemp was also grown. The effect of this programme was to increase average yields of maize from less than 3–4 bags per acre in 1940 to over 13 bags per acre in 1944 (Timson, 1945).

In a bimodal rainfall area, especially where the first rains are relatively light and unreliable as in parts of Sudan and Ethiopia, there is scope for growing a catch-crop on the first rains. In parts of Sudan an abundance of wild sorghum (*S. arundinaceum*) may make it unnecessary even to plant. Otherwise, a thick planting of Sudan grass at 20–25 kg seed per ha should be sown and ploughed in, or perhaps harvested for forage, at 6–8 weeks, before there is any chance of *Striga* seeding. The main crop is then planted on the main rains.

Land preparation

Minimum or zero tillage was one of the main components of a *Striga*-control package tested by the UN Food and Agriculture Organization (FAO) over a number of years in West Africa. Matthews (1989) pointed out that theoretically, if tillage was avoided, the seeds of *Striga* would remain at the soil surface where the extremes of temperature and moisture would substantially eliminate their chances of attacking the crop, and infestations would then steadily decline. Although plausible in theory, and supported by some results in Niger, where it resulted in a modest reduction of infestation of *S. hermonthica* (Hess and Ejeta, 1987), the method did not work well in most of the trials that were done and it is no longer part of the FAO package. On certain light soils, where zero tillage is most practical, the technique should still be considered, though it may require the use of herbicide.

Time of planting

Where there is a monomodal rainfall pattern, the intensity of *S. hermonthica* emergence is usually highest in the earliest plantings. Delayed planting can therefore often result in reduced numbers of *Striga*. Any benefits in crop yield resulting from reduced *Striga* will almost certainly be counterbalanced by the tendency to reduced yield from delay in planting in such rainfall conditions. It cannot, for this reason, be regarded as a solution for extensive use. It can, however, be a useful way to reduce *Striga* numbers on the more seriously infested fields, or parts of fields, where good yields are in any case not expected and it is vital to use all possible methods to reduce numbers.

Experience with *S. asiatica* in South Africa is somewhat different and early plantings are often less severely attacked because of lower soil temperatures at this time.

Where rainfall is bimodal, *S. hermonthica* is normally worst in the crops sown on the less intensive phase of the cycle (e.g. on the early or 'belg' rains in Ethiopia) (see Plate 5B). This is because the weed is favoured by intermittent rather than continuous wetness. Hence farmers should be advised to plant in June/July, immediately before the more continuous rain. Varieties suitable for such later planting should of course be used. The first rains might be used for catch-cropping (see above).

Method of planting

There is conflicting evidence on the ideal density of planting for *Striga*-infested situations. Bebawi (1987) compared different planting densities and concluded that wide spacing of sorghum rows was advantageous in terms of reducing numbers of *S. hermonthica* per ha but his results showed that numbers of *Striga* **per host plant** were higher under wide spacing. Other observations and reports suggest that numbers of *Striga* per ha can be reduced at high crop densities, perhaps associated with the extra shading that then occurs (see discussion of shading and other effects of **mixed cropping** below). Provided the soil and climate are suitable, high planting density should probably be the aim.

Whatever the density of planting of a cereal it is highly desirable that it is planted in rows so that emerging parasites can later be systematically found and destroyed whether by herbicide or hand-pulling.

Nitrogen fertilizer

As noted above, it is a general observation that *Striga* on cereals is most abundant and damaging on soils of low fertility, and that the essential element for reversing this effect is nitrogen. As yet, however, there is no full understanding of the means by which high levels of nitrogen suppress *Striga*. The main effects are believed to be via: (i) reduction of stimulant exudation; (ii) direct damage to *Striga* seeds and seedlings in the soil; (iii) alteration of host root : shoot balance; (iv) reduced osmotic pressure in the parasite relative to the host; (v) a toxic effect of nitrogen on the developing *Striga*; (vi) increased shading by the crop.

Reduction of stimulant exudation from sorghum roots by moderate levels of N has been clearly confirmed *in vitro* by Taye Teferedegn (1973), Sherif and Parker (1988), Raju *et al.* (1990) and Cechin and Press (1993). Although there is evidence for damaging effects of urea and ammonium forms of N on *Striga* seedlings *in vitro* (Pesch and Pieterse, 1982; Pieterse, 1991) there is no clear confirmation that this occurs in the field, nor that there is a better effect of reduced N (ammonium or urea) compared with nitrate as these authors found. Nitrogen is known to modify the root : shoot balance in the host in favour of the shoots, which is quite opposite to the effect of the *Striga* itself as already discussed under 'Effects on the host'. This modification must presumably result in less resources being available to the parasite on the root system and lead to slower parasite growth. Reduced osmotic pressure in the parasite at the optimum N level has been noted in one study (Gworgwor and Weber, 1991), presumably reducing the flux of water and nutrition from host to parasite. This effect will warrant further study. Toxic effects of N on the developing *Striga* would seem to be quite likely, but there is no published evidence on how it might occur. Increased shading as a result of more luxuriant growth of crop shoots will result in higher humidity, lower temperature and reduced transpiration, again reducing the flux from host to parasite. The relative importance of these and perhaps other mechanisms awaits further study. Meanwhile, the reduction in stimulant exudation remains the best documented.

Some examples of the successful use of nitrogen to suppress *Striga* in the field include that of Hess and Ejeta (1987), working in Niger, who reduced numbers and weight of *S. hermonthica* on sorghum by 55–82% using 47 kg N ha^{-1} as urea. Mumera (1983) obtained at least 50% reduction of *S. hermonthica* emergence in maize in Kenya using 39 kg N ha^{-1} as calcium ammonium nitrate. Kabambe (1991) similarly reported 50–75% reduction of *S. asiatica* in maize in Malawi using 112 kg N ha^{-1} as ammonium sulphate. In Togo, Agbobli (1991a) achieved 53% reduction of *S. asiatica* and 132% increase in yield of maize with 120 kg N ha^{-1} as urea. Kim and Winslow (1992) record how the standard fertilizer applications on their maize/*S. hermonthica* research sites had to be modified, from 120 kg down to 60 kg N ha^{-1}, because the higher rate was resulting in a steady decline in *Striga* infestation. Last (1960a,b) obtained varying results with added N on sorghum in Sudan, depending on both variety and on the fertility of the soil. In general he obtained good reductions of *S. hermonthica* except where initial soil fertility was extremely low. Results were better the sooner the N was applied to the crop.

The results of field trials with basal applications of N fertilizer over many years in many countries have not, however, been consistent in terms either of crop yields or *Striga* numbers. As summarized by Pieterse and Verkleij (1991), moderate levels of N have often resulted in an increase in emergence of *Striga*. This has usually been apparent on the most nutrient-deficient soils and need not be surprising if we consider that, given a certain number of parasites attached to the root system, a greater proportion are likely to emerge on the more healthy crop which results from higher nitrogen. A more healthy root system may also tend to favour a larger number of *Striga* attachments, especially later in the season when the levels of N in the soil and/or the crop have declined and are no longer supra-optimal for the parasite. Hence moderate levels of applied N may not always be beneficial in reducing *Striga* numbers, especially those emerging late in the season, but they should normally result in some delay in *Striga* emergence and be beneficial to crop yield. Where 'high' levels of N have been applied there is usually a more distinct and consistent beneficial effect in reducing parasite numbers. Defining the necessary 'high' level, however, is not easy and such levels are often somewhat uneconomic, or even undesirable, if they involve the use of ammonium sulphate and the risk of increasing soil acidity.

The benefits of N can be judged in several ways. The most important from the viewpoint of an integrated *Striga* control system is the reduction in numbers of emerged and maturing *Striga* and hence a reduction in the seed produced, or the effort required to prevent seeding. The farmers' most immediate reason for investing in N, however, will be the prospect of increased yield. In general, it is rare not to achieve an increased yield where *Striga* is reduced. Conversely, however, *Striga* is not necessarily decreased when yields are increased. Apart from the intrinsic benefit of N on yield independently of the *Striga*, there can be two further beneficial effects: (i) reduced *Striga* attack; (ii) increased tolerance (decreased sensitivity) to the effects of *Striga* attack. Many workers have suggested that the latter is important (e.g. Osman *et al.*, 1991) but it is difficult to prove.

There is no clearly optimal method or time of application of nitrogen; consequently there can be no clear-cut recommendation for the control of *Striga* by nitrogen application. It is to be hoped that research will help define more clearly the mechanism by which it works and so lead to more refined recommendations. Meanwhile, on the assumption that most if not all the benefit from N arises **after** it has been absorbed and utilized by the host, it can only be suggested that nitrogen be applied in the form, by the method, and at the level, at which it is expected to have the greatest beneficial effect on host crop growth.

While it is hoped that farmers will have access to nitrogen fertilizers through subsidy or credit arrangements, it would be better still if means can be developed and implemented to restore the fertility of *Striga*-infested soils. All possibilities for the introduction of soil-improving legume species and perhaps agroforestry and alley-cropping techniques need to be explored.

Mixed cropping

Considering the frequency and importance of mixed cropping in the traditional farming systems in many *Striga*-infested regions, it is surprising that there was relatively little interest in the possible benefit of mixed cropping or intercropping since Saunders (1933) showed a considerable reduction in *S. asiatica* from intercropping maize with cowpea in the row. Now there have been several studies showing that *S. hermonthica* is significantly reduced in mixed crops in Gambia (Carson, 1989b) and Mali (Sallé *et al.*, 1987a); also in Cameroon and Kenya (see Plate 6A). No detailed study has yet been published to show just how mixed cropping results in reduced *Striga*, but several of the reports make it clear that intercropping with alternate rows of cereal and legume may have little beneficial effect, whereas if the intercrop is within the cereal crop row, it can be highly effective. It seems that there has to be a close association of the intercrop with the cereal, but why? Several possibilities have been proposed:

1. The intercrop is acting as a trap-crop and stimulating germination of the *Striga* ahead of the cereal roots.
2. The intercrop interferes with the stimulant exudation of the host.
3. The intercrop exudes an inhibitor to reduce *Striga* germination.
4. The shading by the intercrop reduces soil temperature, and hence the germination of the *Striga*.
5. The legume intercrop is fixing nitrogen and leaking sufficient into the soil to have a 'nitrogen effect'.
6. The intercrop is reducing air temperature and raising humidity over the newly emerged parasites, reducing the transpiration, and hence the supply of nutrition from the host.

Few of these possibilities have as yet been directly tested, except for the effect on soil temperature, which Carson (1989b) has shown may be reduced by 2–3°C. Unfortunately, related work does not support the idea that this reduction would be at all critical for *Striga* germination. In the laboratory, germination of *S. hermonthica* is only reduced quite gradually as temperatures fall from 30°C down to 23°C, not sufficient to explain the striking differences seen in the field. Perhaps under field conditions there is some interaction with other soil factors such as gas levels which makes temperature more critical, but this is only supposition at this stage. Of the other theories we currently favour the last; that is, the one concerned with humidity and transpiration rate. The importance of the permanently open stomata, and hence continuous transpiration in *Striga*, in keeping the weed cool under hot conditions has been well recognized and there have been attempts to exploit this using anti-transpirants (see below), but it has not been generally realized perhaps how important this continuous transpiration is to the nutrition and hence vigour and survival of the parasite. This would help to explain how the 'mixed crop' effect is shown by a range of crops and is also apparent with some especially leafy weeds, such as *Xanthium pungens* in Kenya, and by general weed growth in other observations. It would also explain how in Kenya it has been noted that numbers of *S. hermonthica* emerging on mixed crop plots were high and dieback occurred only after emergence. If this explanation can be confirmed it suggests that choice of intercrop species should not be critical provided they are leafy; also that some of the disadvantages of mixed cropping within the row, in terms of competition with the cereal, might be avoided by delaying the planting of the intercrop for a week or two, sufficient to allow the establishment of the cereal but still in time to cast a shade and raise humidity for the emerging *Striga* at 5–6 weeks. It also suggests the possibility of using other techniques to raise humidity, including inert mulches, and dense planting as referred to earlier.

Solarization

The technique of solarization is discussed in more detail under *Orobanche* (see Chapter 3). Osman *et al.* (1991) obtained complete kill of *S. asiatica* seeds at 2 cm depth in the soil, following 34 days under a polyethylene cover, when temperatures repeatedly reached 60°C. However, the crop was still infested by seeds from a deeper soil layer and it seems unlikely that this technique will be of interest in any but very exceptional situations, as costs would be high.

Fire

Burning the stubble of tropical cereals is not a common practice, often because the stover is of importance for animal feed. There is also rarely enough volume of material in a *Striga*-infested situation to create a really hot fire. It does, however, present one more option for reducing seed numbers in a badly infested situation. *S. hermonthica* often continues to grow on the stubble of the cereal for some weeks after harvest, and the burning of the *Striga* plants themselves will greatly reduce seed shed. Many of the seeds already shed onto the soil surface may also be killed. As *Striga* infestations are usually very patchy there is the option of collecting the available material onto the more infested spots before burning.

Resistant and Tolerant Varieties

Resistant varieties have long been seen as the main hope for control of *Striga*. If only we had such varieties of sorghum, maize or millet, all we would have to do would be to give or sell them to the farmer and he would need no further help. Unfortunately, after a great many years of searching, especially for resistant sorghum, even partially resistant varieties are available to relatively few farmers; but some progress has been made and we should see where we now stand.

Sorghum

Early developments Some of the earliest work was in South Africa where Saunders (1933) conducted a remarkably comprehensive study of *S. asiatica* on sorghum, including the identification of a number of different types of apparently mechanical resistance, leading to the development of the resistant variety Radar (Saunders, 1942). The variety Framida is also understood to have been selected in South Africa, though originating from West Africa (Williams, 1959).

Work in India (reviewed by Ramaiah, 1987a) also began in the 1930s and varieties of sorghum were selected with resistance based on low stimulant production, including the local Bilichigan (AS-4003) and Bonganhilo (AS-4693) from Africa, and the resistance incorporated into improved varieties by crossing with Co-1, AS-90, N-1 and T-6. One outcome was the variety Co-20. These were developed for resistance to *S. asiatica*. Bonganhilo was noted also to be resistant to *S. densiflora* but was later shown to be susceptible to *S. hermonthica*. A further programme by the Agricultural Station at Nandyal in Andhra Pradesh in the late 1940s was based on single plant collections of apparently resistant individuals. One of these, culture 109, was subsequently released as variety N-13.

In West Africa, 8000 sorghum lines were screened in the field by King (1975) and Zummo (1974) and resistant/tolerant lines selected by their lack of characteristic *S. hermonthica* symptoms, though none proved completely resistant in subsequent trials. The best of these, SRN-4841, had brown seed and is thought to have been close to, if not identical with, Framida. Further selection and breeding work was done by Obilana

(1983), especially with the variety L-187, and a number of varieties were promoted for distribution to farmers in *Striga*-infested localities. In East Africa, Doggett (1965) identified Dobbs as a partially resistant variety, supporting emergence of only 6–36% as many *S. hermonthica* plants as a susceptible line, and giving correspondingly higher yields. Whether these are still proving useful is not certain.

More recent work on *S. asiatica* A new phase of work on sorghum was begun by ICRISAT (International Crops Research Institute for the Semi-arid Tropics) in the 1970s, first in India and then in West Africa. A systematic screening of the world collection of sorghum germ plasm was initiated and eventually covered over 15,000 entries. This screening was based on the only clearly defined and readily detected resistance mechanism, namely low stimulant exudation. As a result, several hundred lines have been selected for their low stimulant character and these have been tested further, and some incorporated into the breeding programmes; but it is not clear from ICRISAT literature that any of these have made a significant impact.

Varieties already known to have resistance were meanwhile brought into ICRISAT's testing and breeding programmes in India. These included the low-stimulant varieties Framida and 555, and others, 'stimulant-positive', such as 148, N-13 and IS-2203. The work in India led to the development of a number of *Striga asiatica*-resistant ('SAR') lines with excellent potential in India, such as SAR-1, also known as ICSV-145, originating from a cross between varieties 555 and 168 (Vasudeva Rao *et al.*, 1989b) (see Plate 6C). The usefulness of others in this series, especially SAR 16, 19, 26, 29 and 35, has been confirmed for both *S. asiatica* and *S. forbesii* in southern Africa (Riches *et al.*, 1987; Obilana *et al.*, 1991).

S. hermonthica Much of ICRISAT's effort has been devoted to the problem of *S. hermonthica* in West Africa, but with less conspicuous success. The low-stimulant variety Framida (and the closely related SRN-4841) proved generally to have the most reliable resistance to *S. hermonthica*, though even this was somewhat inconsistent, especially in Sudan. Many of those lines and varieties with relatively good resistance to *S. asiatica* in India proved to be significantly more susceptible to *S. hermonthica*. N-13 shows good resistance to *S. hermonthica* in some trials, especially in Ethiopia, but is far from reliable. A number of breeding lines were developed, based on Framida, but with more acceptable white grain. One of these released in West Africa, ICSV-1002-HV, had good agronomic characters and some reduced susceptibility to *S. hermonthica* but was far from fully resistant (Ramaiah, 1985b). Work has continued with many other of the breeding lines and locally selected 'resistant' types, and promising local selections are being tested further. These include ICSV-1019-BF (Seguetana 5153) and ICSV-1020-BF (Seguetana niarabougou) in Mali ('seguetana' means *Striga*-resistant or *Striga*-tolerant locally) and ICSV-1018-BF (CSV-122) in Burkina Faso (Ramaiah, 1987a,b, 1991). SAR-24 (a derivative from Framida and 148) has shown relatively good resistance in Ethiopia (Fasil Reda, 1991) and a recent report suggests that SAR 29 has some useful degree of resistance to *S. hermonthica* in Tanzania (Obilana *et al.*, 1991) but in neither case has the resistance been fully acceptable. In Sudan, ICSV-1007-BF (also known as SRN-39) and IS-9830 have now been released as Mugawim Buda 1 and Mugawim Buda 2, respectively.

An especially interesting new line is IS-7777, which was selected for its low stimulation of *S. asiatica* and tested further in West Africa and found to have a very high degree of resistance to *S. hermonthica* also (Ramaiah, 1991; Olivier *et al.*, 1991b). The level of resistance of IS-7777 to *S. hermonthica* appears from published reports so far to be uniquely high and perhaps to offer a source from which more fully resistant sorghum varieties might be developed. The line itself is agronomically very poor and breeding will be needed. Nothing is known yet of the genetics of this new source of resistance. Other lines noted by Ramaiah (1991) to have been selected for low stimulant character and to have promising resistance to *S. hermonthica* include IS 7739 from Nigeria; IS-14825, IS 14928 and IS 16184 from Cameroon; IS-1840 from Uganda; and IS 6961 and IS 9830 from Sudan. Most have poor agronomic characters.

Resistance mechanisms in sorghum Saunders (1933) described a number of different ways in which different sorghum varieties resisted the full and effective penetration of the parasite to the host stele. The first involved an apparent failure of the parasite to readily penetrate even the cortex, perhaps because of enzymic problems in separating the host cells. Saunders did not regard this as a mechanical resistance. Occasional individuals reaching the stele developed normally. In another type the parasite had particular difficulty penetrating the endodermis, while in the third there appeared to be a reaction within the stele, which prevented further growth. The variety Radar had a form of resistance in which germination and attachment of *Striga* occurred but full penetration and development was prevented. Doggett (1988) concluded the same was true for Dobbs.

In India, Kumar (1940) was able to explain the resistance of the local variety Bilichigan to *S. asiatica* in terms of low stimulant production, and Williams (1959) later showed that the same was true for the variety Framida. When ICRISAT started its systematic screening of the world sorghum collection, this was the only mechanism which could be readily sought. But it was also known quite early that the variety N-13 produced abundant stimulant and must have some other explanation of its widely confirmed field resistance. In the first of a number of studies on this variety it was shown that the primary root would allow penetration of *S. asiatica* only up to 7 days, while that of the highly susceptible hybrid CSH-1 would continue to allow penetration up to 15 days (ICRISAT, 1979). Further work by Maiti *et al.* (1984) presented evidence of a number of anatomical characteristics with which the resistance of N-13 and several other varieties to *S. asiatica* was apparently associated. These included relatively thick inner tangential walls in the endodermis, layers of thick, lignified cells in the pericycle and large silica crystals in the endodermis. Other possibilities for the resistance of N-13 to *S. hermonthica* were explored by Dixon and Parker (1984) without conclusive result. The resistance of N-13 to *S. hermonthica* is in any case much less reliable than the resistance to *S. asiatica* and appeared to be very much influenced by other factors, including nitrogen fertilization and perhaps the rhizosphere microflora. A possibility suggested by Dixon and Parker was that some varieties may avoid heavy attack through having a much simpler, less branched root system. This was explored further by Cherif-Ali *et al.* (1990), whose results showed some support for this idea for the variety P-967083, but not for N-13.

Other possible mechanisms of resistance which were explored by Dixon and Parker included the lack of chemotropic or other directional stimulus by which the parasite locates the host root. They showed that theoretically, without such a stimulus, relatively few *Striga* seedlings would make contact with a host root, and the host would have effective partial resistance. They could not, however, find any evidence for differences in chemotropism between varieties; nor could they show that there were differences in ability of different varieties to stimulate haustorial initiation. It is still possible, however, that resistance based on such mechanisms could yet be found.

The most recent work has been with the variety IS-7777, which was selected as low-stimulant for *S. asiatica* from the routine screening at ICRISAT and passed on to WRO for further testing. In spite of being low-stimulant for *S. asiatica*, it was found to cause normal stimulation of *S. hermonthica*. Olivier *et al.* (1991a) have now shown that the resistance of this variety to *S. hermonthica* involves the development of cellulosic layers in the cortex which prevent penetration to the stele.

Genetics of resistance in sorghum In detailed studies with *S. asiatica* Saunders (1933) used several stimulant-positive resistant lines of sorghum in a breeding programme but apparently came to no clear conclusion concerning the genetics of resistance. There was an indication that resistance was incompletely dominant in one case, but susceptibility dominant in two others. He comments that 'the inheritance of resistance is decidedly complex'. There have been a number of similarly inconclusive studies on sorghum and *S. asiatica* in India, but it has now been concluded that the resistance based on low stimulant production is associated with a single recessive gene (Ramaiah *et al.*, 1990).

Sorghum – the current situation The outcome of great efforts on *Striga*-resistance in sorghum over the past decades has so far been somewhat disappointing. For *S. asiatica*, the SAR lines are proving valuable but for *S. hermonthica* there are still no varieties with immunity, or even a high level of resistance reliable over any wide area. Locally, as in Sudan, new releases are reported to be performing well, so it is to be hoped that the investment by ICRISAT and others is just now beginning to bear fruit. Meanwhile, there are usually varieties which have been identified as being much less susceptible and which should be made available to farmers. Some examples include the varieties CS-54, CS-95 and S-35 in Cameroon, Gambella-1107 in Ethiopia and a range of 'seguetana' types in Mali. The use of such less susceptible varieties, apart from giving improved yield, will usually help to reduce the numbers of *Striga* emerging, and hence the task of removing them to stop seeding. Further useful varieties should emerge from conventional breeding and selection programmes in the next few years but none is likely to be fully immune, and they will need to be introduced as components of integrated packages if there is to be minimum risk of 'breakdown' in their resistance. Fully immune varieties, if they are achievable at all, will almost certainly depend on new techniques of genetic engineering and will take many years to develop.

Millets

We are not aware of any work on varietal resistance in any of the millets other than pearl or bullrush millet (*Pennisetum americanum*). Hence any use of the term 'millet' in this and most other sections of this book refers to this species.

Some reports from India suggested that there was a high level of resistance or even immunity in the inbred TF-23A and hybrid lines of pearl millet derived from it (Krishnaiah *et al.*, 1968; Mathur and Bhargava, 1971). It is believed that this line did not prove resistant to *S. hermonthica* in Africa but other work has repeatedly shown some apparently useful variation in susceptibility among available varieties. Attempts have been made to purify and fix the resistance character, but to date there has been no practically useful outcome, perhaps largely because of the out-crossing breeding system and the corresponding difficulty of maintaining purity of lines. Parker and Wilson (1983) demonstrated in pot experiments a high level of resistance in a range of semi-wild 'shibra' lines selected from the field in Niger. This resistance was near complete for a number of millet-attacking populations of *S. hermonthica* from West Africa, but subsequent testing in the field failed to confirm their promise. The source of this resistance in the shibra types is not explained as the presumed wild parent of these out-crossed rogues, *P. violaceum*, proved to show susceptibility. Other common wild *Pennisetum* species, including *P. ramosum*, *P. polystachion* and *P. pedicellatum* are generally immune, but do not cross readily with the crop species. ICRISAT is continuing its search for useful resistance to *S. hermonthica* in West Africa and we look forward to hearing of good progress.

Maize

It has often been assumed and stated that, because maize evolved in the New World and was not exposed to *Striga* until recent centuries, there was no likelihood of any useful resistance having developed. The validity of this assumption might be challenged and could only apply to gene-for-gene resistance of the wheat rust type; but until the past 10 years, there was indeed little evidence of any useful level of resistance in spite of intensive efforts by Saunders (1942) in South Africa and later, largely unpublished, work in Nigeria. Langston and English (1981) first showed high levels of resistance to *S. asiatica* in some maize lines in the USA, but results were not readily confirmed from one year to another. The International Institute for Tropical Agriculture (IITA) then, by chance, observed interesting differences in the susceptibility of maize lines being tested for resistance to streak disease and identified a number of white- and yellow-grained inbreds and hybrids, showing much reduced damage from *S. hermonthica* (Kim *et al.*, 1985, 1987; Efron *et al.*, 1989). It was not clear at first whether the reduced damage was due to reduced attack (resistance) or reduced sensitivity to damage (tolerance). It now seems that most of the difference is in tolerance, with numbers of *Striga* emerging often comparable with those on standard susceptible varieties; but there are indications of reduced *Striga* numbers in some trials. The most widely tested hybrid is the white-grained 8322-13 (from inbreds 1368 × 9030, the latter, also designated TZi-12 from TZSR × N-28, being the 'resistant' parent). Tolerance is far from complete but when yields of a susceptible hybrid are reduced by 90%, those of the selected hybrids are reduced by only 60% (Kim and Winslow, 1992). Ransom *et al.* (1990) have confirmed that one of the inbreds identified in the IITA programme, TZi-30, has genuine resistance to *S. asiatica* in the USA, *Striga* emergence counts reaching only about 10% of those on a susceptible variety. A hybrid Pioneer 3181 also shows high resistance. Further work is now in progress to refine our understanding of these materials and the genetic basis for tolerance, which so far appears complex (Kim, 1991b). It is hoped that the physiological basis for tolerance might also be identified so that it can be enhanced, or introduced into other cereal varieties.

IITA's work has provided maize varieties with tolerance and/or resistance which will in due course offer useful options to the farmer. Any degree of resistance will help to reduce the numbers of *Striga* emerging, while tolerance, wisely used, also has value as discussed in Box 1.4.

Box 1.4. The value and limitations of tolerance versus resistance.

The term **tolerance** is here used for the reaction of varieties that are parasitized to the same extent but suffer less damage than standard varieties. The converse of tolerance is **sensitivity**. The term **resistance** applies to varieties showing less attack, usually in terms of numbers of parasite attached or emerged. Resistance is rarely complete and, where necessary, the term *partial* **resistance** implies significantly less attack compared with standard varieties ('tolerance' is sometimes used wrongly for this reaction), while total resistance is usually referred to as **immunity**. The converse of resistance is **susceptibility**. Tolerance may or may not be combined with a degree of resistance which results also in reduced numbers of parasite attaching and/or emerging.

The idea of screening sorghum for tolerance to *Striga* was discussed on occasions by staff of ICRISAT but the conclusion was that it was undesirable, on the assumption that tolerant varieties would encourage prolific emergence and seeding of *Striga*. Now a high degree of *Striga*-tolerance has been identified in maize, and the potential for such varieties can be assessed. The indiscriminate growing of tolerant varieties could be disastrous if it allows as many or more *Striga* to emerge and seed normally; even the most tolerant maize varieties are damaged by very high populations of *Striga*. However, combined with other methods to prevent seeding, they offer the farmer a valuable option in the form of a crop that can be grown economically in infested land and support the cost of other methods such as directed 2,4-D spraying. Furthermore, a tolerant variety (plus prevention of seeding) that stimulates many parasite seeds to germinate is advantageous in terms of reducing the seed bank in the soil. In contrast, a resistant variety in which the resistance is due to low stimulant exudation will tend to preserve the seed bank.

Tolerance has been recognized to some extent in cowpea varieties, in relation to *S. gesnerioides*, and in faba bean, in response to *Orobanche crenata*, but in neither case have tolerant varieties been deliberately developed or promoted. Tolerance to *Cuscuta* species or to mistletoes has not been clearly identified.

Mechanisms by which varieties show greater tolerance than others have been little studied but are likely to receive greater attention in future, especially where there seems little chance of finding immunity, and a combination of partial resistance and tolerance will be most desirable. In one of the few discussions of the topic, Wegmann *et al.* (1991) suggest that tolerance of faba bean to *O. crenata* may be associated with high osmotic pressure in the host. In the case of *Striga*, there are suspicions that high nitrogen increases the tolerance of the host as well as reducing attack. As part of the damage to the host is thought to result from the drain of nitrogenous compounds from host roots, this possibility deserves further study.

Rice

Some unpublished observations on pot experiments at the Weed Research Organization suggested that there might be considerable variation in susceptibility to *Striga* among rice species and varieties. This has now been confirmed in a striking manner by Harahap *et al.* (1992) who compared a range of rice lines (all *Oryza sativa*) in a series of field experiments in Kenya and identified seven with a very high degree of resistance to *S. hermonthica* (less than 2% of *Striga* dry weight compared with susceptible checks). Two of these appeared to show complete immunity, namely IR-38547-B-B-B-7-2-2 and IR-47697-4-3-1. Further reports are awaited with interest.

Sugarcane

There have been few studies of *Striga*-resistance in sugarcane, perhaps because the crop is not very commonly subject to serious attack, but it is clear that there is very useful variation in susceptibility to both the main *Striga* species. Robinson and Stokes (1960) tested a range of sugarcane varieties in the USA and found wide variation in response to *S. asiatica*, Co-290 proving the most 'tolerant' and Co-281, CP-33-224, CP-50-28, CP-53-19 and F-31-762 showing moderate tolerance. Visser (1987) compared five South African sugarcane varieties for their response to *S. asiatica* and found N-17, 52/219 and 75-F-2573 much less susceptible than N-14 and N-18. The only reported study on *S. hermonthica*, in Kenya, shows Co-617 to be highly resistant, although still allowing some infestation. Others with nearly equal resistance included KEN-83-1228, KEN-83-538 and KEN-83-1161 (Mbogo and Osoro, 1991).

Stability of resistance

Once a relatively resistant crop variety has been identified, and assuming that it is acceptable in agronomic and other characteristics, two further questions have to be answered. Firstly, will this resistance be stable in space (geographically) – i.e. will the variety prove resistant in other localities, to other *Striga* species, and to other 'strains' of the species first tested? Secondly, will it be stable in time (durable) – i.e. over a period of years when grown repeatedly on the same site? The first of these questions has been addressed already and though there is no clear answer, the evidence so far is encouraging, and the answer should become apparent relatively quickly from routine multi-location testing, before wide-scale release.

Cross resistance to different *Striga* species In general, the sorghum varieties with best resistance to *S. asiatica* also prove to be best for *S. hermonthica*. Chromatographic separation of root exudates at the Weed Research Organization showed very similar patterns of peaks for *S. asiatica* and *S. hermonthica* when they were used to assay HPLC series from sorghum, maize and millet, and it could be generalized that varieties with resistance based on low-stimulant exudation are most likely to show cross-resistance to the other main species. However, there is a strong tendency for *S. hermonthica* to be more virulent so that a variety highly resistant to *S. asiatica* may be only partially resistant to *S. hermonthica*. Occasionally, the reverse seems to apply and a variety may prove susceptible to *S. asiatica* while resistant to *S. hermonthica*; but this is less common (Ramaiah, 1987a). Varieties resistant to *S. asiatica* also appear to show cross-resistance to *S. forbesii*. Obilana *et al.* (1991) found five SAR lines to have dual resistance to both *S. asiatica* and *S. forbesii* in Zimbabwe. Two of these, SAR-19 and SAR-29, also showed resistance to *S. hermonthica*, while SAR-16, SAR-26 and SAR-35 did not.

When other resistance mechanisms are involved, good resistance to *S. asiatica* may also be reflected in relatively good resistance to *S. hermonthica*, as in the variety N-13. But this is not always so as was noted by Doughty in 1941. He tested Saunders *S. asiatica*-resistant sorghum varieties from South Africa (all stimulant-positive), and found them all susceptible to *S. hermonthica* in Kenya.

Evidence for the behaviour of *S. densiflora* is conflicting. In pot experiments at WRO the ranking of a range of sorghum varieties (including Framida) for resistance to *S. densiflora* was remarkably similar to that for *S. asiatica* or *S. hermonthica*. Hence it appears that, in general, varieties resistant to *S. asiatica or S. hermonthica* may also show resistance to *S. densiflora*, but the latter has sometimes shown differences from *S. asiatica* in the range of varieties attacked (Parker *et al.*, 1977) and it would be unwise to assume that it would always be so. Among wild para-sorghum species with a chromosome number (2n) of 10, *S. versicolor*, *S. purpureo-sericeum* and *S. nitidum* are resistant to both *S. asiatica* and *S. densiflora*, while *S. conspicuum* (2n = 20) is resistant to *S. densiflora* but not to *S. asiatica* (Deodikar, 1951). Perhaps it will be shown that there are important differences between different populations within this species.

The cross-tolerance of the promising maize lines has not yet been clearly established. In the USA Ransom *et al.* (1990) found the maize inbred TZi-30 resistant to *S. asiatica*, as reported for *S. hermonthica* (Kim *et al.*, 1987). However, TZi-1, also listed as 'resistant' to *S. hermonthica*, supported normal emergence of *S. asiatica* in the USA. This could result from different response to the two species but it could also be that this line has strictly tolerance only, which was not detected so readily in the US study. There is now some evidence for tolerance of these lines to *S. asiatica* in Africa, but it is known that they are sensitive to *S. aspera* (Kim and Winslow, 1991, 1992).

Cross-resistance to different *Striga* strains Apart from differences in the reaction of resistant varieties to different species of *Striga*, there is the extremely important question of their different response to different 'strains' or populations within a single *Striga* species. As the resistance in sorghum varieties is never complete and varies greatly from trial to trial, it has often been suspected that such variation in response is due to intraspecific variation in the *Striga* and that there was thus the danger of 'resistant' varieties failing when moved from one locality to another. While some pot experiments comparing samples of *Striga* from different localities on a range of sorghum varieties have suggested this to be true, it has proved difficult to confirm statistically that the ranking of varieties according to their relative resistance differs according to the *Striga* sample used. On the basis of evidence available up to that time, Vasudeva Rao and Musselman (1987) concluded there was no good evidence for such 'intracrop-specific strains', though their existence was not ruled out. Unpublished work from the Weed Research Organization and Long Ashton Research Station also suggested that sorghum varieties selected for relatively good resistance to a single sorghum-attacking population of *S. hermonthica* or *S. asiatica* should have relatively good resistance to most, if not all, other populations within each of those species, though not necessarily to those of the other species. It may yet be established that the relative resistance of a variety depends on the particular population of parasite to which it is exposed. However, most variation in the behaviour of 'resistant' varieties from site to site, we believe, is attributable to the very substantial, and as yet poorly understood, influence of environmental conditions on the expression of resistance.

Durability of resistance over time An answer to the question of durability with time has rarely been sought and should be given a great deal more attention than it has received in the past. Loss of resistance in the sorghum variety Radar in South Africa, reported by Grobbelaar (1952), may have been due only to occasional out-crossing of the host, leading to the loss of purity of the variety. It may, therefore, have nothing to do with any change in the virulence of the parasite; but the variety does still show resistance in parts of Botswana, where it has presumably not been grown before (Riches 1989a). We have to consider the possibility that where we have incomplete resistance to a highly variable parasite such as *Striga*, there must surely be a risk of selecting a more virulent population. Why is a small number of parasites able to attack the 'resistant' variety? There are three possibilities. If the resistance is based on low stimulant exudation, the successful parasites may represent that small proportion of the population close enough to the root to receive a critical dose of stimulant. This should not create any selection pressure for change in the parasite. If the successful individuals are able to respond to a lower threshold level of stimulant this could create an undesirable selection pressure towards a population better able to attack the 'resistant' variety. (They will then also be more susceptible to the use of trap-crops and the risk of germinating too far from the root when a standard cereal variety is grown, but the overall result is undesirable). Thirdly, the successful individuals may be responding to a 'new' or different stimulant from the variety, which would lead to a serious selection pressure so long as that variety continues to be grown. In the case of 'stimulant-positive' varieties, we still have inadequate understanding of the resistance mechanism. If it is purely mechanical in the older roots, then successful individuals may only be those closest to the root, which have the best opportunity to penetrate before mechanical barriers develop. This might not result in selection pressure for change. But with most other conceivable resistance mechanisms, we must assume that successful individuals differ in their ability to establish and grow on the resistant variety, and there must consequently be a dangerous selection pressure.

It is long overdue that *Striga*-resistant varieties intended for release or already released should be deliberately tested in simple long-term trials for the risk of selecting more virulent *Striga* populations. It may prove that, especially with the genetically less variable autogamous species, such as *S. asiatica*, any change in virulence is very slow or negligible; but with *S. hermonthica* it could become apparent quite quickly that a change was occurring. This would not necessarily invalidate the usefulness of the variety but would emphasize the importance of an integrated package involving supplementary treatments to eliminate or minimize *Striga* seed production.

The risk of build-up of more virulent parasites should in any case be explained to all farmers so that, whenever it is at all feasible, seeding of the *Striga* is prevented. This principle will apply not only to the introduction of resistant cereal varieties, but also to the change-over of one cereal species to another that is initially resistant, especially from sorghum to maize. Initial infestations of *S. hermonthica* in maize in northern Cameroon were very low, and if only it had been controlled by relatively easy hand-pulling, the problem would have taken many more years to develop, or might have been kept under control indefinitely.

Chemical Control Methods

Fumigants

Although fumigation will rarely have any place in practical field-scale control, because of its very high cost, it has proved an important part of the eradication programme in the USA, where it has been used for cleaning up small infestations in situations where other methods would be less easily applied (Eplee, 1992). The compound of choice has been **methyl bromide**, applied at about 400 kg ha^{-1} under plastic covers to retain the gas for at least 24 hours. Soil has then to be ventilated for 7 days before planting crops. **Dazomet** has also been used as a simpler though slightly less reliable alternative, applied at 320 kg ha^{-1} as a dry formulation. This can be used without plastic covers but it is then important that it receives at least 20 mm of rain or irrigation within a few days to wash it into the soil. Planting of crops has to be delayed for 30–60 days.

Germination stimulants

Ethylene This was first shown by Egley and Dale (1970) to stimulate germination of *S. asiatica* and this gas was soon put to practical field use in the USA as a means of causing 'suicidal' germination of *Striga* and so accelerating the depletion of the seed bank in the soil. The development of this important component of the US *Striga*-eradication campaign is well summarized by Egley *et al.* (1990). The dose needed was found to be not more than 1–2 kg ha^{-1} and in the light sandy soils of the Carolinas the gas could spread up to 1 m laterally and downwards from the point of injection. Hence tines on large scale equipment are spaced 1 m apart, and set to inject at about 20 cm depth. Back-pack kits for injection have also been developed (see Fig. 1.11) which require one injection per m^2. The timing of application has to be when the *Striga* seeds have had the opportunity for normal conditioning (i.e. at least one week of moisture) and temperatures are at least 25°C.

Although *S. hermonthica* was soon shown also to be stimulated by ethylene, the possibilities of it being effectively controlled in the field have not yet been fully confirmed. The limited published results suggest that it may be less well controlled than *S. asiatica* (Bebawi and Eplee, 1986; Ransom and Njoroge, 1991), but whether this is due to an intrinsic difference between the species, or to the heavier soils under which the African tests have been conducted, is not certain. It deserves further experimentation as it is a treatment which, although too expensive for most individual farmers, might be subsidized by governments or aid projects.

Fig. 1.11. Back-pack kit for manual injection of ethylene.

Ethephon The ethylene-releasing compound ethephon (2-chloroethylphosphonic acid) was shown to be effective in stimulating germination of *S. asiatica* (Egley and Dale, 1970), and also *S. hermonthica* (Chancellor *et al.*, 1971). However, it was shown to be much less effective than ethylene in the field, needing high doses of up to 35 kg ha^{-1} and also requiring soil pH to be raised above the natural levels of 5–6. Ethephon is stable under acid conditions and fails to break down and release ethylene unless conditions are alkaline. Field testing of ethephon against *S. hermonthica* has occasionally shown some effectiveness (Babiker and Hamdoun, 1983) but as soil pH in most affected areas is usually acidic, this is unlikely to be a widely useful treatment.

Strigol analogues Following the identification of strigol as the natural stimulant from cotton, there was interest in the possibility of using this or related substances for inducing suicidal germination in the field. Strigol itself proved extremely expensive to synthesize, so the International Development Research Centre sponsored a project aimed at the synthesis of simpler analogues; this quite quickly yielded a series of substances known as GR compounds (after the chemist involved, Dr G. Rosebery) with very high activity on several *Striga* and *Orobanche* species (Johnson *et al.*, 1976, 1981). These compounds have subsequently been tested in pot and field experiments on *S. asiatica* in the USA (Norris and Eplee, 1983) and on *S. hermonthica* in Sudan (Babiker and Hamdoun, 1982) but without fully satisfactory results. To be effective the compound must reach the seeds at depth in the soil, which is not easily achieved with a very insoluble material. A further problem has been the instability of the compounds in soil, especially at high pH. Attempts have been made to look for alternative structures (see e.g. Vail *et al.*, 1985, 1990a,b) or to modify structures to achieve more stability (see e.g. Zwanenburg *et al.*, 1986), but latest studies (Mangnus, 1992; Mangnus and Zwanenburg, 1992) suggest that the instability of the strigol analogues may be an intrinsic characteristic of structures with stimulant activity. As a result of these difficulties and because the compounds would have a market limited to control of parasitic weeds, there has so far been no commercial development; and at the time of writing none is anticipated.

Natural germination stimulants These have been detected in a range of plant materials, including sorghum root residues (Chidley and Drennan, 1987b) and in a range of *Euphorbia* species (Ibrahim *et al.*, 1985). No way has yet been devised for making practical use of these but the possibility deserves further exploration.

Seed hardening

Bharathalakshmi and Jayachandra (1980) reported that alternate wetting and drying of sorghum seeds in solutions of certain phenolic acids would confer resistance to *S. asiatica*, but recent attempts to confirm this result have been unsuccessful (Awad, 1989; Friesen and Korwar, 1991).

Herbicides

Herbicides to prevent attack Ideally, a herbicide should control *Striga* before or immediately after attachment so that damage is prevented. Many herbicides have been screened for this type of action but even with good soil incorporation to maximize the chances of success, very few have shown any promise at all. **Chlorfenac** looked possible in maize but the selectivity was too narrow for practical usefulness. Among large numbers of herbicides screened in the USA (Dowler and Robinson, 1990; Langston and English, 1990a), and in the UK (Kasasian and Parker, 1971), the phenoxyalkanoic acids, including **2,4-D**, had high activity at the germination stage. Both 2,4-D and **MCPA** have occasionally been shown to prevent *Striga* attack and reduce damage to the crop, but success seems to depend on irrigation or rainfall washing the chemical into the soil at the critical time of *Striga* germination, 2–3 weeks after crop planting, and many trials have shown no benefit at all. Parker (1974) showed that in addition to some non-commercial benzoic acid compounds, **benazolin** at about 2 kg ha^{-1} had promising selectivity in both maize and sorghum, but was likely to be too expensive for practical use. Perhaps this compound now deserves reappraisal. Recent work with the sulfonyl ureas has not always shown adequate selectivity in sorghum, but Babiker and Fasil Reda (1991) report selective control of *S. hermonthica* with **chlorsulfuron** at 1–2 g ha^{-1} in Sudan. Two related compounds, **nicosulfuron** and **DPX-E9636,** have given very promising results in maize (Adu-Tutu and Drennan, 1991) and deserve further testing.

An alternative to a soil-acting herbicide which has to reach the parasite directly from the soil, is one applied to the host which is then translocated to the roots and exuded to damage the parasite at the point of attachment. 2,4-D and a few other compounds gave disappointing results when applied only to the foliage of sorghum (Parker, 1974). **Dicamba** has sometimes been used successfully in this way to control *S. asiatica* on both sorghum and maize in the USA (Eplee and Norris, 1987a,b) but results have not been consistent, even on *S. asiatica*, and crop yields have not been much improved. Results on *S. hermonthica* have been even less promising and where some success has been achieved it has apparently been from leaching of the chemical into the soil rather than a translocated effect.

The principle of such a translocated herbicide remains attractive as it is known to work for *Orobanche*, but Samb and Chamel (1992) have shown how translocation of dicamba from millet into *S. hermonthica* is very low compared with that from cowpea into *S. gesnerioides*. They suggest that this is associated with the lack of any phloem bridge in the cereal/parasite haustorium and its presence in the cowpea/*S. gesnerioides* connection. This may mean that results can never be as good with *Striga* on cereals as with *Orobanche* (which has a good phloem bridge with its host), but it will still be extremely interesting to see if **glyphosate** or other herbicides will provide a good technical solution, once genetically modified cereals such as maize are available with suitable herbicide-resistance. Such modified crops may in any case be useful if they allow freer use of potent herbicides for the knock-down of emerged *Striga*, but the advantage will be greatest if they allow early use of translocated herbicide such as glyphosate or imidazolinones, which will kill *Striga* at an early stage and so prevent damage altogether.

Herbicides to prevent *Striga* emergence Although no soil-acting herbicides for genuine prevention of *Striga* attack are being currently used, even in the USA, several compounds are being used to prevent *S. asiatica* emergence from the soil (Eplee, 1992). **Trifluralin** is effective when incorporated shallowly in a layer above the cereal seed. *Striga* attaches to roots below this level but shoot growth is inhibited when the parasite grows up into the treated layer. **Pendimethalin** provides control in the same way following pre-emergence application, while **oxyfluorfen** is applied later as a directed post-emergence treatment with the same effect. These compounds are useful for preventing *Striga* emergence and so reducing the need for other methods to prevent seeding, but they may do relatively little to protect the crop from damage. Furthermore, with the possible exception of oxyfluorfen, there has been no confirmation of their effectiveness on *S. hermonthica*.

Post-emergence knock-down herbicides The simplest use for herbicide is as an alternative to hand-pulling for destruction of the emerged weed. The cheapest and most widely used has been **2,4-D**, an important component of the routine eradication campaign in the USA. This is applied at about 1 kg ha^{-1} plus non-ionic surfactant as a directed spray between maize rows, repeated as necessary through the season. Its effectiveness has been confirmed in trials against *S. asiatica* and *S. hermonthica* in India and Africa, though there has been no extensive routine use outside the USA. Some reports suggest a yield benefit even when the weed is being destroyed long after emergence, but this may depend on special conditions and cannot be relied upon. For scattered, light infestations, the spray can be applied by a pistol-grip hand sprayer rather than a conventional sprayer (Ogborn, 1972b). The hand-sprayer is adjusted to spray about 1 ml of dilute herbicide per shot and can be used to spray plants up to 2 or 3 m away. It is not, however, practical or economic for populations much over 10,000 ha^{-1} (1 m^{-2}.)

Paraquat is also used in the USA at 0.25 kg ha^{-1} either alone or in mixture with 2,4-D for quicker and more complete control. The same combination has been shown to have advantages over 2,4-D alone for control of *S. hermonthica* in the Gambia (Carson, 1988), but the use of this highly toxic herbicide must depend on total confidence in the ability and understanding of the local workers to use it safely. **Glyphosate** has also been shown to be effective as an alternative or additive to 2,4-D but high cost and greater risk of crop damage make this a questionable practice.

One of the serious disadvantages of 2,4-D is its lack of safety on non-cereal crops, precluding its use in most mixed crop situations and even in sole-crop cereal in cotton-growing areas. Ogborn (1987) tested a number of alternatives and suggested **ametryne, linuron, cyanazine** and **bromoxynil**. Bromoxynil is no longer regarded as safe enough for application by hand-operated equipment and the others have not always been confirmed effective by other workers. Further screening may be needed in the African context, but some compounds promising in USA work (apart from oxyfluorfen which can also be used for knock-down) have included **fomesafen, acifluorfen** and **lactofen** which are safe enough to use in soyabean and some other legume crops (Langston and English, 1990a). These may, therefore, be suitable for mixed cropping as well as for control of *Striga* on grass weeds in rotational crops.

In the USA herbicides are used intensively in rotational crops, as well as in maize, for control of grass weeds on which *S. asiatica* might grow and seed. These are usefully summarized, together with those used in maize itself, by Langston *et al.* (1991).

Most farmers will not for the time being be able to take advantage of herbicides, but the cost of 2,4-D is low and there should be many situations where it is feasible for it to be introduced, especially as an aid to prevent seeding. Other more expensive compounds will have less widespread application but are likely to prove useful at least locally.

Antitranspirants

In common with most parasitic plants, *Striga* species have high transpiration rates, associated with stomata which remain open under most if not all conditions (Shah *et al.*, 1987). In exploring this characteristic in *S. hermonthica* it was by chance found that the high rate of transpiration is important not only for ensuring a strong flow of sap and associated nutrients from the host xylem, but was also important for cooling the parasite under hot conditions. Press *et al.* (1989) showed that *S. hermonthica* is less resistant to heat stress than sorghum, and when transpiration was prevented by use of a commercial antitranspirant, the weed collapsed and died. Field trials in Sudan confirmed that *S. hermonthica* could thus be severely damaged by the application of **di-1-*p*-menthene** ('Wiltpruf') at normal recommended dose, and yields of sorghum were, rather surprisingly, significantly increased. In subsequent trials in Nigeria and Kenya the treatment has not been so effective but this remains a potentially interesting method for areas with suitably high temperatures, especially as an alternative to 2,4-D for use in mixed cropping. Work has continued to find a cheaper alternative to 'Wiltpruf' but no reports have emerged as yet.

Biological Control Methods

By insects

The genus of greatest interest for biological control is *Smicronyx*, of which several species are thought to be highly specific to *Striga* (as are other species to *Cuscuta*). Other potentially important species are *Ophiomyia strigalis* (an agromyzid fly) in Africa and *Eulocastra argentisparsa* and *E. undulata* (moths) in India. Most of these insects suffer from their own parasites and rarely cause any high degree of control in their indigenous areas. However, galling of fruits of *S. hermonthica* by *Smycronyx* in West Africa is often very heavy, and seed production of the weed may be reduced by 80% or more. The possibility of augmentation of *Smicronyx* to enhance this quite significant natural control within West Africa has not yet been attempted; but it can be pointed out that when flowering *Striga* plants are collected, as in recommended hand-weeding practice, there can be a large population of *Smicronyx* larvae emerging from the *Striga* plants as they dry. It should be worth considering how these larvae might be preserved rather than destroyed with the *Striga* plants, so that they are available for re-infesting *Striga* the following season.

The possibilities for deliberate use of these insects for 'classical' biocontrol, by introduction from one region to another, were reviewed by Greathead (1984) who concluded that there is potential for the Indian species to be effective in Africa and vice-versa. Only one practical attempt has already been made, with the introduction from India to Ethiopia, in 1974 and again in 1978, of *S. albovariegatus* and *E. argentisparsa*. The second of these releases, involving only the *Smicronyx*, was thought to have resulted in establishment, but in 1988 there was no trace and it is now thought there was confusion with the indigenous, unnamed *Smicronyx* species.

Further work is awaited on the taxonomy and biology of *Smicronyx* species especially, and although it cannot be expected that any of these insects are likely to be capable of a high degree of control on their own, their contribution to an integrated package could be significant. Conversely, the use of insecticides for control of insect pests in the cereals can be expected to result in interference with the natural control balance and cause some increase of the *Striga* problem. Further study of both augmentation and classical biocontrol would seem to be justified.

By pathogens

A number of fungal pathogens are known from emerged *Striga* plants (Nag Raj, 1966; Greathead, 1984; Zummo, 1977) including the powdery mildew *Sphaerotheca fuliginea*, *Phoma* species, *Cercospora* species, *Fusarium equiseti*, *Drechslera longirostrata* and *Macrophomina phaseolina*. Evans (1987) considers only *Cercospora strigae* to have any serious potential as a biocontrol agent but several of these, other than *M. phaseolina*, could be host-specific and could be studied further. The report by Zummo (1977) that there is commonly an unidentified bacterium associated with *S. hermonthica* which suppresses the development of fungal pathogens also deserves confirmation.

There is suspicion that soil-borne pathogens may be important in suppressing *Striga* and there has been interest in locating sites where *Striga* has declined for otherwise unexplained reasons. So far there is no confirmation of such organisms, but it is a topic of active interest and could perhaps lead to the development of an effective mycoherbicide. Such a weapon could be immensely valuable.

Integrated Control of *Striga*

The selection and integration of control measures must depend on a great many factors, including the combination and predominance of the different cereal crops, opportunities for fallowing and rotational cropping, the economic status of the farm, labour availability, access to fertilizer and pesticide supplies. Decisions on the optimum package of control measures is likely to be helped in due course by the development of mathematical modelling techniques. Two models have been published, by Kunisch *et al.* (1991) and by Cardwell *et al.* (1991). In each case, the importance of the *Striga* seed production and of the seed bank in the soil are of paramount importance. A combination of very slow dissipation of the seeds in the soil (whether by germination or natural mortality) and very high potential seed production means that the number of plants which can be allowed to survive and shed seed must be minimized. The models help to show just how complete any control system must be to achieve a steady decline in the infestation. An additional model devised by Cardwell *et al.* (1991) attempts to assess the likelihood of adoption of different control measures in different farming systems, according to socio-economic criteria. Many more data are needed before the models will be of predictive value, but they meanwhile serve a valuable purpose in pinpointing gaps in our knowledge and the need for specific research information.

For purposes of our consideration here of control options, the main factor to be considered is the intensity of *Striga* infestation.

Areas with severe Striga *infestation*

Given severely infested fields on which cereal yields are seriously reduced, prevention of *Striga* seeding by hand-pulling is usually out of the question.

Option A (in the absence of a highly resistant cereal variety) would be to fallow or rotate out of susceptible cereal for at least three years, though this could usually be considered for only a small part of the farm area. If the infestation is associated with loss of soil fertility, then every effort should be made to use crops which are likely to improve soil fertility, such as *Crotalaria* or other green manure crops. The latest techniques of agroforestry and alley-cropping may be relevant. Otherwise, if there is any choice of rotational crops, those known to act as trap-crops should be selected, and planted early or at whatever time *Striga* is likely to germinate most readily. When returning to cereal growing, it may still be necessary to treat the field as though heavily infested and hence follow Option B below, otherwise one of the options described for moderate infestations.

Option B would be to grow the most tolerant available cereal, such as the ITA *Striga*-tolerant maize hybrids, and use 2,4-D or other herbicide known to prevent *Striga* emergence and/or seeding. Several applications may be needed, up to and beyond the time of harvest as necessary. The use of mixed cropping and/or additional fertilizer may not be economic or practical where herbicide is being used to prevent seeding. Early planting will ensure the greatest cleaning effect of the crop, but a moderate delay (planting as last field rather than first) should ensure slightly less severe infestation and better cereal yield, even with a 'tolerant' variety.

It can hardly be emphasized too strongly that an intensive programme of *Striga* control should be pursued **every** year without let-up, if the problem is to be permanently reduced, and it is only after a number of years that any striking benefit can be expected.

Areas with moderate Striga *infestation*

Option A: Rotate or fallow as for severe infestations, for one or more years, and then follow one or other of the options below.

Option B: Plant a tolerant cereal and use herbicide to prevent *Striga* seeding as for severe infestations.

Option C: Do everything possible to reduce *Striga* emergence and flowering, so that hand-weeding becomes feasible. This might entail any combination of: 'resistant' or at least less susceptible cereal variety; dense planting; highest economic rate of N fertilizer plus P and K as required; moderately delayed planting (plant these areas last rather than first); intercropping within the row, with cowpea or other convenient broadleaved crop, possibly delaying the planting of intercrop until after the first weeding (2–3 weeks after sowing) for minimum interference with cereal yield. Hand-pulling should then begin not more than three weeks after first *Striga* flowering, to be repeated every three weeks thereafter through harvest time as necessary; pulled *Striga* to be removed to a safe site and eventually burnt. Spot-spraying with herbicide may be used as an alternative to hand-pulling, but this should begin before the first flowering of *Striga*, when plants are not more than 25 cm high. An alternative to a resistant or less susceptible variety could be a very early maturing variety on which the parasite fails to set seed before harvest. The whole field can then be ploughed, or sprayed, immediately after harvest to destroy the *Striga* before seeds mature. As the *Striga* plants are not being removed from the field, they should be destroyed not later than one week after flowering.

Some variation of this option to be repeated in successive years, whenever the cereal crop is grown.

Areas with light infestation

Option A: Hand-pulling may be all that is required, without other input, but it should be rigorous and continued through time of harvest. Spot-spraying with hand-held 'water-pistol' type sprayer is an alternative. If the cereal is not already being grown in rows, this practice should be encouraged.

Option B: For all but the lightest infestations, it will almost certainly be worth reducing *Striga* numbers by any combination of practices as suggested under Option C in the previous section.

Once again it will be important to sustain a systematic control programme in successive years.

Prevention and quarantine

It is important also to consider those areas on which *Striga* does not already occur, especially those within farms already infested, but also on farms where it does not yet occur but where *Striga* is already in the district or is known to pose a threat because of suitable crop and climatic factors.

Prevention of spread locally Local spread of *Striga* can occur by wind, water, animals (including man), in manure and forage and on implements.

Although the wind itself cannot be controlled, the availability of seed to be blown can be, and adequate attention to prevention of seeding of *Striga* could greatly reduce the risk. Where seed production and shed have already occurred, burning the weed material after harvest and/or ploughing in could be of value.

Movement by water cannot be completely controlled, especially when heavy storms result in local flooding, but some precautions can be taken to prevent parasite seed reaching normal watercourses; disposal of pulled weed should be such that rains or flooding will not result in spread.

Movement of animals, workers, produce, vehicles and implements cannot be completely controlled, but farmers can be made aware of the risks and encouraged to clean their implements before they are moved from infested to uninfested fields. Care may be needed in the use of farmyard manure if stock are known to have had access to mature *Striga* plants. Bebawi and El-Hag (1983) show that seeds of *S. hermonthica* may be devitalized in the rumen juices of sheep, but as the seeds of *Orobanche* are known to survive passage through this animal (Jacobsohn *et al.*, 1987), it would seem wise to take precautions.

The most important advice must be always to prevent the shedding of mature seed, by pulling or other means, so that there is no seed to be spread.

Long-distance spread The main means by which long-distance spread may occur include: (i) contaminated crop produce (seed, planting material, forage, etc.); (ii) contaminated bags and containers; (iii) contaminated vehicles and agricultural machinery such as combines. In the USA, strict quarantine procedures have been followed in the infested areas of North and South Carolina ever since 1957 (Sand, 1987; Sand and Manley, 1990; Eplee, 1992). These procedures have included the designation of three zones – the inner infested core, the suppressive or eradication zone, and the uninfested zone outside. Movement of produce, equipment and materials within the infested core is restricted to a small degree only. Movement from the core zone to the eradication zone involves stricter regulatory procedures, while movement out to uninfested areas is permitted only under the most sanitary procedures. Regulations are applied and monitored by Federal Inspectors assigned to a block of farms and advice is given on approved treatments such as washing, fumigation, method of harvest and heat treatment.

Contamination may occur in the seed or produce of crops other than those normally attacked by *Striga*, especially if they have been threshed on contaminated ground or are otherwise contaminated with soil (e.g. root crops like potato, carrot, groundnut, or transplants of trees or shrubs).

Cleaning of crop seed to remove the minute seeds of *Striga* will not normally be feasible, but where absolutely necessary, risks could be greatly reduced by very thorough washing, while in the case of groundnut, seed should at least be hulled (removed from the shells) before transport.

Any bag or other container that is known or thought to have been contaminated with seed of *Striga* should not be used for the transport of crop seed or other produce from infested to uninfested areas. Metal or wooden containers can probably be cleaned satisfactorily, but bags of jute, hessian, sisal or synthetic fibre must be regarded as uncleanable unless sophisticated bag-cleaning equipment is available. Where necessary, only new bags should be used.

Risks of spread on vehicles cannot be completely eliminated without very rigorous precautions, but the risks can be very much reduced by e.g. scheduling field visits to go to uninfested areas before infested ones, and where combine harvesters are used, harvesting uninfested crops before infested ones. Where necessary, vehicles should be thoroughly washed by high-pressure hose-pipe before leaving the infested area, or at some intermediate station where there is no risk of the washings reaching land in which the parasite could develop.

Resource allocation in the whole-farm context

Many decisions may need to be made in the whole-farm context, in terms of the allocation of resources, to maximize the alleviation of the *Striga* problem while minimizing the costs of control and crop losses due to the parasite. No detailed economic analysis of this problem is attempted here, but some generalizations are suggested. For instance, the siting of fallow or non-cereal crops should normally be on the most *Striga*-infested fields, while cereals are grown mainly on the least infested. The allocation of limited labour for hand-pulling should normally be on the *least-infested* fields, where it has a chance of being fully effective and so contribute to a reduction of the problem areas, or at least to minimizing the risk of increased intensity. Allocation of limited nitrogen fertilizer is more problematic but should probably be applied first to the moderately infested land on which it is most vital to reduce infestations to a level feasible for hand-pulling.

Conclusion

We believe the prospects for practical reduction of the *Striga* problem, in Africa especially, are seriously damaged by the tendency to pessimism on behalf of researchers, extension workers and farmers alike. There is a feeling that the available solutions are quite inadequate and that the only hope is some dramatic research breakthrough, involving resistant varieties, genetic engineering or a miracle herbicide. While it has to be admitted that progress in the search for resistant varieties has been disappointing and the available methods are far from perfect, there is much that can be achieved with these methods if they are applied in a determined, persistent manner. The priority in all field projects should be to generate information from which farmers can be helped towards the optimum decisions on rotation, fallowing, choice of cereal species and variety, time and method of planting, mixed cropping, manuring, inexpensive herbicide and hand-pulling, as relevant to their farming system and the intensity of the problem in different fields on their farms. Long-term integrated systems have to developed, based on sound agronomic principles and involving especially all options for maintaining and improving soil fertility. A single 'package' for all situations will **not** be sensible. For maize in wetter ecologies, with greater flexibility in cropping and better economic conditions for the use of inputs, such systems should not be difficult to achieve. On densely infested sorghum and millet farms they will be more difficult and more dependent on assistance from government or aid agencies in the form of subsidy or other help in the use of more expensive inputs such as fertilizer or herbicides (or ethylene?). These agencies have to be persuaded of the importance of supporting such *Striga*-control interventions. Meanwhile, the search for the ultimate solution, truly resistant varieties, should go on, using whatever sophisticated methods are available, including genetic engineering. If this approach succeeds it will be a welcome bonus but we should not be waiting for it. We shall be failing the farmer if we rely too much on the prospect of a golden egg which may or may not be laid in ten years' time.

2

STRIGA AND *ALECTRA* ON COWPEA AND OTHER BROADLEAVED CROPS

Knowledge of the biology and control of the cereal witchweeds has been developed over many decades. In contrast, it is only relatively recently that the extent of the infestation of grain legumes by *Striga gesnerioides* or *Alectra vogelii* in Africa has been recognized. As the biology of these and other *Alectra* species is similar they are treated together in this chapter. The control of *S. gesnerioides* and *A. vogelii* on cowpea, their most important host crop, is based on similar principles. In particular, the same germplasm pool is being exploited in what is likely to be, in the near future, a successful attempt to develop cowpea varieties which are resistant to both species.

COWPEA *STRIGA*: SPECIES DESCRIPTION

Striga gesnerioides (Willd.) Vatke (= *S. orobanchoides* Benth., *Buchnera gesnerioides* Willd.)

Striga gesnerioides, included in the key to the genus in Chapter 1, is unlike other *Striga* species in forming a very large tuberous haustorium which may reach up to 3 cm in diameter (see Plate 7C). On crop hosts the plant is usually extensively branched from just below ground level with stems growing 15 to 20 cm high. The reduced scale-like leaves, ranging from 0.5 to 0.7 cm long, are appressed to the stem, and plants have a rather succulent appearance compared to other witchweeds. Leaves and stems turn blue-black if they are wounded. The flowers, which are borne from just above ground level, most commonly vary from light pink to a dark purple or mauve but more rarely may be white or yellow. They are closely packed together up the stem so that with two or three flowers open on each stem at the same time plants appear very showy (see Plate 7A). The corolla, which is 0.3 to 0.5 cm across, turns blue-black as it withers and remains dried on the developing capsule. The basic number of calyx nerves is five, one running to the tip of each calyx lobe. As with other species this is a variable character and plants in one population in Botswana have six or seven nerves (Ralston *et al.*, 1987). The chromosome number (2n) is 40.

Distribution *S. gesnerioides* is widely distributed in Africa in a range extending from Morocco and Sudan to around 30°S in South Africa. Although this species is best known from infestations in cowpea fields in the dry Sahelian and savannah zones of West and Central Africa stretching from Senegal to Chad and to southern Togo and Benin, it is perhaps the commonest *Striga* species on the continent in natural vegetation. In semi-arid areas of Southern Africa it is frequently seen in undisturbed grassland. *S. gesnerioides* is also found in India, particularly in Maharastra and Tamil Nadu, north central Sri Lanka, and Saudi Arabia. It was accidentally introduced to central Florida, USA, where the species was first seen in 1979.

Host range Despite an extensive host range including species of Acanthaceae, Convolvulaceae, Euphorbiaceae, Leguminosae, Pedaliaceae and Solanaceae, *S. gesnerioides* is only of economic significance as a parasite of cowpea in West Africa, sporadically on tobacco in Southern Africa and rarely on sweet potato in South Africa (Doidge, 1924) and Ethiopia (Parker, 1991b). Although sunflower has been observed to support the growth and flowering of the parasite from the Florida population in pot culture (Herbaugh *et al.*, 1980) there have been no reports of attack of the crop anywhere in the field. Attack of cowpea extends from Senegal eastwards to Chad through the Guinea savannah of Mali, Burkina Faso, Niger, Benin and Kano State, northern Nigeria, where the crop is the main legume intercropped with pearl millet, sorghum or maize. Infestations extend almost to the coast in southern Togo (Agbobli, 1991b) and into northern Cameroon. The east and southerly limits of the occurrence of *S. gesnerioides* attack of cowpea have not been clearly defined. The problem may extend into the Central African Republic and is largely restricted to regions north of the equator, though there has been a recent report of an infestation on the crop in Kasai Oriental province of southern Zaire (Pieterse, 1990). Other important grain legume crops including bambara and groundnut are not hosts.

A number of wild legumes are common hosts including *Tephrosia pedicellata*, *T. lupinifolia*, *Indigofera diphylla* and *I. obtusifolia* in West Africa;

T. purpurea, I. costata, I. schimperi and *Rhyncosia subulata* in Southern Africa; and *I. nummulariifolia* in Sudan. *Jacquemontia tamnifolia* and *Merremia emarginata* (both Convolvulaceae) are frequent hosts in North and West Africa and the parasite attacks *Ipomea magnusiana* in Botswana and *I. pes-caprae* in Senegal. *Euphorbia abyssinica* is parasitized in Sudan and *Euphorbia* species are hosts elsewhere in Africa. A *Pterodiscus* species (Pedaliaceae) is a common host in northern Botswana. *Indigofera hirsuta* is the main host in Florida where the parasite is also seen growing on *J. tamnifolia* and *Alysicarpus vaginalis*, the one leaf clover. In India the host range of *S. gesnerioides* includes *Euphorbia antiquorum* and *E. nivulia* (Euphorbiaceae), *Lepidagathis* and *Hygrophila* species (Acanthaceae).

Variation Considerable morphological variation is found within *S. gesnerioides* and there are a number of strains within this species with a much narrower host range than the species as a whole. Field observations suggest that certain morphotypes may be host specific. A number of morphotypes can be recognized with distinct combinations of stem succulence, internode length, flower colour and size (see Plate 7B). Parker and Reid (1979) reported that the much-branched, white- and blue-flowered form which attacks cowpea in West Africa did not parasitize other host species. A more slender branched, often red-stemmed and more reddish-flowered form was found growing in the same region on *Tephrosia, Jacquemontia* and *Merremia* species, but never on cowpea. A number of apparently distinct, host specific, morphotypic variants occur in Botswana (Ralston *et al.*, 1987). These range from a yellow, small-flowered type with short internodes and succulent stems growing up to 18 cm tall, to a light-pink, large-flowered form with long internodes, and occasionally red-pigmented non-succulent stems growing up to 30 cm tall. At some sites contrasting forms are found on different hosts within a few metres of each other. *S. gesnerioides* is strongly autogamous (Musselman *et al.*, 1981) which accounts for the maintenance of genetically isolated host specific morphotypes whose distributions overlap.

The extreme host specificity of populations of *S. gesnerioides* has been demonstrated on a number of occasions in pot experiments. The strain which attacked tobacco in Zimbabwe did not parasitize cowpea (Wild, 1948), though it was stimulated to germinate by cowpea and other leguminous non-host crops including pigeon pea and velvet bean. *S. gesnerioides* can be observed in cowpea fields in Botswana growing on *Indigofera daleoides* but never cowpea. Musselman and Parker (1981b) demonstrated clearly that four African strains of the parasite would only attack and emerge on the host species from which they had been collected (Table 2.1). The Florida population, which was also included in this study, can develop successfully on two species of *Indigofera* and

Table 2.1. The host range of *Striga gesnerioides* strains from Africa and Florida.

	Source of *S. gesnerioides*			
Test host	*Indigofera* (Florida)	Cowpea (Niger)	Tobacco (S. Africa)	*Jacquemontia* (Niger)
Indigofera	H	NH	NH	NH
Cowpea	NH	H	NH	NH
Tobacco	NH	NH	H	NH
Jacquemontia	NH	NH	NH	H

Source: Musselman and Parker (1981b).
H = host; NH = non-host.

Jacquemontia tamnifolia so is apparently related to the *Jacquemontia* strain from Niger. The narrow host range of the American strain makes it unlikely that this population will pose a problem to American agriculture.

There has been a widely accepted view that the strains which are specific to wild legumes will not give rise to infestations of cowpea or other crops. Indeed it was assumed that the tobacco strain in Southern Africa arose by mutation (Wild, 1951), though a more recent experiment demonstrated that the form collected from a *Tephrosia* species in natural grassland in Botswana could emerge on tobacco (Ralston *et al.*, 1987). It would therefore appear possible that crop infestations may occasionally originate from parasite populations on wild species.

A further narrowing of host range is seen among those populations which attack only cowpea in their ability to parasitize some cowpea varieties but not others which are susceptible elsewhere (Parker and Polniaszek, 1990). This phenomenon has important implications for resistance breeding programmes.

Importance *S. gesnerioides* is a major, indeed according to some observers, the most important biotic constraint to cowpea production in the Sahel, the Sudan savannah and the northern Guinea savannahs of Mali, Burkina Faso, Niger, Nigeria and Cameroon. The problem appears to have increased in importance over the past decade particularly in Nigeria. Obilana (1983) described the rapid spread of infestations southward to the northern Guinea savannah, a trend which Aggarwal (1985) predicted would continue due to an increase in the area and frequency of cowpea production. Sole crop cowpea has been encouraged at the expense of intercropping by greater availability and use of chemical pest control. Erratic rainfall, insect pests, diseases and parasitic weeds are all factors associated with low yields of cowpea which is commonly grown at low populations as an intercrop with the staple cereal. It is extremely difficult to measure yield loss due to one component of the pest complex under these conditions. It is not surprising, therefore, that the information which has been collected on the impact of *S. gesnerioides* comes from trials with sole crops. In trials utilizing artificially infested soil in Burkina Faso the yield loss of a number of susceptible cowpea varieties due to parasite infestation averaged 30%. The greatest damage of 56% was recorded on a local variety of which 83% of plants were attacked (Aggarwal and Ouedraogo, 1989). At Kano, in Nigeria, *S. gesnerioides* has caused cowpea trials to be abandoned. Many farmers' fields in Kano, Sokoto, Katsina, Bauchi and Borno states have been described as completely blighted by the parasite with losses as high as 100% (Emechebe, 1981; Emechebe *et al.*, 1991).

Attack of tobacco is occasionally reported from the Rustenberg and Britts areas of Transvaal in South Africa but other localized infestations of the crop in southern Africa, including those on a few farms in southern Zimbabwe reported in the late 1940s (Wild, 1948), no longer appear to be of concern.

ALECTRA: THE MAIN SPECIES

The majority of species in the hemi-parasitic genus *Alectra*, which is closely related to *Striga* in the Rhinantheae tribe of the family Scrophulariaceae, are found in tropical Africa and subtropical Southern Africa. One species is also found in subtropical Asia, to west China and the Philippines, and one occurs in tropical and subtropical South America. The most recent monograph on the genus (Melchior, 1941) described some 43 species but more recent collection and detailed study has led to the combination of a number of the older species names. Some 30 species are now recognized (Philcox, 1990) of which only five are known to attack crops.

Alectra species are morphologically quite distinct from other parasitic weeds which share their distribution and crop hosts. The five-lobed, sulphur yellow to pale orange, bell-shaped flowers which have a large horseshoe shaped stigma are characteristic of the genus. Identification of individual species, however, can be difficult as this relies on the presence or absence of hairs on, and the shape of, the stamens. In order to distinguish between the species which have well developed leaves it is necessary to examine the flowers with a hand lens.

The following key covers those *Alectra* species which may be found parasitizing broadleaved crops in Africa. An additional species, *A. fluminensis*, occurs in South America and is described below.

Key to the Main *Alectra* Species

A Plants with well developed leaves

Plants 15–50 cm tall, **sharply toothed leaves** 14–25 mm long by 8–18 mm wide, calyx 6–8 mm long with **hairs on nerves and margins only, anther lobes apiculate**. Parasitic on grasses and broadleaf species, particularly Compositae. Recorded from niger seed in Ethiopia . . . . ***A. sessiliflora***

Plants 15–30 cm tall, leaves 15–35 mm long by 3–15 mm wide with two to five widely spaced blunt teeth, calyx 7–8.5 mm long, **covered with short hairs, anther lobes not apiculate, stamen filaments glabrous**. Parasitic on cowpea, groundnut and other cultivated legumes . . . . ***A. vogelii***

Plants 15–45 cm tall, leaves 8–20 mm long by 2–8 mm wide with up to two shallow teeth on each margin, calyx 5–7 mm long with obtuse lobes and **covered with short hairs, anther lobes not apiculate, stamen filaments hairy**, sometimes densely so. Parasitic on cowpea and groundnut . ***A. picta***

B Plants without well developed leaves

Plant 10–45 cm tall, **leaves scale-like** up to 2 mm wide and often appressed to the stem. Parasitic on tobacco and sunflower in Southern Africa only . . . . ***A. orobanchoides***

Alectra sessiliflora (Vahl) Kuntze

Two floral characters can be used to distinguish *A. sessiliflora* (see Fig. 2.1) from *A. vogelii* and *A. picta*. The anther lobes of *A. sessiliflora*, having pointed tips, are termed apiculate. The anther lobes of *A. picta* and *A. vogelii* on the other hand, do not have pointed tips. Easier to distinguish is the hairiness of the calyx. Those of *A. vogelii* and *A. picta* are covered evenly in short hairs while the calyx of *A. sessiliflora* has long hairs on nerves and margins only (see Fig. 2.2). In practice the identification of *A. sessiliflora* compared to the other two species should present little problem. In contrast to the two species which attack legume crops *A. sessiliflora* tends to have sessile leaves which are more sharply toothed and pointed, while the flowers are less closely packed along the stem. These characters are, however, variable throughout the range of these species. The presence or absence of a beard of hairs on the stamen filaments is also an important, though again variable, character which is used to distinguish *Alectra* species in published keys. Two of the four filaments of *A. sessiliflora* are bearded except in the case of specimens found in South Africa which have no hairs (Hepper, 1960).

Fig. 2.1. *Alectra sessiliflora.*

Fig. 2.2. The calyces of (l.) *Alectra vogelii* and (r.) *A. sessiliflora* showing different types of hairiness.

Distribution The species occurs most commonly in wet or moist grassland and woodland throughout West, Central, East and Southern Africa, Madagascar and Mauritius to India, Burma, China, and the Philippines.

Host range This is almost entirely non-cultivated species, especially the Compositae (e.g. *Tridax procumbens* and *Ageratum houstonianum*) and some Gramineae. It has been collected from niger seed in Ethiopia (Parker, 1988). *A. picta* and *A. vogelii* are found almost always in association with grain legune crops so the only confusion may arise when *A. sessiliflora* attacks weeds in a legume crop. With careful excavation it is relatively easy to expose the connection between an *Alectra* plant and the host root.

Alectra vogelii Benth.

Those not familiar with root hemi-parasites are not likely to recognize that *A. vogelii* (Plate 8A) is parasitic when observing the species for the first time. Flowering specimens are particularly leafy and have a similar habit to many free-living weed species. Individual plants grow up to 30–45 cm tall, often as a single stem, but sometimes branch from near to ground level. As with other *Alectra* species the stems are bright orange below ground. The stems and leaves, which are 1.5 to 3.5 cm long by 0.3 to 1.5 cm wide, are conspicuously hairy. The shape of the leaves, particularly the extent of toothing along the edge of the lamina, varies considerably. Plants from Kenya have five or six sharp teeth, each up to 3 mm long, along each margin. In Central and Southern Africa leaves may have two to five large, widely spaced teeth along each edge, while in parts of West Africa leaf margins are almost entire. The flowers are borne individually on a short stem in the axils of the upper leaves or bracts. Up to 10 flowers may be open on one day on individual plants depending on the extent of branch development and age. The buds are enclosed in a densely hairy calyx whose five lobes each have a triangular tip with an obtuse apex. The corolla, formed of five petals which are fused into a tube for the bottom half, is bell-shaped when open, 0.6 to 1 cm in diameter, and somewhat longer than the calyx. Petals are generally pale yellow and may or may not have three deep red veins. Both flower forms can be found in the same stand of *A. vogelii*. Anthers and filaments are glabrous (see Plate 8B). After flowering the corolla withers and remains covering the developing globose seed capsule which eventually swells to approximately 5 mm in diameter. The chromosome number (2n) is 38.

Distribution *A. vogelii* is distributed throughout semi-arid areas of tropical Africa and subtropical Southern Africa. From Swaziland and Transvaal in South Africa in the south, the range of this species extends through Central Africa to Burkina Faso and Mali in the west and Kenya in the east of the continent. This species is closely associated with cropping and has rarely been collected from undisturbed savannah.

Host range *A. vogelii* is likely to be found wherever cowpea is grown in semi-arid savannah areas of sub-Saharan Africa and this is indeed the major leguminous crop host. Bambara, groundnuts, mung bean, tepary, and common beans are also common hosts in some parts of the parasite's range. There have also been occasional reports of infestation of chickpea, soyabean and runner bean. Pigeon pea appears to be the only widely grown pulse crop which is not parasitized. Many other legumes are hosts and this presents a problem for the introduction of alternative pulse or fodder break crops into an arable rotation. Pot trials and field observations in Southern Africa have shown that *Dolichos lablab*, siratro, velvet bean and *Stilozobium deeringianum* can be attacked (Riches, 1989b). Leguminous weeds including *Indigofera daleoides* and *Tephrosia purpurea* may also be hosts. Reports of the attack of non-legume crops almost always involve weeds on fallow or cropped arable land. *A. vogelii* has a wide host range having been occasionally found on *Acanthospermum australe*, *A. hispidum* and *Vernonia poskeana* (Compositae), *Euphorbia chamaesyce* and *E. inaequilatera* (Euphorbiaceae), *Acrotome inflata* (Labiateae), *Hibiscus* species (Malvaceae) and *Sesamum grandiflorum* (Pedaliaceae).

Variation Although *A. vogelii* can attack the grain legume crops listed above there is clear variation in the host range in different regions of Africa (Table 2.2). Among some populations which are able to attack both cowpea and bambara differential attack is seen of at least one selection of each crop which is resistant in other areas. The distribution of particular

Table 2.2. Geographical variations in the host range of *Alectra vogelii* strains from Africa.

	Host species			
Alectra origin	Cowpea	Groundnut	Mung bean	Bambara
S. Africa: E. Transvaal	H	H	H	H
S. Africa: N. Transvaal	H	H	H	NH
Botswana	H	H	H	NH
W. Africa and Cameroon	H	H	NH	NH
Malawi	H	H	H	H
Kenya	H	H	H	H

H = host; NH = non-host.

strains of the parasite has been determined by growing differential host species in soil infested with seed collected from different parts of its range (Polniaszek and Parker, 1987; Riches *et al.*, 1992). Populations collected from Mali and Nigeria in West Africa and Cameroon only attack groundnut and cowpea. Seed from infested sites in Botswana and Northern Transvaal in South Africa attack mung bean in addition to cowpea and groundnut. Populations sampled from Kenya, Malawi and eastern Transvaal parasitize bambara as well as the crops which are susceptible elsewhere. This distribution is based on well watered pot trials. Groundnuts are not attacked at infested field sites in Botswana. Further collecting and testing is likely to reveal additional local variation in host range. No correlation has been observed between the morphological variation, particularly leaf shape, found in various parts of the range of *A. vogelii*, and host preference.

Importance *A. vogelii* replaces *S. gesnerioides* as an important constraint to cowpea production in East, Central and particularly Southern Africa. The true extent of the problem has only been recognized since national research and extension services began to focus on staple rather than cash crops of smallholder cropping systems in post-independence Africa. The parasite has, however, been a cause of concern for many decades. In notes accompanying a specimen of *A. vogelii* in the Kew herbarium the collector McDonald in 1929 estimated a yield loss of 20% for cowpea crops attacked by the parasite in the Emberre District of Kenya. In 1966 the Agriculture Department in Botswana reported that 'an estimated 24,000 acres was planted to cowpea . . . "yellow witchweed" was widespread in all areas and some fields in the south east of the country were so badly affected as not to be worth harvesting'. *A. vogelii* occurs on local landrace cowpeas in Botswana which have traditionally been grown at low populations in mixed crops with sorghum or millet. The variety Blackeye was introduced from the USA during the late 1950s and has proved to be highly susceptible to the parasite. Blackeye attracts a premium over most local varieties on the market and farmers have responded by planting Blackeye monocrops. When the variety is grown for just one season on lightly infested land which had previously been growing mixed crops including local cowpeas, the soil parasite seed bank is massively increased. Seed production levels up to 600,000 seeds per plant have been estimated for *A. vogelii* (Botha, 1946). In on-farm trials in southeastern Botswana in 1977/78 season six out of 25 Blackeye crops produced no yield due to *A. vogelii* infestation (Plate 8E). While Blackeye yields averaged 602 kg ha^{-1} in 1979/80 a heavily infested plot yielded less than 100 kg ha^{-1} (Riches, 1989b). Total crop loss has also been observed in recent years in the Embu district of Kenya (Bagnall-Oakeley *et al.*, 1991). Most farmers in this area consider the incidence of the parasite to be on the increase.

A. vogelii is also of importance in the Nigerian savannahs where it can be found in cowpea crops which are also attacked by *S. gesnerioides*. It is reported to be the major parasite of the crop in the northern Guinea savannah (Lagoke, 1989). Elsewhere in West Africa *A. vogelii* infestations tend to be more localized, as in southern Mali. The species is widespread in groundnut producing areas of Nigeria where yield losses due to the parasite of up to 15% have been recorded (Salako, 1984) and the extent of the problem has increased in recent years. Complete crop failure of some groundnut varieties due to infestation occurs in the KaNgwane area in the east of South Africa. Between 30 and 50% reduction in the yield of bambara nuts has been recorded in this area (Beck, 1987). *A. vogelii* has also been reported to cause serious damage to bambara in Zimbabwe (Mabasa and Lagoke, 1991). Late-sown soyabean crops may be completely destroyed by the parasite in northern Nigeria (Lagoke, 1989), though there have been few reports of the extent of infestation of this crop.

Alectra picta (Hiern) Hemsl.

The presence of so called 'beards' or hairs on the stamen filaments of *A. picta* compared to glabrous filaments in *A. vogelii* is the character used by current taxonomic works to maintain these as separate species (see Plate 8C). In other respects, plants of both species collected from parasitized crops are morphologically similar. Leaves of *A. picta* have an entire margin or in some localities have one small tooth on either edge of the lamina. These are quite different to the obviously toothed leaves of

A. vogelii in Central and Southern Africa but similar to specimens of this species in parts of West Africa. Plants with bearded filaments have recently been found parasitizing groundnut in eastern Hararghe, Ethiopia; cowpea near Moutouroua, northern Cameroon (Riches *et al.*, 1992); and groundnut in southern Malawi. The Cameroon and Malawi populations occur in areas where *A. vogelii* is also found. In Malawi both species have been observed side by side in the same crop row. These are the only reports of *A. picta* parasitizing crops. This has previously been considered a southern African species parasitic on wild genera of Convolvulaceae and Leguminosae over a range which includes southern Angola, Botswana, Mozambique and Tanzania (Hemsley and Skan, 1906; Philcox, 1990).

A. picta has a similar host range to West African populations of *A. vogelii* parasitizing cowpea and groundnut but not bambara or mung bean in pot trials (Riches *et al.*, 1992). In view also of morphological similarities with *A. vogelii* there have to be some doubts whether *A. picta* is a distinct species.

Alectra orobanchoides Benth. (= *A. kirkii* Benth.)

This rather variable species has scale-like leaves 0.35 to 0.9 cm long by 0.15 to 0.2 cm wide. Growing up to 45 cm tall, the stems are extensively branched from or near to ground level (see Plate 8D). Although containing chlorophyll, *A. orobanchoides* plants tend to be purplish in colour. The flowers are yellow to orange with up to three dark purple veins on each of the five corolla lobes. The stamen filaments are at least slightly and often heavily bearded. Found throughout savannah areas of Southern Africa *A. orobanchoides* extends northwards into Tanzania, Uganda and Kenya. Known as Kirk's tobacco witchweed, the species was of limited importance in the early 1940s on tobacco crops on a few farms in the area which is now eastern Zambia (Wild, 1948) but is no longer thought to be a problem. It is occasionally found on sunflower in the Transvaal of South Africa (Visser and Johnson, 1982). The host range otherwise consists of wild species and, like other members of the genus, *A. orobanchoides* can parasitize plants from a wide range of genera. According to Visser (1981) these include *Dicliptera, Hypoestes* and *Justicia* (Acanthaceae); *Commiphora* (Burseraceae); *Tagetes* (Compositae); *Cyperus* (Cyperaceae); *Becium* and *Salvia* (Labiatae); *Ceratotheca, Dicerocaryum* and *Sesamum* (Pedaliaceae); and *Nenax* (Rubiaceae). *Thunbergia annua* (Acanthaceae) is a common host in northern Botswana.

***Alectra fluminensis* (Vell.) Stearn (= *A. brasiliensis* Benth., *A. melampyroides* (L.C. Rich.) Kuntze)**

A. fluminensis is a robust species growing up to 1 m tall. The conspicuously toothed leaves are up to 5.5 cm long by 2.5 cm wide. Flowers are approximately 1 cm long. This South American species is of local importance as a parasite of sugarcane in Venezuela (Trujillo and Ordosgoitti, 1983). It is also found in Brazil, Guyana, Panama, Nicaragua, Guatemala and Trinidad.

BIOLOGY AND PHYSIOLOGY

In comparison with the cereal witchweeds, relatively little research effort has been committed to understanding the biology, physiology and ecology of *Striga gesnerioides* and *Alectra* species. These are obligate root parasites with a similar life cycle to the cereal *Striga* species which have been discussed in detail in the previous chapter. The following comments refer largely to the association between *A. vogelii* or *S. gesnerioides* and cowpea.

The Seeds, Dispersal and Germination

The seeds of *S. gesnerioides* are similar in size to those of *S. asiatica* (see Table 1.2, p. 21). *A. vogelii* seeds (see Fig. 2.3) have a more complex structure. An outer sclerified cell layer of the testa is modified into a cone or trumpet-like structure about 1 mm long in which the 'kernel' of the seed, measuring about 0.15 mm by 0.25 mm, is suspended (Visser, 1978). In both species the surface of the seed coat has characteristic reticulations. It has been suggested that these indentations allow air bubbles to be trapped when the seeds are floating on water, making them buoyant and easily dispersed, at least for short distances, in rainwater run-off. Wind dispersal of these tiny seeds over longer distances is also clearly important, especially for *Alectra* species whose trumpet-like structure is aerodynamically suited to being wind-blown even on the lightest breeze. Windy conditions, including 'dust-devils' (localized whirlwinds) are common in the dry season of regions where these parasites are found.

Each capsule of *A. vogelii* produces 2000 to 3000 seeds, and with as many as 200 to 350 capsules some 400,000 to 600,000 seeds may be produced by a single plant (Botha, 1946; Visser, 1978). Seed production by *S. gesnerioides* is also massive with a single plant producing 400 to 500 seeds per capsule and a total of 50,000 to 500,000 seeds per plant (Aggarwal, 1985).

The germination requirements of these species are broadly similar to

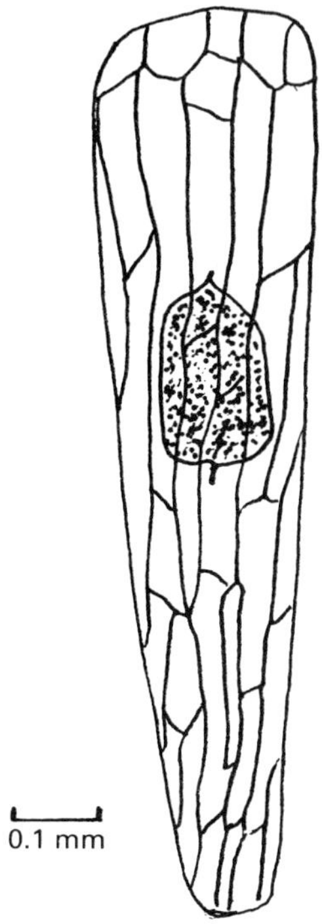

Fig. 2.3. Seed of *Alectra vogelii.*

those already described for the cereal witchweeds. Germination follows exposure to the root exudate from potential host plants. *A. vogelii* has no after-ripening requirement, and seed taken from mature dry capsules will germinate immediately after collection from the field, following exposure to a suitable germination stimulant at a suitable temperature (Riches, 1989b). Experience at Long Ashton has shown that seed of *S. gesnerioides* harvested from glasshouse-raised plants will give low levels of germination immediately after shedding but maximum germination levels are reached following three to six months' storage (Lane, personal communication). As both species produce seed at the end of the rainy season in semi-arid areas the long dry season imposes a natural period of dormancy.

Natural germination stimulants True hosts, for example cowpea, and some so-called 'false-host' species which are not parasitized themselves, produce root exudates which contain chemicals capable of stimulating the germination of these parasite species. A number of chromatographically different *A. vogelii* stimulants, isolated from the root exudates of leguminous hosts including bambara and cowpea, are also produced by the roots of the false-host sunflower (Visser, 1975a). Four substances which stimulate the germination of both *A. vogelii* and *S. gesnerioides* have been isolated from the root exudate of cowpea (Herb *et al.*, 1987). One of these has structural similarities to the cytokinin group of plant growth hormones. A chemical stimulant, termed 'alectrol', which is active in promoting the germination of *A. vogelii*, has recently been isolated from cowpea root exudates (Müller *et al.*, 1992). This is chemically very similar to strigol (see Fig. 1.5c, p. 23) isolated from cotton roots. Strigol is a potent germination stimulant of *A. vogelii* as well as the cereal witchweed species, but has lower activity on *S. gesnerioides*.

Other stimulants Fungi of the genus *Fusarium*, isolated from the surface of *A. vogelii* seeds, when grown in a liquid malt extract medium, also produce a germination stimulant which is active on this species. Chromatographically this is similar to a stimulant which is present in the root exudate of bambara (Visser, 1975b).

A number of chemicals which are not related to the natural stimulants promote the germination of *S. gesnerioides* in laboratory tests. These include cytokinins, ethylene and sodium hypochlorite (Igbinnosa and Okonkwo, 1992). Hypochlorite is also highly active on *A. vogelii* (Okonkwo and Nwoke, 1975) but this species is apparently unresponsive to ethylene, when the gas is provided as the sole germination stimulant (Jackson and Parker, 1991). Synthetic analogues of strigol including GR7 and GR24, effective stimulators of the germination of the cereal witchweeds, are not active on *S. gesnerioides* but are effective on the seeds of a number of *Alectra* species (Visser and Johnson, 1982; Okonkwo and Okafor, 1988). Little is known about the role of stimulant substances in the germination process of these species. It has been suggested that hypochlorite may act via the disruption of seed coat permeability or the removal of germination inhibitors.

Conditioning Maximum germination of *S. gesnerioides* has been reported to follow relatively long periods of conditioning. Reid and Parker (1979) found highest germination following 5–6 weeks conditioning in distilled water at either 23°C or 33°C for parasite seed samples which were only a few months old. With 11 year old seed samples 70–80% germination was achieved following 10 days' conditioning at 25°C (Igbinnosa and Okonkwo, 1992). For these older samples, only a short conditioning period if any was needed and up to 30% germination was recorded some six days after incubation of dry seed in stimulant solutions at 30°C. *S. gesnerioides* then germinates after two to three days. For routine germination assays, 14 days' conditioning at 33°C gives reliable results for seed samples collected from a wide range of sites. *A. vogelii* and *A. orobanchoides* will also germinate within five days of adding a suitable stimulant to dry unconditioned seed incubated at 28°C or 30°C (Visser and Johnson, 1982; Riches 1989b). In earlier studies of the germination biology of *A. vogelii*, Botha (1948, 1950) demonstrated that germination is enhanced when the seed is conditioned prior to exposure to a stimulant. Maximum germination of 85% followed nine days conditioning at 30°C to 35°C. Almost 30% germination occurred four days after the addition of a stimulant following only 24 hours conditioning, but no germination was noted when no conditioning was used. *Alectra* species do then appear to require a very short period of conditioning which will take place once dry seed has been added to a stimulant solution and before germination is assessed four to five days later. Little attention has been paid to possible effects of conditioning *S. gesnerioides* in stimulant solutions. Igbinnosa and Okonkwo (1992), however, have reported that germination in response to a range of stimulants is enhanced following conditioning in low concentrations of hypochlorite. Higher concentrations inhibit subsequent germination.

Prolonged periods of conditioning do not appear to induce a 'wet dormancy' of *A. vogelii*, at least for periods up to 40 days (Botha, 1948). There is some evidence that this phenomenon may occur in *S. gesnerioides*; a marked decline in germination following conditioning periods of 35–70 days has been reported in one study (Reid and Parker, 1979).

Penetration of the Host and the Development of the Parasite

Growth occurs largely at the root meristem during the germination of the seeds of both *A. vogelii* and *S. gesnerioides*, resulting in seedlings with elongated radicles (Okonkwo and Raghavan, 1982). These grow to a maximum of 3 mm and 2 mm respectively and need to penetrate a compatible host root within about 8 to 10 days to survive. Radicles which are pointing away from a root following germination quickly curve round towards the root. For *A. vogelii* this has been shown to be a chemotropic response to a concentration gradient of host root exudate (Visser, 1978).

Although the seedlings of both species are capable of forming normal roots in culture on nutrient agar the root cap is lost during post-germination growth of the radicle in host root exudate. The radicular apex is then adapted to penetrate the host root and, as the primary haustorium, this grows through the host root tissue to make contact and form a union with the host conductive tissue in the stele. A prominent feature of the initial attachment of these parasite species to the roots of some hosts is the development of so-called 'haustorial hairs' on the tip of the radicle (Visser *et al.*, 1990; Lane *et al.*, 1991a). These may allow the radicle to adhere to the root while the epidermis is breached.

The anatomical development of the primary haustorium, which forms as the parasite radicle penetrates and ramifies within the host root tissue, and structural changes within the host itself have been described for *A. vogelii* by Visser and co-workers (Visser *et al.*, 1977; Visser *et al.*, 1990) and for *S. gesnerioides* by Okonkwo and Nwoke (1978). The actual physiological mechanisms and biochemical processes involved with the development of a successful host–parasite association have not, as yet, been identified and remain a matter of considerable conjecture. Contact between the *A. vogelii* radicle and the host root is accompanied by a local increase in size of the outer cell layers of the root. Cell walls of the advancing front of parasite cells are thickened and appear to push between host cells. As *S. gesnerioides* penetrates the host the root surface is split around the site of infection (Lane *et al.*, 1991a). It is assumed, but not proven, that enzymes are involved in the breakdown of host cell walls as the primary haustorium extends both inter- and intra-cellularly within the host root. While it has been claimed that secretions of penetrating radicular cells diffuse into the surrounding host cells and also accumulate in cortical intercellular spaces (Okonkwo and Nwoke, 1978) it may be that the radicle stimulates the production of hydrolytic enzymes in the host cells. However it is achieved, the breakdown of host cell walls is clearly crucial to the further development of the parasite. Following the initial penetration of the roots of pea by *A. vogelii* the cell walls of this non-host remain relatively intact. The difficulty the parasite encounters in breaking down the cell walls may contribute to the incompatibility between the two species (Visser *et al.*, 1990).

In both species the parasite endophyte grows until contact is made with xylem vessels in the host stele, and subsequently vessel elements differentiate within the haustorium to link the host xylem with parasite radicular xylem. Penetration by *A. vogelii* induces the host to proliferate and subsequently to differentiate phloem sieve tube elements towards the intruded haustorium. Simultaneously, sieve tube elements are differentiated within and towards the advancing front of the haustorium. It is thought that the transfer of organic compounds from host to parasite is effected by a bridge of parenchyma cells between the phloem systems of host and parasite (Dörr *et al.*, 1979). Although phloem is known to occur in the haustorium of *S. gesnerioides* it has not been possible to distinguish with certainty the relationship between this and phloem of the host root (Ba, 1984). The movement of the foliar applied herbicide *glyphosate* from cowpea to *S. gesnerioides* suggests an efficient phloem link between host and parasite (Samb and Chamel, 1992). Both *A. vogelii* and *S. gesnerioides* stimulate cell division in the host root, a proliferation which, together with expansion of the parasite haustorial tissue, eventually leads to the formation of the large tuberous primary haustoria which are so characteristic of these species. The internal structure of these haustoria is a complex intermeshing of both parasite and host tissue (Ozenda and Capdepon, 1972; Visser *et al.*, 1977).

The structure of the fully developed haustorium of *A. vogelii* is further complicated by the production of numerous 'haustorial roots'. About 12 days after a seed has germinated, and the parasite radicle has made contact with the host, the host root develops a localized swelling from which about 12–15 host lateral roots develop around the point of attachment (see Fig. 2.4). At this stage the parasite shoot begins to break out of the seed coat following differentiation of the first leaves. The proliferation of host lateral roots may be induced by an extract made from parasite seedlings indicating that growth regulators produced by the

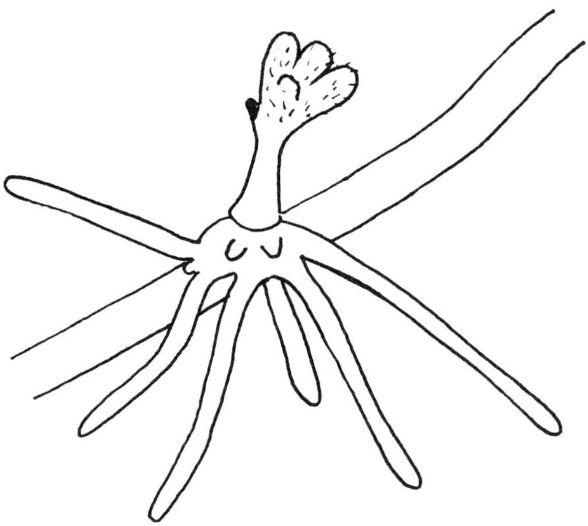

Fig. 2.4. Proliferation of host roots around young *Alectra vogelii* on cowpea.

parasite are involved (Dörr *et al.*, 1977). The parasite effectively stimulates an increased root volume which increases the ability of the host to supply water and mineral salts to the parasite, perhaps independently of the main root system. Nwoke and Okonkwo (1978) suggest that this may account for the large number of infections on one host plant and the ability of one host root to support the development of a chain of primary haustoria. *A. vogelii* also produces adventitious roots at the base of the developing stem. These are able to develop secondary haustoria through attack of host roots including parasite induced-lateral roots. Secondary haustoria arising from adventitious roots are also sometimes seen on *S. gesnerioides* but these are not such an obvious feature as those associated with *A. vogelii*. The adventitious roots of *S. gesnerioides* reach up to 6–8 mm long but soon wither if no secondary haustorium is formed (Hussain and Aigbokhan, 1991).

The life cycle of *S. gesnerioides* parasitic on cowpea, from host root penetration until the first flowers open, lasts from 5–6 weeks. The development of the haustorium and below ground growth of the primary shoot is completed within three weeks. Shoots are in full bloom by two weeks following emergence (Hussain and Aigbokhan, 1991). *A. vogelii* emerges above ground about four weeks after the parasite radicle has penetrated a cowpea root. The first flowers are produced some two weeks later (Visser, 1978).

Parasite Nutrition

Both *A. vogelii* and *S. gesnerioides* can be grown successfully in the absence of a host plant on a culture medium containing mineral salts and sucrose. Normal seedlings with green hairy shoots and leaves develop with normal hairless roots, with a structure typical of non-parasitic species. Although *A. vogelii* did not flower when maintained in culture, *S. gesnerioides* did so 8–10 weeks after planting onto the nutrient medium (Okonkwo, 1975, 1982). A minimum of mineral salts and sugar is then all that is required to stimulate shoot/root morphogenesis so apparently the parasite is self-sufficient for growth regulator requirements. Much stronger shoot growth of *A. vogelii* does, however, follow the addition of cytokinin to the growth medium. Strongest growth of *S. gesnerioides* is seen when vitamins are supplied. In nature these are apparently obtained along the host–parasite xylem stream. As these parasites have no direct access to soil they are also dependent on the host for their nitrogen requirements. *S. gesnerioides* is unable to assimilate nitrate ions, the form of nitrogen usually obtained from soil solution and utilized by higher plants. High levels of the enzyme glutamine synthetase in the tissues of this species suggest that in common with the cereal witchweeds, organic nitrogen compounds are obtained by the parasite via the xylem connection to the host and these are the raw materials for ammonia assimilation by the parasite (Stewart *et al.*, 1984).

The chlorophyll concentration of leaves of *Alectra* species varies from 2.5% of the level found in sunflower for *A. orobanchoides* to 18% for *A. vogelii*. Chlorophyll concentration of leaves and stems of *S. gesnerioides*, which is often regarded as holoparasitic, is only 9% of the level found in cowpea and only one third of the level present in *S. hermonthica* (Samb, 1992). Indeed, in contrast to the chloroplasts of *S. hermonthica* which have a structural similarity to those found in non-parasitic plants, those of *S. gesnerioides* are less well developed (Thalouarn *et al.*, 1991). The photosynthetic activity of these parasites is consequently low. The rate of carbon dioxide assimilation by *A. orobanchoides* and *A. vogelii* is just 11% and 21% respectively of that in sunflower (de la Harpe *et al.*, 1979). The CO_2 fixation capacity of *S. gesnerioides* as estimated from carboxylase enzyme activity is about 20 times lower than is found in tobacco leaves and is comparable to that found in the holo-parasite *Lathraea clandestina* (Thalouarn *et al.*, 1991). Very low rates of photosynthesis coupled with very high rates of respiration in the shoot tissues prevent the parasite making any net gain of carbon (Graves *et al.*, 1992). These species are therefore highly dependent on their hosts for carbon, and a considerable accumulation of host derived metabolite has been shown to occur in *A. vogelii* parasitic on bambara and in *S. gesnerioides* parasitic on cowpea (Gouws *et al.*, 1980; Graves *et al.*, 1992).

Effects on the Host

Prior to *A. vogelii* emergence, affected cowpea plants appear wilted compared to those which are not attacked, particularly when the crop is short of moisture. Adverse effects of parasitism also include delayed onset of flowering of the host, a reduced number of flowers and pods and hence a reduced yield of dry pods and cowpea seed even under well watered conditions (Mugabe, 1983). A common symptom of *S. gesnerioides* attack of cowpea is the appearence of pale patches or chlorosis on the leaves which may be seen before the parasite emerges above ground. The leaves of heavily infested plants develop necrotic straw coloured dry patches which coalesce until the entire lamina and petiole becomes desiccated and the leaf is lost from the plant. Infested plants are often stunted and have a smaller leaf area than parasite free plants. The reduction in leaf area is proportional to the degree of infestation (Aigbokhan and Hussain, 1991). More detailed studies of pot grown cowpea have demonstrated that there is a rapid reduction in the total leaf area of infested plants some 40 days following cowpea emergence. At this stage the canopy of uninfested plants is still expanding (Graves *et al.*, 1992). This study demonstrated that the severe reduction in the growth of *S. gesnerioides* infested cowpea plants is primarily caused by this transfer of carbon from host to parasite. The reduction of host photosynthetic capacity is also important. *S. gesnerioides* reduced the growth of the host by 75% and pod production was virtually absent. Host root mass was also reduced considerably in contrast to promotion of root growth of cereal hosts when parasitized by *Striga hermonthica*.

Pollination and the Role of Insects

Flowering commences 7–10 days after the emergence of *S. gesnerioides*, which is highly autogamous (Musselman and Parker, 1981b; Ralston *et al.*, 1987). The flowers are well adapted for self-pollination, shedding a very sticky pollen mass inside the unopened flower bud. Upward pointing hairs which line the corolla tube serve to position the pollen so that it is picked up by the elongating style as the flower develops and lodges in the forked stigma. The pollen persists as a dried mass on the stigma and can be clearly seen even on the developing capsule. While the structural barrier to cross-pollination is strong, preliminary evidence from hand-pollinations suggests that host specific strains of *S. gesnerioides* may be inter-fertile (Musselman, 1979).

A. vogelii can set seed as a result of either self- or cross-pollination (Riches, 1989b). As may be expected for a species in which individuals produce very showy flowers over a period of 20–30 days insects are essential for pollination. Field observations in Botswana have shown that the anthers only release pollen when stimulated to do so by insect floral visitors (Riches, 1989b). A number of flies which feed on pollen, including *Cosmia, Ischiodon* and *Rhyncomya* species, visit the newly open flowers during the early morning. As these forage over the anthers a cloud of powdery pollen is released on to the adjacent large stigma resulting in self-pollination. Various bees including *Amegilla* and *Lasioglossum* species also liberate pollen as they 'buzz' during floral visits. As these carry loads of pollen while foraging throughout a flowering stand it is possible that cross-pollination may also result from insect activity. The relative importance of self- or cross-pollination has not been studied. Other *Alectra* species of economic importance also probably require insects for pollen release to be stimulated. The beards on the filaments of *A. orobanchoides, A. picta* and *A. sessiliflora* may have an important role in pollination. It has been suggested that hairs on the stamen filaments of other parasitic Scrophulariaceae, by mutual interlocking, ensure that the anthers shed pollen as a single unit when stimulated by a pollinating insect (Pennell, 1935). The contrasting red veins on otherwise yellow petals may guide insects to either pollen or nectar (Visser, 1981).

CONTROL OF *STRIGA* AND *ALECTRA* IN LEGUME CROPS

Recognition of the importance of *Alectra* and *Striga* as constraints to pulse crop production in Africa has occurred at a time when the exploitation of host-plant resistance has been gaining prominence as a crop protection strategy. Much of the effort committed to the control of these species has therefore been within crop breeding departments of national research programmes and at the International Institute of Tropical Agriculture (IITA).

Although there has been considerable progress towards the goal of the first improved cowpea varieties with *Alectra* and *Striga* resistance it is likely to be some years before adapted lines are available for all infested areas. For *A. vogelii* in groundnut such an objective is even further in the future. It is therefore important to consider which other methods of control may have an impact on populations of these parasites. In comparison to the cereal witchweeds there have been few demonstrations of control measures for *A. vogelii* and *S. gesnerioides* in the field. Many of the principles covered in Chapter 1 on the control of *Striga* in cereals are no doubt relevant and of potential importance also for the control of *Alectra* and *S. gesnerioides*. Any practice which leads to a reduction in the amount of parasite seed returned to the soil and the associated exhaustion of the 'seed bank' should always be encouraged. It is, however, often difficult to persuade farmers of the benefit of this approach, particularly when labour-intensive operations are involved, as yield gains in the short term are unlikely. In this section the likely effect of various agronomic practices on parasite populations will be considered. These include a few control measures which can be of practical value to the farmer now, and topics on which further research may lead to the development of additional strategies. A range of methods should be used in combination with resistant varieties to lessen the chance of virulent adapted strains of the parasite building up. Unless crop varieties are effectively immune, so that no emergence of the parasite occurs, there is likely to be some seed production by the parasite.

Cultural and Mechanical Control Methods

Land preparation and time of planting

Little information is available on the effect of soil conditions on *A. vogelii* and *S. gesnerioides* populations. It is claimed that shallow tillage to only 6–8 cm prior to planting redistributes *S. gesnerioides* seeds and encourages infestation of the crop (Sallé, 1991). Whether or not the occasional use of deep ploughing, on land which is usually shallow ploughed with animal draught, will reduce parasite populations has not been confirmed. Reduced infestation of cowpea by *A. vogelii* followed later planting of the crop in Kenya (Bagnall-Oakeley *et al.*, 1991), though late-sown groundnut and soyabean crops have been completely destroyed by the parasite in northern Nigeria (Lagoke, 1989). If it is beneficial at all, late planting, which is perhaps associated with lower soil temperatures and therefore less germination of the parasite, could, however, be very risky where the length of the growing season is unpredictable. In Botswana *A. vogelii* damages susceptible cowpea varieties planted either at the beginning of the rains in mid-November or at the end of the planting period in late January.

Hand-pulling

Removal of emerged parasite plants before these flower and seed should always be encouraged where infestations are limited in area or exist at low populations. It is particularly important to prevent the build-up of the problem where a new infestation is encountered. This is equally the case for previously clean fields or portions of a field which the farmer may have considered to be parasite free. Small stands of the parasite emerging in fields sown to resistant varieties should also be pulled and destroyed. Removal of *A. vogelii* plants after emergence did not improve the yield of bambara (Beck, 1987). There have been no other reports on whether hand-pulling of either *A. vogelii* or *S. gesnerioides* can benefit yields of the crop from which the parasite is removed. As the effect of these species on the host is often seen prior to their emergence it is probable that subsequent removal will not be directly beneficial. Long-term trials are needed to determine how many seasons hand-pulling must be continued to make an impact on the parasite population and, equally importantly from the farmers' perspective, to ascertain how much labour is needed.

Timely destruction of crop residues

Stands of *A. vogelii* and *S. gesnerioides* flower and set seed over many weeks and may continue to do so after the crop has been harvested. In Botswana it has been observed that where cowpea residues have been removed from the field for conservation as animal feed, *A. vogelii* continues to emerge and seed as crop roots remain. When the soil is sufficiently moist and draught power is available, prompt post-harvest tillage can reduce the period of parasite seed production. Fast-maturing varieties of cowpea, particularly the '60-day' types, may offer the farmer the opportunity to harvest cowpea seed and then plough in the residues, destroying the parasite stand before many plants have been able to set seed. In order to do this the crop must be planted as a mono-crop or in a strip arrangement as an intercrop. Studies of the phenology of *A. vogelii* and *S. gesnerioides* in relation to that of early maturing crop varieties are needed. Post-harvest tillage is likely to have other benefits including the reduction of non-parasitic weed populations, and moisture conservation.

Rotation, trap-cropping and catch-cropping

As has been discussed in Chapter 1, population pressure now rarely allows fallows of sufficient duration to allow the seed bank to become depleted naturally before land is returned to cultivation. Anecdotal evidence suggests that *A. vogelii* seed may remain viable in the soil for as long as 12 years. The longevity of *S. gesnerioides* is thought to be similar. Rotations which include non-susceptible crops do provide an opportunity to manage populations of these parasites by preventing the further build up, and/or effecting a reduction, of the seed bank. This management option may in many situations prove unattractive, particularly to subsistence producers and where the susceptible legume is produced as an intercrop and accounts for a relatively minor component of the arable enterprise. In many parts of Africa the subsistence pulse crop tends to be predominantly cowpea or groundnut. Farmer preference, seed availability and long-established marketing arrangements for any surplus would make a switch into another pulse difficult to introduce. In Botswana and West Africa, bambara may seem a suitable alternative to cowpea on *A. vogelii* infested land while, where *S. gesnerioides* is the problem in West Africa, groundnut would seem the obvious alternative to cowpea. For many farmers, however, the continued production of the multi-purpose cowpea crop is felt to be essential as this provides leaf spinach and livestock fodder in addition to seed.

Trap-crops If switching to a non-susceptible alternative legume is not likely to provide an answer, is it possible to modify the farmers existing rotation so as to encourage a decline in the parasite seed bank? Inclusion in the rotation of a so called *trap-crop* is possible in many situations (see Box 3.1, p. 136). Trap-crops, also referred to as false hosts, produce *Alectra-* or *Striga*-germination stimulants but are not susceptible to attack. Cowpea, pigeon pea and velvet bean stimulate the germination of *S. gesnerioides* in Southern Africa and were suggested for trap-cropping of infestations of the parasite in tobacco (Wild, 1948). The existing species grown as components of the crop mix in infested areas may already include trap-crops. In Botswana, both bambara and pearl millet, which are commonly grown at *A. vogelii* infested sites, are potent stimulators of germination of the parasite (Table 2.3).

In well watered pot trials both those species and also sunflower, a further false-host, were shown to be potential trap-crops (Table 2.4). Two consecutive seasons of trap-cropping with bambara or millet reduced the *A. vogelii* population in a following cowpea crop by 79% and 68%, respectively, of that following two seasons of fallow (Riches, 1989b). Further long-term field trials would be needed to define how many seasons' trap-cropping would be necessary before the parasite seed bank

Table 2.3. Germination of a Botswana sample of *Alectra vogelii* in the root zone of some hosts and non-hosts.

Species	*A. vogelii* % germination
Cowpea cv Blackeye	95
Bambara 'local'	68
Pearl millet cv Babala	59
Pearl millet cv Serere 6A	18
Sunflower cv Russian No. 4	18
Sorghum cv Segaolane	0
Fallow	0

had been exhausted. In many areas, including Botswana, sparse populations of cowpea are often grown scattered throughout the cereal crop. As the total area sown to cowpea is relatively small the crop could, if grown at a higher density, be rotated around the field following a number of seasons' millet. In this particular situation millet is a useful rotational crop because it is not susceptible to *S. asiatica* which commonly occurs in association with *A. vogelii*. In mixed cropping the trap-crop will be planted in association with the susceptible species, as in the case of a millet/cowpea mixture. Whether or not the additional level of stimulant in the soil leads to greater or less infestation of the susceptible crop is not known. In mixed arable/livestock farming systems it is helpful if fodder break crops are also trap-crops. The fodder millet variety Babala is useful in this respect for *A. vogelii*-infested sites, at least in Southern Africa. A much needed useful line of research is screening of varieties of known

Table 2.4. The trap-cropping effects of the non-hosts pearl millet, bambara and sunflower on *Alectra vogelii* in pot culture.

Previous crop	% reduction in *A. vogelii* on following cowpea crop
Bambara 'local'	67
Millet cv Serere 6A	48
Sunflower cv Russian No. 4	48
No crop	0

trap-crop species for germination stimulant production. Variation in the level of germination of *A. vogelii* caused by different varieties of the trap-crop guar have been observed (Visser and Beck, 1989).

Soyabean, a crop which has been promoted by a number of national programmes throughout the savannahs of West Africa, produces *S. gesnerioides* germination stimulant in laboratory tests (Igbinnosa and Okonkwo, 1991). Choice of variety of this potential trap-crop would be critical to successful reduction of cowpea strains of the parasite. While one variety stimulated 68% of the parasite seed to germinate, a number of others produced no stimulant. Similar variation occurs in the ability of pigeon pea varieties to stimulate *S. gesnerioides* germination. This is a potentially useful species for incorporation into rotations as a trap-crop for sites where both *A. vogelii* and *S. gesnerioides* occur. Pigeon pea was found to be a potent stimulator of the germination of samples of *A. vogelii* from Botswana (Riches, 1989b). Sorghum, the staple cereal at many locations where *S. gesnerioides* infests cowpea in West Africa, also causes some cowpea witchweed germination (Parker and Reid, 1979). These areas, however, tend to be infested with the cereal parasites *S. hermonthica* and *S. aspera* to which any *S. gesnerioides* trap-crop should not be susceptible.

Catch-crops These are susceptible species which are ploughed in or harvested after parasite attachment, offering a further modification to rotations. The principle is to plant a catch-crop as early as possible in the season and then plough under the host and parasite while the crop is still young when there is sufficient time to replant and take the susceptible crop through to yield. In theory the catch-crop will reduce the infestation of the field for the subsequent crop. Risk-reducing strategies are of prime importance to smallholders who grow crops under the extremely variable rainfall patterns of many of the *A. vogelii* and *S. gesnerioides* infested areas. Planting opportunities are already limited under these conditions, so any practice which requires changing the time of planting or forgoing periods of good soil moisture are unlikely to be adopted by smallholders. A catch-cropping regime was recommended to commercial farmers for the management of *A. vogelii* infestations in South Africa (Hattingh, 1954). This involved planting cowpea, soya or sugar bean in the latter half of November. Towards the beginning of January when the parasites emerged the crop was cut and land ploughed to prevent *A. vogelii* from seeding. At this stage of the season a quick maturing sunflower variety could then be planted and subsequently harvested, at least in seasons of good rainfall. The prompt destruction of cowpea or groundnut residues after harvest, described earlier, will have the same effect as a purpose sown catch-crop. Some weed species may also act in this way to reduce parasite infestations. *Acanthospermum hispidum* becomes dominant towards harvest in eastern Botswana and stands become parasitized by *A. vogelii* (Riches, 1989b). The parasite develops very poorly on *A. hispidum* following attachment. Subterranean stems generally grow little more than a few millimetres long. A combination of drought, frost or grazing kills the host–parasite system before parasite emergence. Flowering stems of *A. vogelii* are rarely seen on the weed. In the traditional farming system poor weed control is the norm. *A. hispidum* stands reach upwards of 70,000 plants per hectare and clearly, acting as a catch-crop, bring about a reduction in the *A. vogelii* population.

Fertilizer use

Few studies have been made of the effect of fertility on *A. vogelii* or *S. gesnerioides*. In sand culture the incidence of *S. gesnerioides* on cowpea decreases with increased nitrogen concentration in the watering solution (Emechebe *et al.*, 1991). Parasite number fell by more than 50% at an N_2 level of 20 ppm. The effect of fertilizer in the field has, however, been inconsistent. Trials in Niger and Nigeria have shown that, in contrast to the situation with the cereal witchweeds, nitrogen application may have no significant effect on *S. gesnerioides* populations or the yield of infested cowpea crops (Emechebe *et al.*, 1991; Laycock, 1989). Rates of 30 to 120 kg ha^{-1} of N applied as urea or calcium ammonium nitrate were ineffective in the Nigerian study. In subsequent studies cowpea varieties responded to nitrogen differently, 30 kg ha^{-1} of either nitrogen source reducing *Striga* incidence on cowpea variety Suvita 2 and of one local type, but not on variety Sampea-7. Substantial yield increases were associated with reduction in *Striga* incidence (Lagoke *et al.*, 1990). Unpublished results from the Weed Research Organization showed that inoculation of cowpea with *Rhizobium* resulted in much greater infestation by *S. gesnerioides*. Phosphate alone has also been reported to increase cowpea yield (Laycock, 1989), but this needs to be confirmed in follow-up trials. Little work has been carried out on the effect of fertility on *A. vogelii*. Observations in Botswana showed that neither single superphosphate nor ammonium nitrate at recommended levels for cowpea and sorghum affected the infestation level of *A. vogelii* on Blackeye cowpea when applied to the current crop or as a residual treatment. Parasite emergence and the number of plants affected was reduced following an application of 60 kg N ha^{-1} in a recent trial in Nigeria, but cowpea yield was depressed (Maganin *et al.*, 1992). In-depth study of the reaction of *Alectra* species and *S. gesnerioides* to fertilizer is clearly overdue.

Chemical Control

Susceptible legume crops are well established by the time *A. vogelii* and *S. gesnerioides* emerge. Achieving selective control of parasite stems among the crop canopy with a post-emergence herbicide would therefore be difficult and would be unlikely to have much effect on crop yield. Problems of selectivity would be even greater in mixed crop situations. The translocation of the herbicides **dicamba** and **glyphosate** from the point of application on cowpea leaves to *S. gesnerioides* has been demonstrated in the laboratory (Samb and Chamel, 1991, 1992) but the potential for using these translocated chemicals for the control of the parasite in cowpea in the field has not as yet been determined. In preliminary tests with some 15 herbicides only **imazaquin** at 0.4 kg ha^{-1}, applied as a soil drench, and **pendimethalin** at 1 kg ha^{-1} applied before emergence of cowpea, showed any potential to reduce *A. vogelii* and *S. gesnerioides* numbers without causing serious damage to the host (Polniaszek and Parker, 1987). Imazaquin reduced *S. gesnerioides* populations in field trials in Niger but severe crop phytotoxicity was reported (Laycock, 1989).

Natural and Biological Control

Although biological agents have often been suggested for the control of parasitic weeds, little critical research has been carried out and there has been no practical application of the approach to suppress *Alectra* or *Striga*. There are a number of insects which feed on these species and also fungal diseases which play a role in the natural regulation of parasite populations. The complex of insect natural enemies of *S. gesnerioides* in West Africa is broadly similar to that affecting *S. hermonthica* (Markham, 1985). Weevils in the genus *Smicronyx*, which cause the formation of stem or seed galls, are widespread and possibly the commonest insect affecting *S. gesnerioides*. If stem galls occur at the base of the stem, vigour and flowering can be reduced substantially; seed galls on the other hand are rarely common enough to account for more than 20% of seed production. Larvae of the moth *Stenoptiloides taprobanes*, found on *S. gesnerioides* in West Africa, also feed on floral parts of *A. vogelii*, preventing seed production of predated flowers. Only a small proportion of parasite plants are attacked in any stand so the overall effect on seed production is probably insignificant. Fungal pathogens of *A. vogelii* are probably more important. At some sites in Botswana as many as 40% of parasite stems fail to produce seed. These die back following infection by a stem wilt apparently caused by *Fusarium oxysporum* and *F. solani* (Riches, 1989b). *S. gesnerioides* is affected at many sites in Niger and Mali by a black rot or blight which can be highly destructive, though seed production in a stand of the parasite is usually not prevented entirely (Markham, 1985).

No research has been reported on possible use of such organisms for biological control. It would be helpful to describe the cropping practices which encourage the survival of existing natural enemies and to ensure that if pesticides are used on cowpea crops these do not reduce the current effectiveness of insects which feed on *Striga* or *Alectra*.

Resistant Varieties

One major cause of the increased importance of *A. vogelii* and perhaps also of *S. gesnerioides* has been the widespread adoption of introduced high yielding cowpea varieties in infested regions. Varieties such as the Blackeye types introduced into southern Africa from California during the 1950s have by and large greater susceptibility than many of the traditional longer-duration lower-yielding landrace types which have been displaced. The relatively high susceptibility of Blackeye was recognized in Botswana over 30 years ago (Anon, 1959) and an important, although simple, piece of advice to farmers remains that this variety should not be planted on *A. vogelii*-infested fields. During the past 15 years a number of workers have focused on the selection of legume crop varieties which are resistant in the field to *A. vogelii* and *S. gesnerioides*, with considerable success.

S. gesnerioides-resistant cowpeas

Attempts to screen cowpea lines for resistance to *S. gesnerioides* began with field trials in Nigeria in 1977. The work was subsequently expanded to include pot screening trials, research groups and sites elsewhere in West Africa, and assistance from *Striga* researchers in the UK. Although many hundred lines were tested in the early years the initial lead on sources of resistance came from IITA scientists based in Burkina Faso. The accessions Goram local (subsequently called Suvita 2) from Burkina Faso and 58-57 from Senegal supported little or no *Striga* emergence in the field and screen house, compared to a large number of emerged *Striga* on other varieties (Aggarwal *et al.*, 1984). When these two resistant lines were tested elsewhere in the region, 58-57 was found to be susceptible in Mali while both were susceptible in Niger and Nigeria (Aggarwal *et al.*, 1986). This variation in response is due to the presence of at least three strains of the parasite with differential ability to parasitize various cowpea genotypes (Parker and Polniaszek, 1990). Multiple strain resistance was subsequently discovered in B301 from Botswana and the IITA line IT82D-849 (Aggarwal, 1991; Parker and Polniaszek, 1990). The value of these lines has now been confirmed in field trials throughout the main *S. gesnerioides*-infested areas of West Africa. B301 has also been examined by comprehensive glasshouse screening at Long Ashton Research Station, UK, involving challenge by *Striga* samples from many parts of the region. To date the resistance has only failed to hold up to one sample of *S. gesnerioides* from southern Benin which apparently represents a further race of the pathogen (Lane and Bailey, 1992). Other sources of resistance, including IT82D-994 selected in Burkina Faso, are also available along with a number of derivatives produced by the breeding programme at IITA. Line TN88-63 selected by the Niger national programme is one of a number of lines which have been reported to be tolerant to *S. gesnerioides*. It is able to produce high yields despite heavy *Striga* infestation (Aggarwal *et al.*, 1984).

Recognizing that improved varieties of cowpea for West Africa must incorporate resistance to both *A. vogelii* and *S. gesnerioides*, IITA began a systematic breeding programme using B301 as a source of resistance to both species in 1987. This became possible following the discovery, described in the following section, that B301 is resistant to West African populations of *A. vogelii*. A number of high yielding lines have been selected which combine the multiple resistance of parent IT84S-2246-4 (to aphids, bruchid, thrips and several diseases) with the **dual parasite resistance** of B301. These are now being used in further crosses to breed a range of varieties differing in plant type, maturity and seed characteristics to suit farmer preferences for cowpeas in different parts of the region (Singh and Emechebe, 1992).

A. vogelii-resistant cowpeas

A number of landrace accessions collected from infested fields in Botswana are considerably less susceptible to Botswana samples of *A. vogelii* than Blackeye. Collected originally at infested sites as plants with no associated emerged parasite stems, this material included B359 which remained parasite free in a number of pot trials (Riches, 1987). Field trials subsequently confirmed that B359 is resistant in the field in Botswana (Riches, 1989a). In preliminary screening of some 358 local accessions from the Botswana cowpea collection, a further 12 were found which appeared to be completely resistant while a further 28 supported no more than two *A. vogelii* per plant compared to a mean of 14 per plant on Blackeye. Of 311 introduced lines which were also screened, only four were not parasitized at all while a further seven supported no more than two parasite stems per plant (Riches, 1989b). There is therefore a considerable range of germplasm which could be used as a source of *A. vogelii* resistance in future breeding programmes for Southern Africa. B359 is an indeterminate late-maturing type of cowpea and as such is not well adapted to seasons of poor rainfall. An initial attempt to combine *A. vogelii* resistance and early maturity has been successful, though breeding work has not yet been followed through to the stage of a resistant variety suitable for release to farmers (Riches, 1989b; Phillips, personal communication, 1992). A number of lines showing resistance or low susceptibility in Botswana are also resistant to collections of *A. vogelii* from West Africa (Polniaszek *et al.*, 1991). B359 has now been shown to be resistant in pot culture at least to *A. picta* from Ethiopia and Cameroon and to *A. vogelii* collected from various locations throughout West Africa, Cameroon, East, Central and Southern Africa (Riches *et al.*, 1992). The resistance of B359 has been confirmed in the field in Kenya (Bagnall-Oakeley *et al.*, 1991). Of other lines showing low susceptibility to the parasite in Botswana, B301 which is being exploited as a source of resistance to *S. gesnerioides* in the IITA breeding programme is also resistant to samples of *A. vogelii* from Mali and Cameroon and has now been shown to be completely resistant in the field in northern Nigeria (Singh and Emechebe, 1992). Although B301 is resistant in the field in Kenya, pot trials have shown that it is susceptible to strains of the parasite from Malawi, Botswana and some areas of South Africa (Riches *et al.*, 1992). Multi-location field trials will therefore be essential to confirm the geographic utility of *A. vogelii*-resistant varieties. As a number of strains of the parasite exist, different sources of resistance may be required for different areas, though B359 does appear to offer multi-strain resistance. A number of other lines from the IITA breeding programme including IT86D-534, IT86D-472, IT86D-371 and IT84D-666 are also highly resistant to *A. vogelii* in Nigeria (Singh and Emechebe, 1991). Resistance to the parasite in determinate high yielding varieties will therefore be available for West Africa in the near future. There is, however, no general cross-resistance in cowpea to both *Alectra* and *Striga*. Varieties Suvita-2 and IT-82D-849, two of the most resistant lines to *S. gesnerioides* in West Africa, are highly susceptible to *A. vogelii* in both Southern and West Africa (Riches, 1987; Polniaszek *et al.*, 1991).

A. vogelii-resistance in other crops

Differences have been demonstrated in the degree of yield loss suffered by a number of bambara selections when grown under infested conditions in South Africa (Beck, 1987). Actual parasite numbers per plant were not reported so it is difficult to judge whether reduced levels of yield were due to fewer parasites (some degree of resistance) or a tolerance of *A. vogelii* by the host. The accession TVSU 870, originally collected in Zambia, is resistant in pot trials to samples of the parasite from Kenya and Malawi but not from Transvaal in South Africa (Riches *et al.*, 1992). The resistance of this accession needs to be tested at a number of sites in East, Central and Southern Africa.

Little work has been carried out to examine the differential susceptibility of groundnuts to *Alectra* species. On the basis of a single pot trial, Fasil Reda and Parker (1990) suggested that the variety ICG-273 may be resistant to *A. picta* in Ethiopia. A wide range of susceptibility to *A. vogelii* among a number of lines has also been observed, suggesting that there is scope for selecting resistance in this crop (Riches *et al.*, 1992).

Mechanism of resistance

The mechanism which limits the host range of *A. vogelii* apparently operates after the parasite seed has germinated. Although samples from Botswana germinate when exposed to root exudate collected from bambara, the crop does not support the development of the parasite (Riches, 1989b). In addition to germination stimulant, the root exudate of the leguminous crop guar also contains inhibitory substances which prevent the post-germination elongation of *A. vogelii* radicles. This effectively prevents the penetration of guar roots (Visser and Beck, 1989). In other non-host species the parasite may penetrate the root but further growth is prevented. Penetration of pea and faba bean root cortical cells is limited to intercellular growth. Invading parasite cells are subsequently encapsulated and isolated from the potential host leading to subsequent starvation. Chemical or biochemical processes appear to be the basis of parasite recognition by these species (Visser *et al.*, 1990) and are probably widespread determinants of host specificity.

The resistance of certain cowpea lines to *Alectra* and *Striga* is also expressed after the parasite has penetrated the cowpea root. At least two mechanisms of resistance to *S. gesnerioides* are involved (Lane and Bailey, 1991). In the variety 58-57, resistance is expressed two to three days after the parasite radicle has penetrated the cowpea root as a necrosis of host tissue around the point of infection. This is associated with the death of the parasite. The necrotic response to infection is also seen in the case of B301. However, most infections of this line are initially successful, small tubercles 1–2 mm in diameter developing after 3–4 days, but no further growth of the parasite occurs. Full development of the parasite is prevented by this host even though connection is made with the host vascular system. Resistance of cowpea B359 to *A. vogelii* is expressed in a similar manner to the response of B301 to *Striga*. Although the radicles of *A. vogelii* penetrate B359 and parasite cells become established in the host vascular tissue, little further development of the parasite takes place (Lane *et al.*, 1991b). In field trials in Botswana less than 5% of B359 had visible parasite attachments at harvest and these had developed stems no more than a few millimetres in length (Riches, 1989b). A number of local accessions in Botswana which show low susceptibility in terms of the number of *A. vogelii* attachments present at cowpea maturity are also poorer hosts for parasite development after attachment. Parasite shoots are smaller on these than highly susceptible lines and usually do not emerge (Riches, 1987). The actual physiological or biochemical basis of resistance in cowpea to *A. vogelii* and *S. gesnerioides* has not yet been determined.

The resistance of cowpea types studied to date is not associated with a reduced ability of the host to stimulate the germination of seed of either parasite species. While preventing the development of parasite attachments, these varieties will in effect act as trap-crops to reduce the parasite seed bank. Another feature of this type of resistance is that, as parasite emergence is prevented, there is no opportunity for plants which begin to develop after germination to flower and set seed. There should therefore be no risk of a more virulent strain of the parasite building up where the resistant types are grown. It will always be sensible, however, for farmers to monitor their crops for any parasite emergence and to rogue accordingly to prevent the build up of a resistance-breaking strain.

Genetics of resistance in cowpea

Monogenic dominant inheritance of resistance to *S. gesnerioides* has been demonstrated in the progeny of crosses between Suvita 2, 58-57 or B301 and susceptible lines (Aggarwal *et al.*, 1984; Aggarwal, 1991; Singh and Emechebe, 1992). Because it is controlled in each case by a single dominant gene the resistance is relatively easy to transfer. If further studies indicate that there are different genes for resistance in the different parents, particularly those with multiple strain resistance, the potential exists for combining a number of resistance genes in one variety. As new varieties are currently being bred with single gene resistance the concern exists that *Striga* strains may arise which can overcome the resistance. Broadening the genetic base of resistance will improve its stability in the field. Resistance to *Alectra* in B301 is controlled by duplicate dominant genes (Singh and Emechebe, 1990).

Prevention of New Infestations

Although there have been no critical investigations of the potential for dissemination of parasitic weed seeds in crop seed, crop improvement and seed distribution programmes should be aware of this possibility. With the regular movement of breeding lines and promising new varieties between research programmes across Africa it is vital that every precaution is taken to ensure that seed batches are not contaminated. Every effort must be made to ensure that biotypes of *S. gesnerioides* which are capable of attacking cowpea are not introduced into Central and Southern Africa. Although cowpea is usually harvested by hand, contamination could occur particularly in late maturing varieties if parasite seed is shed onto pods and then onto crop seed during threshing. *A. vogelii* seed in soil could theoretically contaminate groundnut pods and this again could get into the seed during processing. *A. picta* infestations now seen in Ethiopia probably originated from imported contaminated seed. To be completely safe any seed due for export to as yet uninfested regions should be produced from uninfested sites or, in the case of cowpea, from fields which are regularly rogued of parasite plants before these flower. At a local level, seed multiplication operations must ensure that only clean seed is offered to the farmer. Once again fields should be regularly inspected and if necessary rogued. Particular care should be taken if the crop is to be mechanically harvested.

Research on parasitic weeds should only be carried out at sites which are already infested. Parasite seed should not be collected in one region to be used at an infested research site elsewhere in the same country until the host range of the sample being moved is fully understood.

Integrated Control

In view of recent progress with the identification and transfer of resistance in current breeding programmes, the prospects for combating the problems of *S. gesnerioides* and *Alectra* species in cowpea appear excellent. Until resistant varieties are released extension advice will need to concentrate on an integration of cultural methods of control with the emphasis on practices which fit in with existing cropping practice and which are affordable. This will also be the case for some time for groundnut, until sources of resistance are found, and minor legumes including mung bean, the cultivation of which in infested areas may be too limited to justify resistance-breeding programmes.

The starting point of an integrated package of control measures should be the choice of the least susceptible variety available. Even where a ranking of variety susceptibility has not been conducted, some information will probably be available within the farming community on farmers' perceptions of which varieties and landraces are least affected. In Botswana, for example, it is generally recognized that certain landraces of cowpea are less susceptible than introduced varieties, including Blackeye. Recommendations of local and probably lower yielding lines will be easier to make for farmers who are producing a legume predominantly for subsistence and who therefore will require stable rather than maximum production.

Where susceptible legumes are grown within multiple cropping systems the farmers will most likely aim to maximize yield of the cereal component, which may limit control options for *S. gesnerioides* or *Alectra* species. Opportunities for including known trap-crops should, however, be exploited wherever possible, taking care that these are not susceptible to other parasites. In many areas fields are infested with one or more of the cereal witchweeds and either or both *S. gesnerioides* and *A. vogelii*. In parts of southeastern Botswana both *A. vogelii* and *S. asiatica* are a problem. In this situation pearl millet which is not susceptible to *Striga* but is a trap-crop for *Alectra* is a useful component of the rotation. If monocropping predominates it is easier to suggest rotational schemes to farmers. In mixed farming systems the incorporation of fodder crops into the arable rotation may also offer the opportunity to use a trap-crop. Fodder types of pearl millet could be useful for trap-cropping *A. vogelii* in Southern Africa. Care is also needed that the introduction of a fodder does not increase parasitic weed problems. Lablab, for example, is susceptible to *A. vogelii* and should not be promoted as a fodder crop in infested areas.

Hand-pulling should always be recommended for light infestations or small patches of parasites, particularly if these are seen in fields which had not previously been affected. Removal of all emerged stems should also be promoted for farmers who are planting varieties which have low susceptibility or are resistant. Timely destruction of crop residues to prevent parasite seed production after harvest will also be sound advice. This practice may be easiest to implement when early maturing varieties are planted. Where mixed cropping is the norm, farmers could be encouraged to plant their susceptible legume between strips of the cereal so that the legume crop may be easily hoed or ploughed out after the pulse harvest which is likely to be a number of weeks before the cereal ripens.

The design of sensible integrated control packages will depend on an in-depth knowledge of the target farming system and particularly of farmers' production goals and resource constraints. It is only with this information that appropriate control measures can be recommended.

Conclusion

Provided research and seed multiplication programmes in affected areas receive adequate support and take full account of farmer preference in terms of plant type and seed quality, the eventual adoption of resistant cowpea varieties should be assured. In parallel with the promotion of resistant material, extension services should be encouraged to operate follow-up monitoring systems so that early warning of any breakdown of resistance is possible. It will also be important to encourage farmers to remove any emerged parasite stems which do emerge in stands of resistant varieties, to guard against the build-up of particularly virulent strains of either *A. vogelii* or *S. gesnerioides*.

3

OROBANCHE SPECIES: THE BROOMRAPES

The name 'broomrape' derives from the association of certain species with 'broom' (*Cytisus* spp.) in Europe. The damage is usually less sinister than that caused by *Striga* in cereals, but high seed production and longevity contribute to the build-up of very damaging populations in cropping systems where there is inadequate rotation, especially in tomato and tobacco growing. It is not so often a problem for subsistence smallholders, but growers of faba bean in the Mediterranean region are affected and even for the more sophisticated farmer or vegetable grower it can prove extremely difficult to bring under control and may lead to enforced changes in cropping.

TAXONOMY AND IDENTIFICATION

The family Orobanchaceae includes about 14 genera, all holo-parasites, lacking chlorophyll, with a fleshy, pale stem bearing a simple spike or raceme of flowers and with small scales in the place of leaves. By far the most important genus is *Orobanche*, known as broomrapes, but *Aeginetia* and *Christisonia* may also occasionally occur as weeds (see Chapter 4).

The genus *Orobanche* has more than 100 species but only five are considered here as significant parasites of crops and are included in the following identification key. A few further species occur more locally and are noted below. It is unlikely that other species will be found on crops in any numbers but many other species could be encountered on wild hosts and for those a local flora or taxonomist should be consulted.

Orobanche species can be difficult to identify when dead, or when dried as herbarium specimens, and are much easier when flowers still show their natural shape and colour (and fragrance, in *O. crenata*). Hence it is important to make good notes when collecting and preparing herbarium specimens. The following key should, however, be suitable even for dried specimens. Branching usually distinguishes the first two species from the rest but the bracteoles may need to be identified (see Fig. 3.1). The main difficulties are between *O. ramosa*, especially the larger forms sometimes known as *O. mutelii*, and *O. aegyptiaca*, for which it is necessary to look closely at the hairiness of the anthers. There can be difficulties also between *O. minor*, *O. cernua* and *O. crenata*. For these the size and shape of the corolla and corolla-lobes are important, also the attachment of the filaments inside the corolla tube (see Fig. 3.2).

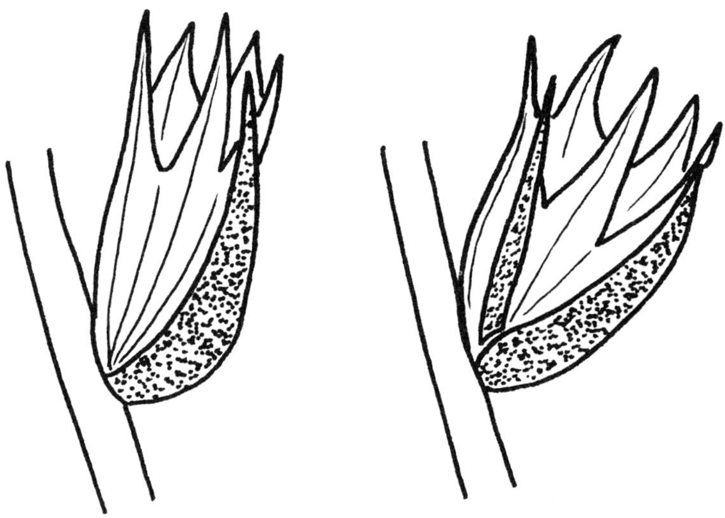

Fig. 3.1. Calyces of *Orobanche* with (l.) bract only and (r.) bract and bracteole (shaded for clarity).

Key to the Main *Orobanche* Species

A Stems usually branched above ground, bracteoles present

Plant branched (except very small individuals), usually 15–20 cm high but up to 30 cm. Flowers mainly uniform pale blue/purple outside but with two white cushions on lower lobes, tube narrowest at base and widening towards the open mouth, length usually **less than 20 mm. Anthers glabrous or sparsely hairy at the base**. Parasitic on a wide range of crops but especially tomato . . . . **_O. ramosa_**

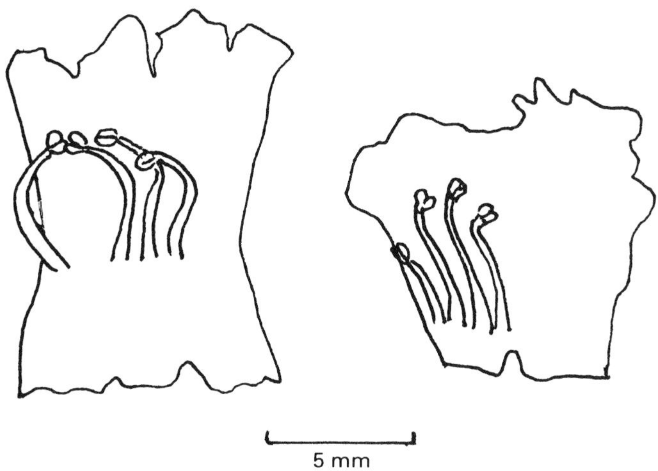

Fig. 3.2. Flowers of (l.) *Orobanche cernua* and (r.) *O. minor* opended to show different levels of insertion of the filaments.

As above but flowers **over 20 mm, anthers glabrous or sparsely hairy**. Host range as above . . . . ***O. mutelii*** or ***O. ramosa*** **sub-species** ***mutelii***

Plant branched, usually 20–30 cm high, sometimes larger. Flowers as above but longer, **always more than 20 mm. Anthers densely hairy**. Parasitic on a wide range of crops, especially tomato, also cucurbits . . . . ***O. aegyptiaca***

B Stems usually unbranched above ground, bracteoles absent

Plants unbranched above ground though sometimes branching from the base below, 20–30 cm high. Corolla up to 20 mm long, tube uniformly pale, becoming inflated below, with a narrow mouth and **small, 1–2 mm lobes with contrasting deep blue or purple coloration**. Stigma whitish. **Stamens inserted well up the corolla tube**, at least 4 mm from the base. Parasitic on Solanaceae (especially tomato, tobacco, eggplant) and Compositae (especially sunflower) . . . . ***O. cernua***

Plants unbranched, up to 50 cm (occasionally 100 cm). Flowers up to 25 mm long, the tube often streaked purplish at least along the veins, opening out into **lobes 2–3 mm long, normally white**. Stigma usually purple. **Stamens inserted low down in the corolla tube**, not more than 3 mm from the base. Parasitic on a wide range of hosts, especially Leguminosae . . . . ***O. minor***

Plants unbranched, robust, commonly up to 100 cm high. Flowers 20–30 mm long, the tube usually white, opening out to **widely divergent lobes 5 mm long**, making total width of the flower 1.5–2 cm. **Distinctly fragrant when fresh**. Stigma white, yellow or pinkish. Stamens inserted 2–4 mm from the base. Parasitic on several crops, especially faba bean and lentil . . . . ***O. crenata***

THE MAIN *OROBANCHE* SPECIES

Orobanche ramosa L.

O.ramosa (see Plate 9A) is perhaps the most widespread of the problem species and occurs in a range of types varying from the very small, unbranched forms only 10 cm high, sometimes known as *O. nana*, to the much larger forms, 30 cm or more, known as *O. mutelii*. Flower size is correspondingly variable, also the colour, varying from deep blue or purple to almost pure white. Confusion of the larger forms with *O. aegyptiaca* can be a problem, resolved to some extent by looking at the hairiness of the anthers. If the anthers are glabrous one can with confidence call the plant *O. ramosa* but Musselman (1991) shows that there are small-flowered forms with densely or moderately hairy anthers which are difficult to ascribe. It is clear therefore that *O. ramosa* and *O. aegyptiaca* are very closely related. They have the same chromosome number and Musselman (1986) has shown them able to cross. Hence hybrid intermediates may well occur. There is a useful key to all the main species in the *Flora Europaea* (Tutin *et al.*, 1972). Chromosome number (2n) = 24.

Distribution *O. ramosa* has a mainly Mediterranean distribution but extends to central Europe, the Middle East and northern Africa. It has been introduced to a number of other regions, including South Africa, Mali, Australia, Cuba and several sites in Central America and the USA.

Host range This is extremely wide. The crops most commonly attacked are indicated in Table 3.1. The most serious economically are in the families Solanaceae, Cruciferae and Leguminosae, but crops in several other families can be affected including a number of ornamentals such as *Chrysanthemum*. There is a long list of other families in which wild plants are hosts, including Chenopodiaceae, Amaranthaceae, Rubiaceae, Oxalidaceae, Euphorbiaceae, Plantaginaceae, Capparidaceae, Labiatae, Linaceae, Malvaceae, Polygonaceae and Rosaceae. In addition to the monocot crop onion, listed in the table, there is a single report of attack on maize, but in general, members of the Gramineae are immune.

Table 3.1. The main host crops of *Orobanche* species

Family/crop	*O. ramosa*	*O. aegyptiaca*	*O. minor*	*O. cernua*	*O. crenata*
Alliaceae					
onion	x	—	—	—	—
Cannabidaceae					
hemp	xx	—	—	—	—
Compositae					
lettuce	xx	—	x	—	x
noog/niger seed	—	—	x	—	—
safflower	—	—	x	—	x
sunflower	x	xx	—	xxx	x
Cruciferae					
cabbage	x	x	—	—	—
mustard	xx	xxx	—	—	—
rapeseed	xxx	x	—	—	—
Cucurbitaceae					
cucumber	x	xx	—	—	—
melon	x	xx	—	—	x
squash	x	xx	—	—	—
water melon	x	xx	—	—	—
Leguminosae					
bird's foot trefoil	—	—	x	—	—
chickpea	x	x	—	—	xx
clovers	x	—	xx	—	—
groundnut	x	x	x	—	x
faba bean	x	x	x	—	xxx
lentil	x	x	—	—	xx
lucerne/alfalfa	—	—	xx	—	—
pea	x	x	—	—	x
Solanaceae					
eggplant	xxx	xxx	—	xx	—
pepper	x	x	—	—	—
potato	x	x	—	—	—
tobacco	xxx	xxx	x	xxx	—
tomato	xxx	xxx	—	xx	x
Umbelliferae					
carrot	x	x	x	—	xx
celery	x	x	x	—	x
fennel	x	x	—	—	—
parsnip	x	x	—	—	—

xxx seriously attacked; xx moderately attacked; x lightly attacked; — attack doubtful.

Variation Size, vigour and flower colour variations have already been noted, and are recognized in the use of different names, treated usually as subspecies but occasionally as full species. The very small forms, known as ssp. *nana* are usually associated with wild hosts (e.g. *Oxalis pes-caprae* in Lebanon) but also occur sometimes on crops, especially late in the season. Whether these forms are genetically distinct, or merely reduced in vigour by growing under suboptimal conditions is not certain. There has also been insufficient study, as yet, of the possible physiological variation as expressed in host range. Musselman and Parker (1982) showed that forms attacking Solanaceae were less virulent on lettuce. Jacobsohn *et al.* (1991a) also showed that a sample of '*O. mutelii*' from potato would attack crops in Leguminosae, Cucurbitaceae and Umbelliferae but attack on some of these was later and less serious than on Solanaceae. And in the former USSR, no cucurbits are among the 55 hosts listed for '*O. mutelii*' by Beilen (1948). There appears therefore to be some small degree of host preference in the populations studied so far, but insufficient to allow any confident recommendation for the inclusion of any of the known host crops in the rotation on an infested site. Much more work is needed on this topic.

Importance *O. ramosa* causes some of the most acute problems of any *Orobanche* species, especially on tomato, on which it is able to build up extremely rapidly, causing serious reductions of yield and precluding the economic production of this and a number of other valuable vegetable crops. It is listed by Holm *et al.* (1979) as a 'principal' or 'serious' weed in Egypt, Jordan, Lebanon, Italy, Turkey, Hungary, Nepal and Cuba. There have been few attempts to quantify the losses, because of the same problem referred to under *Striga*, namely the difficulty of creating weed-free plots for comparison with infested (see Box 1.1, p. 8). Emiroglu *et al.* (1987) used methyl bromide to create such plots and arrived at an estimate of 33% yield loss due to an infestation of 1.5 parasites emerged per host plant (tobacco). This is likely to be an overestimate, given the other pests and diseases which may have been incidentally controlled by the fumigant, but suggests that losses from the commonly much higher infestations are likely to be at least 33–50%. Tobacco quality can also be reduced and in Cuba, losses of 16–32% in yield due to *O. ramosa* resulted in 25–50% losses of crop value. Cordas (1973) reported 21–29% loss in yield of tomato in California. *O. ramosa* causes no characteristic symptoms in infested hosts, rather a general weakening of growth. It is suspected that it interferes especially with the host water balance, hence causing more serious damage under conditions of water stress, but damage can still be serious in well irrigated crops.

Orobanche aegyptiaca Pers.

This species, closely related to *O. ramosa* as noted above, is usually more robust, with flowers almost always over 20 mm long (see Plate 9B), but it is also highly variable. Plants with both large flowers and densely hairy anthers can be confidently ascribed to *O. aegyptiaca*. Until further work has been done, the naming of types with smaller flowers and varying hairiness of the anthers may need to follow local convention. Chromosome number (2n) = 24.

Distribution *O. aegyptiaca* is distributed less extensively than *O. ramosa*. It occurs mainly in southeastern Europe, northeastern Africa and the Middle East, but extends also to southern areas of the former USSR, Afghanistan, Pakistan, northern India and Nepal.

Host range The host range of *O. aegyptiaca* is not known to differ significantly from that of *O. ramosa* but there is some indication that it is the more important species on Cucurbitaceae.

Variation As for *O. ramosa*, there has been inadequate study of variation within *O. aegyptiaca*. Morphological differences have been noted and some subspecies have been recognized, but there is limited information on physiological variation in terms of host range. Jacobsohn *et al*. (1991a) show that samples of *O. aegyptiaca* from winter carrots are much less successful on summer-grown Solanaceae than a sample from summer-grown tomato. Hence there are again indications of some degree of host preference but no encouragement to think that rotation would be safe into any but a crop not known to be attacked at all.

Importance The seriousness of *O. aegyptiaca* as a crop pest is not as well documented as that of *O. ramosa* but it is noted by Holm *et al*. (1979) to be a major weed in Jordan, Iran, Arabia, Italy and Afghanistan and common in Iraq, Pakistan and Turkey. As for *O. ramosa*, no distinctive symptoms are caused, other than some yellowing or paleness of foliage. One report from the former USSR showed losses of yield up to 50% in water melon (Panchenko, 1974), 13–52% in musk-melon and 15% in tomato (Kabulov and Tashpulatova, 1974), and there is no reason to believe that *O. aegyptiaca* is any less damaging than *O. ramosa*. It is certainly regarded as a major problem in Lebanon, Jordan, Iran, Arabia, Afghanistan and Israel.

Orobanche minor Sm.

This and the following species differ quite clearly from the previous two in having a simple, stout, unbranched stem (see Plate 9C), often more than 1 cm in diameter. The difference in lack of bracteoles is less obvious but can be checked by reference to Fig. 3.2. Flowers are somewhat similar to those of *O. crenata* but smaller, while they differ from *O. cernua* in that any purple colour occurs in streaks down the veins, into the corolla tube, whereas in *O. cernua* it is a uniform localized coloration of the corolla lobes. Chromosome number (2n) = 38.

Distribution *O. minor* is more temperate than that of the two preceding species, extending to northern Europe (other than Scandinavia). It also occurs further south in Africa down to the highlands of East Africa, and further afield, presumably by introduction, in South Africa, Australia, New Zealand and the eastern USA.

Host range *O. minor* has an extremely wide range of hosts, including the crops indicated in Table 3.1, especially the forage legumes, clovers, lucerne and trefoil in Leguminosae but also occasionally others in Compositae, Solanaceae and Umbelliferae, and non-crop hosts in many other families.

Variation *O. minor* is a highly variable species and a number of subspecies are recognized in Europe, varying in stature and in the colour of the stem, corolla and stigma. These are to some extent associated with particular host species or genera and some of this variation could be due to an influence of the host on which they are growing but there has been little experimental exploration of this possibility. Musselman and Parker (1982) provide evidence for the existence of some differences in host preference, a sample from lettuce having a wider range of hosts, including clovers and tobacco as well as lettuce, compared with samples from clover hosts which were unable to attack lettuce. As with the preceding species, therefore, there seems to be some degree of host preference but this is rarely at all narrow and in the rotation of crops as a control measure it is necessary to assume that most populations of the species may be able to attack all of the recorded hosts.

Importance Although occurring on a range of crops and occasionally conspicuous, especially in forage legumes in Europe and on tobacco, as in Ethiopia and New Zealand, it is rarely recorded as a major or intractable problem. Populations are rarely as dense as those of *O. ramosa* and the damage caused is also apparently less serious, though Evans (1960) recorded 25% yield reductions in subterranean clover infested by 125

O. minor per square metre and 50% yield reduction by 430 m^{-2}. In the past it was regarded as a significant problem in clovers and lucerne in the UK but is now relatively uncommon. This change in status is mainly because these crops are no longer grown in the UK for seed, when the weed has the opportunity to flower. When they are grown only for grazing or for hay or silage, the repeated harvesting interrupts the maturation of the parasite, and build-up is prevented. The subspecies known sometimes as *O. griesbachii* Reat. is recorded attacking carrot (severely) and tomato (lightly) in Jordan (Abu-Irmaileh, 1979a).

Orobanche cernua Loefl. (including *O. cumana* Wallr.)

The status of the two names *O. cernua* and *O. cumana* will be discussed further under 'Variation' below. Both forms are superficially similar to each other and to *O. minor* and separation of the two species from herbarium specimens is not easy, unless there are good notes on the colour of the fresh flowers (see Plate 9D), or if there are mature capsules present, when it is noted in most keys that the corolla tube of *O. cernua* (in the broad sense) is swollen and shiny below and constricted above the capsule. After dissection of flowers, the point at which the filaments are attached to the inside of the corolla tube can usually provide a positive identification, being at least 4 mm in *O. cernua* and less than 3 mm in *O. minor* (see Fig. 3.2). Chromosome number is apparently variable; (2n) = 24 or 38.

Distribution The distribution of *O. cernua* (in the broad sense) is similar to that of *O. ramosa*, mainly in the Middle East, southern Europe and northern Africa but extending further southwards in Africa, to Niger (Agadez) in the west and Tanzania and Uganda in the east, also further eastwards in Asia, into India and many of the states of the former USSR. Musselman (1986) believes the plant known as *O. australiana*, present in Australia, may also belong to this species complex.

Host range The host range of *O. cernua* is much more restricted than that of the other weedy species, and the crops affected are in only two families: Compositae (especially sunflower) and Solanaceae (especially tomato, tobacco and eggplant). Wild hosts may occasionally be found in other families.

Variation There is a very important distinction between the two main forms of this species in its broadest sense, i.e. that which attacks the Solanaceae and is almost always referred to as *O. cernua* and that which attacks Compositae, still referred to by many as *O. cumana*. The two forms appear to have strict host preference as confirmed by Jacobsohn *et al.*

(1991a), which is apparently **not** based on differences in germination stimulants, as Krishnamurthy *et al.* (1977a) show that the form attacking tobacco is stimulated to some extent by sunflower. There are also physical differences, at least in Israel, both in seed characters (Joel, 1987a,b, 1988a) and in shoot and flower morphology (Joel, 1988b). The seeds have a series of pits separated by 'bars'. In the seeds of *O. cernua* (on Solanaceae) the diameter of pits is larger than the width of the bars between them, while in '*O. cumana*' (on sunflower) the pit diameter is smaller than the distance between pits. The shape is also described as 'longish' in *O. cernua* and 'pear-shaped' in '*O. cumana*'. In the seedling stage, *O. cernua* has a 'smooth apex' while that of '*O. cumana*' has a 'papillate apex'. In the inflorescence, *O. cernua* has a split in the calyx on the side facing the stem, and the corolla tube is long and bent, the opening facing down; while in '*O. cumana*', the calyx is not split and the corolla tube is shorter, not facing down. So far there have been no reports of important variation within the *O. cernua* attacking Solanaceae, but '*O. cumana*' is known to exist in a range of physiological 'races' varying in their ability to attack different sunflower cultivars. These races can also show some morphological differences (see Cubero, 1986). This will be discussed further in the section on varietal resistance.

The specificity of the two forms to their respective families appears to break down occasionally. In highly infested sites in Ethiopia, *O. cernua*, normally attacking only Solanaceae, is seen to attack *Xanthium strumarium* (Compositae) and also *Cynoglossum* (Boraginaceae).

Importance *O. cernua* (in its '*O. cumana*' form) has been of great importance as a pest of sunflower. Pustovoit (1976) describes how, even in 1912, the problem was sufficiently serious in Russia to prompt a resistance-breeding programme which was quite quickly successful but which has required further work over the years to cope with different races of the parasite (see below). It has also caused problems for sunflower growers in Yugoslavia, Hungary, Bulgaria, Turkey, Israel and more recently, in Spain. In the former USSR, yield losses from four parasites per sunflower host plant were estimated at 20% (Malykhin, 1974) and in Israel infestations of 6, 8 and 25 parasite shoots per sunflower plant have caused 52, 82 and 90% yield reduction respectively (Shalom *et al.*, 1988). As a problem on tobacco and tomato, *O. cernua* (in the strict sense) has been especially troublesome in Pakistan, India, Nepal, Iran, Arabia, Egypt and Italy. Krishnamurthy *et al.* (1977a) estimated yield losses of 24–52% in tobacco, depending on the time and intensity of infestation and the availability of soil moisture. Quality of tobacco is also seriously reduced.

Orobanche crenata Forsk. (= *O. speciosa* DC.)

This is the largest of the weedy *Orobanche* species, commonly reaching 100 cm high when growing on faba bean, and having a stem 1–2 cm thick (see Plates 9B, 10A,B,C). Small specimens could be mistaken for large *O. minor* and vice versa, but the widely flared corolla and the carnation-like fragrance of the *O. crenata* should be diagnostic. Chromosome number (2n) = 38.

Distribution *O. crenata* is restricted mainly to the Mediterranean but extends eastwards through the Middle East into Iran. Only isolated infestations have been reported in northern Europe as in the UK and Sweden, but the spring conditions in such places are not unlike the Mediterranean winter conditions where it thrives and it is not clear why it does not occur more commonly in northern Europe.

Host range Crop hosts of *O. crenata* occur in at least four families including Leguminosae (especially faba bean), Umbelliferae (especially carrot), Cucurbitaceae and Compositae, though occurrence on the latter two families is relatively unimportant. Tomato is not normally attacked but may be if planted during the winter period. Non-crop hosts are mainly in these same families but there can be sporadic occurrences on wild plants in others.

Variation Morphologically, *O. crenata* shows a moderate range of variation in stature and flower colour but perhaps less than is seen in other weedy species, and there are no generally recognized subspecies. Radwan *et al.* (1988a) claims to detect some differences between populations in Egypt, in terms of their virulence on a range of faba bean varieties, but detailed isoenzyme studies on this species in Syria and Spain (Verkleij *et al.*, 1991a,b) suggest that variation between populations is much less important than the high genetic variation within populations, which is to be expected from its predominantly out-crossing behaviour (Musselman *et al.*, 1981).

Importance *O. crenata* is especially important for its widespread occurrence and devastating effect on faba bean, and to a lesser extent on lentil and chickpea all around the Mediterranean. From a range of sources, Sauerborn (1991a) estimates that in just four countries – Morocco, Portugal, Spain and Syria – the area of infested faba bean totals nearly 200,000 ha or about 63% of the planted area. This area might be even larger but for farmers abandoning the growing of faba bean on land they know to be seriously infested. The problem is almost equally serious in a number of other countries including Egypt, Lebanon, Jordan,

Tunisia, Italy, Turkey and Malta. The symptoms of attack by *O. crenata* are no more distinctive than from other species but they can be a great deal more severe, resulting in total collapse and death of faba bean, perhaps most commonly when there is a degree of moisture stress (see Plate 10A). By comparing yields of large numbers of individual adjacent host plants, Mesa-García and García-Torres (1984) established that each parasite per host plant causes approximately 12% loss. By a different method (see Box 1.1, p. 8), Zaitoun *et al.* (1991a) show that yield loss of faba bean can be predicted from the percentage of plants infected (% yield loss = approximately % infection minus 5). On a regional basis, the overall loss has been estimated at 5.4% for Andalusia, Spain (Mesa-García *et al.*, 1984) and 12% for certain parts of Morocco (Aber, 1984a).

MINOR *OROBANCHE* SPECIES

Orobanche coerulescens Steph. ex Willd.

This species is 10–40 cm high, has no bracteoles, and has a pale violet blue corolla 14–23 cm long and pale yellow stigmas. Its main distinction is that bracts, calyx and corolla are covered in arachnoid hairs. There are reports of this species attacking crops in widely scattered locations, including eggplant in Israel (Joel, 1987b) and sunflowers in China (Liu and Li, 1988), but little more is known of its range or importance. In Nepal, Vasudeva Rao *et al.* (1988) reported attack on *Brassica* crops but there may have been confusion with *O. aegyptiaca*.

Orobanche foetida Poir.

This species has no bracteoles, a strikingly deep red corolla, 12–23 mm long, with a shiny red surface inside, purple stigmas, and a fetid smell. Distribution of *O. foetida* includes southern Spain and northern Morocco, where it occurs on lucerne and on bird's foot trefoil; also Tunisia where it attacks faba bean.

Orobanche gracilis Sm.

This species is closely related to *O. foetida* but the corolla is partially yellow, and less uniformly red outside, though similarly shining red inside. Stigmas are white or yellow. *O. gracilis* occurs throughout southwestern Europe and in extreme northeastern Africa, occasionally infesting forage legumes such as bird's foot trefoil.

Orobanche lutea Baumg.

This species also has no bracteoles but is distinguished by bright yellow or purplish yellow corollas, 20–30 mm long. Stigmas are also usually yellow. Distribution is mainly in central and southern Europe where it occasionally occurs on clovers and lucerne.

Orobanche solmsii Hook.f.

This species is characterized by simple stout shoots and lack of bracteoles, as in *O. cernua*. The calyx is 4-lobed, corolla tube broad, curved and yellow, 15–20 mm long. It occurs locally on tobacco, tomato and eggplant in Nepal, causing serious damage, especially on tobacco (Bharati, 1991; Khatri *et al.*, 1991).

Biology and Physiology

The Seeds

The seeds of *Orobanche* are similar in size to those of *Striga*, mostly about 0.3 × 0.2 mm and weighing 3–6 μg. They also have a pattern of raised ridges, but the pattern is coarser (see Fig. 3.3).

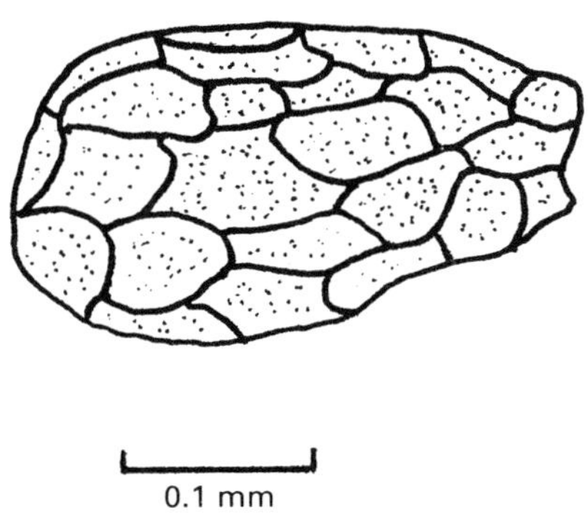

Fig. 3.3. Seed of *Orobanche aegyptiaca*.

The number of seeds per capsule varies from 600–800 in *O. ramosa*, about 900 in *O. minor*, to more than 4000 in *O. crenata*. As the number of capsules in *O. crenata* averages about 50, the total seeds per plant can be over 200,000.

Longevity of seeds in the field has been reported as 13 years for *O. ramosa* and ten years for *O. crenata*. Linke and Saxena (1991a) studying seeds stored in the laboratory found samples of *O. ramosa* retaining viability for 12 years, and of *O. crenata* and *O. aegyptiaca* still viable after

8–10 years, while other samples had lost viability within 3–5 years. They comment on the importance of low humidity for long storage, and the difficulty of relating longevity in the laboratory to longevity in the field. In Arabia it is often said that an *Orobanche* problem has reappeared after seven years or more of uncropped fallow, yet it is also said, of *O. cernua* in India, that three years of thorough weeding to prevent new seeding can largely eliminate the problem. These differing reports are presumably a reflection of different species and differing environmental conditions. It seems probable that, at least under conditions of low moisture and lack of cropping, longevity in the field can exceed five years, making it very difficult to control by fallowing or rotation.

Germination

As an obligate parasite with extremely limited seed reserves, *Orobanche* species have evolved the same requirement as *Striga*, i.e. for a chemical signal from a host root before the seeds will germinate. Thus it is normally ensured that a suitable root is available for attachment within a few millimetres of the germinated seed.

Again, like *Striga*, there is a need for a conditioning period of 1–2 weeks before seed will respond to a germination stimulant. The ideal temperatures for conditioning, however, are somewhat lower than those for *Striga*. Kasasian (1973a) showed that optimum temperatures for both conditioning and germination were about 18°C for *O. crenata* and about 23°C for *O. ramosa* (reported as *O. aegyptiaca* but since redetermined) but temperatures down to 13°C were also effective. Temperatures over 28°C were unfavourable, especially during the conditioning phase. Other reports of optimum temperature for germination of *O. ramosa* have varied from 18°C in the USA to 20°C in Germany and 25°C in Lebanon (ter Borg, 1986b). In Israel the optimum germination temperatures for *O. crenata* and *O. cernua* were about 20°C, while that for *O. aegyptiaca* was somewhat higher, about 25°C and showing less serious decline at 30°C (Foy *et al.*, 1991). Several authors suggest some enhancement of germination when conditioning temperatures differ from those for germination, also when there is some fluctuation of temperature at the germination stage. Extending the conditioning period does not result in any clear 'wet dormancy' as in *Striga*. After-ripening is reported to be necessary for *O. crenata* (Edwards, 1972). At favourable temperatures, germination of the main species normally takes 2–3 days.

Light is inhibitory to germination of *O. crenata* seeds, whether they are exposed during conditioning or germination phases (Hiron, 1973). *O. aegyptiaca* is also reportedly inhibited by light (Jain and Foy, 1987), while *O. ramosa* may not be (ter Borg, 1986b).

There has been even less success in identification of the natural germination stimulants of *Orobanche* than there has for *Striga*. Attempts by several teams working with linseed and *O. minor* (Brown *et al.*, 1951a,b, 1952), maize and *O. minor* (Sunderland, 1960), and faba bean and *O. crenata* (Mallett, 1973) all resulted in only partial purification and no reliable information on chemical structure. This was largely due to the dual problems of very low yield from their separation processes and the serious instability of the active substances. As for *Striga*, it is not only true hosts which exude effective germination stimulants. Two of the sources of stimulant used in the above studies, namely linseed and maize, are non-hosts and can be referred to as trap-crops or false hosts. As for *Striga*, a wide range of potential trap-crops have been identified, which have potential for inducing germination and reducing field infestations without themselves being parasitized. These will be discussed further under control measures. Although a crop such as maize stimulates germination of both *Striga* and *Orobanche*, Sunderland (1960) concluded that different substances in the maize root exudate were involved in the germination of the two parasite genera. Conversely, it is known that the *Striga*-germinating substance strigol, from cotton roots, and also its analogues such as GR-7 and GR-24, can trigger germination of *Orobanche* as well as *Striga* (Johnson *et al.*, 1976).

Although the root exudates from linseed and maize used in the attempts at isolation of the natural stimulant, noted above, were obtained from young seedlings, there has often been difficulty obtaining exudates from young faba bean plants with activity on *O. crenata*. Good germination may not be obtained until the bean plants are about 45 days old (Kadry and Tewfic, 1956; Khalaf, 1982, 1992; El-Ghamrawy *et al.*, 1990). One possible explanation for this, which has still not been positively confirmed, is that production of the stimulant is dependent on, or at least enhanced by, the infection of the roots with *Rhizobium* nodules which only occurs after some weeks. Petzoldt (1979) has even proposed that the *Rhizobium* nodules on roots of faba bean are the main site of attachment for *O. crenata*, but this has not been confirmed by other workers.

The germination response of *Orobanche* species to ethylene is still a controversial matter. Studies on *O. crenata* (Edwards *et al.*, 1976; Linke, 1987) and on *O. ramosa* (Kasasian, 1973a; Eplee, 1984; Linke, 1987) each failed to demonstrate any stimulation of germination by either ethylene or the ethylene-releasing substance ethephon. However, in two separate studies in the USA, significant response was obtained with *O. ramosa* (Chun *et al.*, 1979; Foy *et al.*, 1989).

Among other substances reported to influence *Orobanche* germination, gibberellic acid (GA) has sometimes been reported to stimulate germination alone and has even been thought to be the main natural stimulant of *O. crenata* in faba bean roots (Abdel Halim *et al.*, 1975;

El-Ghamrawy *et al.*, 1990). More commonly it is reported that GA causes only very limited germination on its own, e.g. 5–8% in *O. ramosa* (Kasasian, 1973a), but does significantly enhance germination when seeds are exposed to it during the conditioning phase, before exposure to natural stimulant. Joel *et al.* (1991a) believe that gibberellin biosynthesis may be an essential part of the conditioning process, and have been able to prevent germination of *Orobanche* seed by exposing it to inhibitors of gibberellin synthesis such as paclobutrazole or uniconazole during conditioning. Cezard (1973) showed that germination was promoted by various substances including calcium hypochlorite and claimed further that germination may also depend on certain unidentified microorganisms. This theory, however, has never been clearly confirmed.

Contact, Attachment and Penetration of the Host

On germination, the seed of *Orobanche* produces a radicle up to 3–4 mm long. Some authors refer to this structure as a 'germ-tube' or (in French) 'pro-caulome'. It does differ from a normal root in lacking a root cap but this probably does not justify the distinct terminology. The elongation of the radicle is thought to result largely from cell extension, though some authors have observed cell division also.

Chemotropism is believed to be involved in the successful location of the host root although the evidence for it is meagre. Whitney and Carsten (1981) proposed that inhibitory components of the root exudate would help to cause a bending towards the host root and, using an agar medium, confirmed a highly significant tendency for seedlings to demonstrate this, but only at 3 cm distance from the root, a distance from which the parasite seedlings would not be able to make contact. In zones closer to the root, there was virtually no chemotropic tendency at all. As with *Striga*, it would seem strange if there were not some mechanism for ensuring directional growth towards the host root.

On contact with the host root sticky papillae are formed, as in *Striga*, presumably with the function of adherence to the root surface (Aber, 1984b, Losner-Goshen *et al.*, 1991). It is assumed that haustorial initiating substances are involved but these have not been isolated or identified.

The penetration of the host root occurs by enzymic action, resulting in separation of the host cells, rather than their destruction or intracellular penetration. The anatomy of the haustorial contact between *O. ramosa* and its host, hemp, has been studied with immense care by Dörr and Kollmann (1974, 1975, 1976) and also summarized by Kuijt (1977, 1991). This differs from that between *Striga* and host in showing an effective link between the phloem of host and parasite, consisting of three different types of cell. Contact cells develop against the host sieve tube.

Next to these are transitional sieve elements, showing unusual characteristics including loss of nucleus, and finally the sieve tubes of the parasite which are more or less normal. Kuijt emphasizes that these studies were done on secondary rather than primary haustoria and the latter may show differences. This comment is surprising in that secondary haustoria in *Orobanche* are much less apparent than in *Striga* and it is not clear whether they are a common feature of all species, or peculiar to some such as *O. ramosa*.

Secondary roots are usually quite rudimentary, though there is often an outgrowth of short roots from the tubercle, which do not form secondary attachments to the host as in *Striga* but may be capable of some absorption of mineral nutrients, especially P, direct from the soil (Ismail and Obeid, 1976).

After the establishment of the haustorium in the host tissue, it is apparent that the radicle of *Orobanche* does have a more limited function than the radicle of *Striga* in that it is only responsible for carrying the active meristem of the embryo from the seed to the surface of the root where it establishes the infection of the host. Kumar (1977) working with *O. aegyptiaca* has observed provascular strands in what he refers to as the radicular tube and also development of a shoot at the plumular end, but it seems these may have been due to the influence of hormones used in his culture medium. More normally it seems that the radicle withers soon after the establishment of the haustorium and does not act as the main initial conducting route from the host to the parasite shoot. Instead, the haustorium swells to form the 'tubercle' or 'nodule' on which, after some 1–2 weeks, a shoot bud differentiates (see Fig. 3.4), eventually elongating and emerging to produce the flowering shoot.

Dependence on the Host

As holo-parasites, lacking chlorophyll, *Orobanche* species are totally dependent on the host for all organic carbon, which is acquired from both xylem and phloem, mainly, it is assumed, in the form of sucrose.

In terms of nitrogen metabolism, *Orobanche* species apparently lack nitrate reductase and hence are presumed not to utilize nitrate. Metabolism of ammonium nitrogen depends on glutamine synthetase which is present only in the form of GS1 (GS2 is very much associated with active chloroplasts). Although there are other minor differences in some enzymes involved in amino acid synthesis there is no firm evidence that *Orobanche* species have any serious limitation in their nitrogen metabolism. Hence, they may be acquiring a range of amino acids and amides direct from the host xylem and phloem but are not dependent on these elaborated forms of N (Press *et al.*, 1986; Stewart and Press, 1990).

Lacking any well developed roots, they are also dependent on the

Fig. 3.4. 'Nodule' of *Orobanche ramosa* on host root.

host for water. The flow of water from host to parasite is enhanced by a high osmotic pressure and also perhaps by a high rate of transpiration. Tuohy *et al.* (1986) recorded a high density of stomata on the scale leaves of *O. crenata* and it is believed there may also be exudation from specialized glands (Rakhimov, 1967). The high osmotic pressure in *O. crenata* can be attributed to invertase activity leading to lower levels of disaccharide such as sucrose and higher levels of monosaccharide in the form of glucose and fructose (Whitney, 1973). *Orobanche* is also found to have high levels of mannitol, especially under conditions of water stress. The exact function of these high levels is still a matter for conjecture (Press *et al.*, 1986) but one suggestion is that the mannitol provides a 'compatible solute' which protects enzymes against denaturation (Nour *et al.*, 1984). Wegmann (1986) shows how mannitol is synthesized in *O. ramosa* and theorizes that an interference with this synthetic pathway could be detrimental to the parasite. As mannitol is relatively rare in autotrophic plants, the host is likely to be much less dependent on it, and a compound inhibiting its synthesis could be the basis for a selective herbicide.

Although *Orobanche* cannot photosynthesize and must therefore depend on the host to increase its carbon resource, it is usually rich in starch and is able to survive and even continue growing for some time after the death of the host. Matthies (1989) showed that after the host faba bean was cut down, most of those *O. crenata* with emerged shoots exceeding 10 cm at the time were able to continue growing slowly for a further six weeks and produce some capsules.

Effects on the Host

The effects of *Orobanche* species on their hosts are not yet as well understood as those of *Striga* species. Various studies have shown changes in the constitution of the host tissues, including a decrease in K and P, while Ca is increased (Abou-Raya *et al.*, 1973). Depletion of K in the host is thought to be directly responsible for the damage from *O. ramosa* on tobacco (Ernst, 1986).

Nitrogen levels were shown to be higher in tomato infected by *O. cernua* but the ratios of different forms of N were unchanged and it was thought there was little disturbance to the host nitrogen metabolism (Singh *et al.*, 1972). In the same study, soluble carbohydrate in the host was higher in the infected host but stored and total carbohydrate were reduced. In carrot, a dramatic reduction in total sugars and especially sucrose due to *O. crenata* infection has been recorded by Schaffer *et al.* (1991). There is no evidence as yet for specific reductions of photosynthesis as a result of infection, but Stewart and Press (1990) consider that the depletion of carbohydrate resulting from *Orobanche* infection may cause a serious reduction in the ability of the host to absorb water from the soil under moisture stress conditions, as illustrated in faba bean by Whitney (1972); hence the commonly seen tendency for damage from *Orobanche* to show itself most acutely at times of moisture stress. Such moisture stress in the host can lead indirectly to reduced photosynthesis via closed stomata and reduced access to carbon dioxide.

The root system of the host is not stimulated as a result of infection by *Orobanche*, as it is by most *Striga* species, and there is often death of the distal part of the root system, beyond the point of attachment of the parasite.

ECOLOGY

Temperature

Optimum temperatures for the conditioning and germination of *Orobanche* species are considerably lower than those for *Striga*, perhaps explaining the absence of *Orobanche* from most warm tropical areas. However, once germinated in a cooler part of the year, several *Orobanche* species thrive under the extremely hot conditions which can prevail in the Middle East, as in the Jordan Valley; and germination of somé species is not apparently restricted to the cool winter season, as there can be severe *Orobanche* problems in many of the important summer-grown crops in hot regions of the Middle East, including tomato, eggplant and sunflower (Jacobsohn *et al.*, 1991a).

Light/Daylength

Kasasian (1973a) showed that reducing light intensity from 170 to 40 W m^{-2} resulted in much reduced attack of *O. crenata* on faba beans, though a further reduction to 9 W m^{-2} still allowed some attack, and this is not thought to be a critical factor in the field.

Daylength is not thought to have any direct influence on the flowering of *Orobanche* species and the observation of Holdsworth and Nutman (1947) that flowering of *O. minor* depended on the prior flowering of its host red clover (*Trifolium pratense*) has not been confirmed by other workers. Jacobsohn (1989), reviewing the topic, concludes from a number of published studies and his own observations that several of the major *Orobanche* species can flower and mature on non-flowering hosts.

Moisture

Orobanche species are associated with both rainfed and irrigated crops. *O. crenata* occurs mainly in unirrigated faba bean, and is observed to be reduced under wet conditions (Cubero, 1983) but it can also occur in irrigated carrots. Ter Borg and van Ast (1991) showed that numbers and dry weight of this species could be increased under wetter conditions when those conditions resulted in a larger host root system. *O. cernua* also occurs on both unirrigated tobacco and irrigated tomatoes, while the '*O. cumana*' form attacks mainly unirrigated sunflower, and *O. minor* is also mostly associated with rainfed crops. *O. ramosa* and *O. aegyptiaca* are commonest in irrigated vegetables but not exclusively so. Although these observations suggest no critical influence of soil moisture, it remains possible that *Orobanche* species are favoured by relatively low atmospheric humidity, which ensures a high rate of transpiration, and hence enhanced transfer of water and solutes from the host. The extreme of flooding, as in rotation with flooded rice, has been known to suppress *Orobanche*, and will be discussed under control methods.

Soil Type

The soils on which *Orobanche* species occur in the Middle East are mainly alkaline, and Bischof and Foroughi (1971) have shown that *O. aegyptiaca* is reduced somewhat at lower pH. Kasasian (1971) refers to other studies which suggest that *O. minor* and *O. ramosa* are also favoured by alkaline soils, and proposes that this could be due to the relative stability of the natural stimulant at lower pH and hence a possible tendency for seeds of *Orobanche* in acid soils to germinate too far from the host root and to die from suicidal germination. Izard (1959) comments that *O. minor* is associated with alkaline soils in England but that the crop tobacco which is commonly attacked by *O. ramosa* is more often grown on acidic soils. Hence there is no very strict association of all *Orobanche* species with high pH.

Fertility

Soil fertility appears to be a less critical factor for *Orobanche* than for *Striga*. There are a few reports of *Orobanche* species being favoured by low soil fertility but none of these appears to be based on any detailed survey work and in one (Southwood, 1971), it is pointed out that although infestation of subterranean clover by *O. minor* is lower on plots receiving fertilizer (mainly P), this could be due to suppression by this treatment of *Chondrilla juncea*, a favoured weed host of the *Orobanche*. Furthermore, the levels of fertility, especially nitrogen levels, required to achieve control are generally higher. This question will be discussed in more detail in the section on control.

Insects

By far the most important insect in the ecology of *Orobanche* species is the agromyzid fly *Phytomyza orobanchia* which occurs throughout most if not all of the regions in which the important *Orobanche* species are native. The insects overwinter as pupae in the seed capsules or the stem. On hatching in the spring, they mate and lay eggs on the developing inflorescence. The young larvae burrow into the developing capsule and/or the stem (see Fig. 3.5). Natural infestations can result in over 90% of capsules being damaged by the larvae in Yugoslavia (Mihajlovic, 1986). In Syria, Linke *et al.* (1990) surveyed in northwestern Syria and recorded 95% of sites with the fly present, a mean 55% of plants infested, 32% of all capsules infested and an overall reduction of 29% in seed production. In Turkey a survey of *O. crenata* in faba bean revealed 94% of the seed capsules destroyed by *Phytomyza* (Giray and Nemli, 1983) and in Bulgaria, the percentage of seed capsules damaged by *Phytomyza* was 20–30% for *O. ramosa* and 85–100% for *O. cernua* (Trenchev, 1981). Infestation in the stem can often result in secondary fungal infection followed by complete collapse and death of the shoots before seeds have matured.

Hyper-parasites commonly reduce the effectiveness of *P. orobanchia*, but in spite of this there has been successful exploitation of the fly as a means of biological control by enhancement. This will be discussed further in the section on control methods.

Further insects associated with *Orobanche* species include a *Psila*

Fig. 3.5. Larvae of *Phytomyza orobanchia* in stems of *Orobanche ramosa*.

species collected in Ethiopia and identified by the International Institute of Entomology. This is a larger insect than *Phytomyza*, the adult being over 1 cm long. The larva, up to 1 cm, was found in the stems of *O. minor* along with *P. orobanchia*. The broomrape root fly, *Chyliza extenuata*, is also known from the Netherlands (van der Goot, 1989). Both of these insects are in the dipteran family Psilidae.

Pathogens

Orobanche species are quite commonly damaged by pathogens, and a recent paper by Linke *et al*. (1992) greatly extends the list of fungi recorded from *O. crenata* and *O. minor*. They isolated and identified a total of 16 species from nine genera which included *Ulocladium* species of considerable interest for bio-control as well as *Fusarium* and *Alternaria* species that have been commonly recorded before. The possibilities for biological control will be discussed below.

CONTROL OF *OROBANCHE* SPECIES

Cultural and Mechanical Control Methods

Hand-pulling

Hand-pulling is the option most readily available to the small farmer, and must always be the first approach for light infestations when populations have not yet got out of hand, and also to supplement other methods which have given incomplete control. The practicality of hand-pulling will vary according to the *Orobanche* species and to the host crop. The larger, unbranched, thick-stemmed species *O. minor, O. cernua* and *O. crenata* are easiest to locate and grip, while the small, much-branched *O. ramosa* can be much more difficult and time-consuming to pull. Wide-spaced crops such as sunflower and tobacco offer better possibilities for seeing and reaching the parasite than densely planted or spreading crops such as faba bean and tomato which present much greater difficulties.

The purpose of hand-pulling may be mainly short-term, to reduce the damage caused, or long-term, to prevent seeding and hence reduce the problem for the future. In tobacco, it is claimed by Krishnamurthy and Rao (1976) that thorough hand-pulling for three years can bring *O. cernua* well under control in tobacco in India, although they also refer to the problem persisting after 10 years of growing non-tobacco crops. Hence it may be difficult to achieve good results under all conditions.

If the pulled plants are not being kept as a source of the insect *Phytomyza* (see Biological control), it is important, as for *Striga*, that pulled plants are removed from the field and destroyed, e.g. by fire. Storing them in a shallow pit until dry enough to burn will reduce the risk of seeds being washed or blown back onto the field. The frequency of hand-pulling, where it is done purely to prevent seeding, will depend on the rate of maturation and shedding of the seed. Seeds of *O. ramosa* mature within one week of flowering and are shed one week later, so the frequency of pulling may need to to be every two weeks.

The short-term benefits of removing emerged *Orobanche* in terms of the yield of the current crop have not often been measured, but the benefits claimed for use of *Phytomyza* in the former USSR, which only damages the parasite after emergence, suggest that good yield increases can be expected.

One disadvantage of hand-pulling can be that, due to the strong connection between host and parasite, the host root system may be disturbed, or the crop even completely uprooted as the *Orobanche* is pulled. This has been reported for eggplant (Misra *et al.*, 1982) and is presumably most seriously a problem on lighter soils. There can also be a problem of damage to tobacco leaves if hand-pulling requires frequent passage between the rows.

Rotational trap- and catch-cropping

Any rotation out of susceptible host crops is beneficial in terms of preventing new seed production, and allowing natural decline of the soil seed bank. The length of such rotation needed to achieve a high degree of control is not, however, well established. From the hand-pulling experience, mentioned above, it may be as little as 3–4 years for *O. cernua*. However, Puzzilli (1983) observed that a 4-year break from tobacco only reduced the problem by about 75%, and Zazzerini *et al.* (1981) claim that infestation was not checked even by rotations of 9 to 12 years' duration, though it is not certain that seeding (on weeds) was completely prevented during the rotation period.

The effectiveness of rotation should be improved by selecting crops which cause germination of the *Orobanche* but are not themselves attacked, usually known as **trap-crops** (see Box 3.1 below). Many such crops have been identified by laboratory tests for various *Orobanche* species, e.g.: for *O. crenata*, sorghum, barley and *Vicia dasycarpa* ssp. *villosa* (Kasasian, 1973a; Linke *et al.*, 1991a); for *O. ramosa, Phaseolus* bean, sorghum, maize and cucumber (Labrada and Perez, 1988); for *O. cernua*, sorghum, cowpea, chilli, hemp, mung bean, linseed/flax, lucerne, soyabean, chickpea (Krishnamurthy and Rao, 1976; Krishnamurthy *et al.*, 1977b).

The potential of some of these has been confirmed in pots, but rarely in the field. Krishnamurthy and Rao observed that infestations of *O. cernua* in tobacco were much less after sorghum than after fallow but Bischof and Koch (1974) reported that sweet pepper, while effective on *O. aegyptiaca* in the laboratory, had no beneficial effect in the field, and Khot *et al.* (1989) failed to demonstrate significant benefits from a range of potential trap-crops over several seasons on *O. cernua* in tobacco. The best result achieved by Linke *et al.* (1991a) was 62.5% reduction of *O. crenata* infestation after 3 years of growing *Vicia dasycarpa* spp. *villosa*.

Catch-crops are rotational crops which are susceptible to attack and usually have economic disadvantages, but have sometimes been technically more successful than trap-crops. One of the most successful results is that reported by Al-Menoufi (1991) who achieved a 98% reduction of *O. crenata* after four years planting to berseem in the winter and maize in the summer, compared with faba bean each winter over that period. Reduction after three years was 92%. The berseem supported some infection by *O. crenata* and success depended on regular harvesting for forage, to prevent seeding of the parasite. Abu-Irmaileh (1984) in pot experiments has shown that flax and sweet pepper can be very effective as catch-crops for *O. ramosa*, i.e. when grown for 4–6 weeks and then destroyed before planting tomatoes. These two crops reduced the subsequent infestation by 95 and 80% compared with a control without catch-crop. They were both superior to tomato itself which caused only about 55% reduction. Sesamum, eggplant and garlic were relatively ineffective. Unfortunately there have been no confirmatory reports from field experiments in Jordan.

Although flax has often been referred to as a trap-crop for *Orobanche* spp., Kleifeld *et al.* (1993) have shown that this crop is often parasitized, at least by *O. aegyptiaca*, and although the parasite does not always emerge and mature, the crop is damaged. Hence flax is to be regarded in this case as a catch-crop rather than a trap-crop. Even if it does not have to be destroyed to prevent *Orobanche* seeding, it is likely to be uneconomic due to damage from the parasite. Zahran (1982) found flax to be the best of several rotational crops tested against *O. crenata* in Egypt, reducing infestation by 50% in each of two years of testing. In the USA, also, it is being used to accelerate depletion of seed reserves of *O. ramosa* in Texas (Eplee, 1984). Flax varieties have been shown to differ in their ability to stimulate *O. ramosa* in Texas (R.E. Eplee, personal communication) and this is a factor to be considered in all work with trap- and catch-crops.

As with *Striga* the theoretical benefits from trap- and catch-crops may not always be realized in the field and possible reasons are noted in Box 3.1. Nevertheless, wherever there is a choice of rotational crops, one known to cause stimulation should be considered in preference to one which does not. When there is a means of ensuring that the parasite does not set seed, e.g. by herbicide, catch-cropping with the normal host crop can be more effective than using non-host trap-crops (Sauerborn, 1991b). The benefit of growing the susceptible crop in this way, compared with a rotational crop, has to be recognized in any long-term integrated programme.

An effective rotational crop for *O. cernua* in tobacco is rice (Krishnamurthy and Rao, 1976), but in this case it appears that the

Box 3.1. Trap-crops and catch-crops.

Where rotation is used as a means of controlling, or avoiding the build-up, of a root parasite such as *Striga* or *Orobanche*, its effectiveness in reducing the parasite seed bank in the soil should be improved by selecting crops which cause germination of the parasite but do not allow its seeding.

Trap-crops These are ideal, as by generally agreed definition they are crops whose roots stimulate parasite germination but do not allow normal attachment and parasite development (they trick the parasite into suicidal germination). They can thus be grown through to normal maturity and harvesting. They are sometimes known as 'false hosts'. Many crops have been identified from laboratory and glasshouse testing as having the capacity to stimulate germination of *Striga* or *Orobanche* but there have been relatively few reports of highly successful use of the technique in the field. Even the best results suggest that three seasons of trap-cropping are necessary for substantial reductions in infestation.

The reason for generally disappointing results from trap-cropping is becoming clearer as more workers go to the trouble of monitoring the parasite seed bank in the soil rather than measuring success only on parasite emergence in subsequent years. Emergence is not necessarily proportional to the seed bank but varies greatly according to climatic conditions each season. Hence substantial reductions of the seed bank may not always be reflected in the visible infestation. It is becoming clear, however, that reductions of the seed bank under field conditions cannot be expected to match the theoretical expectations from laboratory, or even pot, experiments. While 80–90% germination can be caused by a potential trap-crop in the laboratory, this level cannot be expected in the field. Linke *et al.* (1991b) assessed the success of a number of crops in terms of the reduction of the *O. crenata* soil seed bank and recorded up to 50% reduction in pots. But in the field the most effective crops caused only 15–20% reduction of a natural infestation per year, compared with a fallow which resulted in only a few percent decrease per year. Work on *S. asiatica* in the USA (Langston and English, 1990b) suggests much greater potential for trap-crops on this species, with depletion rates in the field of up to 90% for cotton and over 50% for a number of other crops, compared with 10–30% for bare fallow. These studies all involved artificial *Striga* infestations in a sandy soil, however, and might not be representative for more natural conditions. Results will always be influenced by the subtleties of *Striga* and *Orobanche* seed germination, especially the interactions of temperature, moisture and gas levels in the soil. The findings in Sudan of a secondary dormancy being induced in *S. hermonthica* seed buried below 20 cm in wet soil (Babiker *et al.*, 1987) seem likely to be due to restrictions in gas diffusion. Under such conditions, a trap-crop could only influence the seed in the upper soil levels and its effectiveness would be much reduced. Increased use of techniques for the separation of *Striga* and *Orobanche* seeds from the soil should soon throw much more light on the exact effectiveness of trap-cropping and other techniques for reducing parasite seed banks. The methods published by Visser and Wentzel (1980) for *Striga* and *Alectra*, by Eplee and Norris (1990) for *S. asiatica*, and by Kachelriess (1988) and Sauerborn *et al.* (1991) for *Orobanche*, are all based on similar principles, involving the use of sieves and high density salt solutions.

One general principle to be considered in trap-cropping is that, for greatest effectiveness, the time of planting of the trap-crop should correspond with the time shortly before that when the parasite is most likely to germinate. There should also be consideration of the planting arrangement and other cultural practices aimed at achieving the healthiest possible development of the trap-crop root system. Crop varieties are also likely to vary in effectiveness.

Catch-crops These are normal hosts which not only stimulate germination but can carry the parasite to maturity. Seeding of the parasite is generally prevented by destroying the crop before any parasite seeds can be formed. Catch-crops are rarely as economically favourable as trap-crops because of the lack of produce, but can sometimes be grown during a short, unreliable, rainy season unsuitable for normal cropping. Forage crops can also be satisfactory where they are relevant to the farming system, and provided the harvesting process results in sure prevention of parasite seed production.

While rotation into trap- and catch-crops needs to be considered where there is no other means for preventing parasite seed production, it should not be overlooked that the normal crop host is likely to be the most effective catch-crop of all, provided the *Striga* or *Orobanche* can be controlled by herbicide or other means.

---

effectiveness may be due more to the flooding of the rice than to a trap-crop effect (see 'flooding', below).

Land preparation

Deep inversion ploughing has been proposed as a means of controlling *Orobanche*. This has often been based only on observation, e.g. that there was little emergence of seeds of *O. cernua* placed at 20 cm depth (Krishnamurthy *et al.*, 1987). But in several cases the practice has been recommended apparently on the basis of field trials, e.g. trench ploughing to 45–50 cm for *O. ramosa* in tobacco in eastern Europe and for *O. minor* in tobacco in New Zealand. Reductions in infestation of 80–90% can apparently be achieved in this way, though it presumably cannot be repeated, otherwise buried seed would be brought up again. This method should be considered as one initial step in an integrated effort for control of a serious infestation, at least where soil and other conditions permit. The fact that most reports are from tobacco growing is probably linked to the greater feasibility of the practice on light soils.

The converse, of minimum or zero tillage, has failed to show any advantage in trials so far reported from Syria and Egypt (Kukula and Masri, 1984; Nassib *et al.*, 1984), though theoretically, if continued for long enough in the absence of new seeding, zero tillage should result in depletion of the parasite seed in the host root zone.

Flooding

Although intermittent irrigation does not usually suppress *Orobanche* species, a period of flooding, as for rice, is believed to reduce the problem of *O. cernua* in a following tobacco crop. This belief is based on the observation that seeds of this species apparently lose their viability after one month storage in water (Marudarajan, 1950), and that the problem is less severe when tobacco is rotated with a flooded rice crop (Krishnamurthy and Rao, 1976). As these same authors failed to show any germination stimulation of *O. cernua* by rice, it is assumed to be a flooding effect rather than a trap-crop effect. In Egypt, Zahran (1982) reported about 65% reduction in *O. crenata* infestation in faba beans following a flooded rice crop compared with fallow, but no decrease in *O. aegyptiaca* infestation of tomato. In Nepal, Vasudeva Rao *et al.* (1988) observed much less *O. aegyptiaca* in *Brassica* crops following rice than following other crops. A good suppression of *O. cernua* was achieved by maintaining a continuous flood for about two months prior to draining and planting tomato in Ethiopia in one year, while the following year results from the same treatment were not successful (Sherif and Mohammed Gelma, 1992). The factors influencing success of this treatment clearly need more study. Apart from differences according to *Orobanche* species, as is apparent from Egypt, it needs to be established whether any suppression of germination after a period of flooding is a result of loss of viability, or merely a temporary induced dormancy. Effects of different length of period, temperature and soil type may also be important, but it remains one of the few direct control measures available and deserves more systematic study and exploitation.

Solarization

Solarization involves the use of plastic sheeting to trap the heat from solar radiation so that soil temperatures are raised to levels lethal to seeds or other organisms. It was first reported effective for the control of fungal pathogens in Israel (Katan *et al.*, 1976) and subsequently shown also to provide excellent control of *O. aegyptiaca* in Israel (Jacobsohn *et al.*, 1980), *O. ramosa* in Sudan (Braun *et al.*, 1984, 1987, 1988), and of *O. crenata* and *O. aegyptiaca* in Syria (Sauerborn and Saxena, 1987; Linke *et al.*, 1991b) and in Egypt (Abdel-Rahim *et al.*, 1988). Abu-Irmaileh (1991) has also confirmed its effectiveness on *O. aegyptiaca*, *O. ramosa* and *O. cernua* in Jordan. In Iraq Elia *et al.* (1987) report best results with a combination of solarization with prior application of chicken manure.

Requirements for success include moist soil, sufficiently high air temperatures and solar radiation, and adequate length of exposure. Moist soil is essential for heat conductivity and for keeping seeds in a more susceptible imbibed state. Drennan and Mohamed-Ahmed (1992) have shown that seeds of *O. ramosa* can tolerate 50°C for 35 days when dry, but are killed in a few days at 45°C when wet. At 40°C they are killed in 20 days. Wetness is usually achieved by pre-irrigation. Critical levels of air temperature and/or solar radiation have not been precisely defined, but it appears that the temperatures under the plastic must exceed 45°C at 5 cm depth and successful treatments are usually associated with temperatures exceeding 50°C for at least part of each day. The higher the temperatures, the shorter can be the time of exposure, down to as little as two or three weeks, though periods of 4–7 weeks are usually recommended.

The type of plastic used has usually been clear polyethylene, the thinner (0.03 mm) better than thicker (0.1 mm) for achieving highest soil temperatures, and UV-absorbing preferable to regular, both for soil temperature and for durability (Horowitz *et al.*, 1983). Black plastic results in somewhat lower soil temperatures and longer periods are needed to achieve effects equal to clear plastic, but under the hottest air temperature conditions it can also be effective, and Abu-Irmaileh (1991) has shown that it has the advantages of: (i) being longer-lasting than regular clear material and hence perhaps reusable; and/or (ii) when left in place, it can usefully provide prolonged control of other weeds after the crop, such as tomato, has been planted through it.

Although many other weeds can be controlled by short-term solarization, there are a number of species that tolerate the treatment, even with clear plastic, especially legume species such as *Scorpiurus muricatus* (Sauerborn and Saxena, 1987). As the depth of soil effectively solarized does not usually exceed 10–15 cm, species which can emerge from those depths, especially perennials such as *Cyperus rotundus* can also survive and immediately be troublesome in the planted crop. Hence the use of black plastic can be economically preferable to the use of clear plastic, reducing the cost of controlling non-parasitic weeds. The limited depth of effectiveness in the soil also means that it is important to minimize the disturbance of soil after removal of the plastic. Hence fertilizer applications and any other operation requiring soil disturbance should be carried out prior to treatment. Reduced emergence of *Orobanche* and other weeds can usually be observed in subsequent crops, planted 6–12 months later, but the benefits are likely to be reduced if cultivation is too deep.

Yield increases of 331%, 441% and 92% have been recorded in faba bean, lentil and pea respectively following solarization to control *O. crenata* in Syria (Linke *et al.*, 1991b). Even in the absence of weeds, yield of pea was increased by 32%, and it is suggested that there is a benefit from increased availability of N and P as well as from control of other soil pathogens including nematodes. *Rhizobium* is also decreased initially, but rapidly recovers, presumably by reinfection from lower soil horizons. An incidental benefit of solarization reported by Satour *et al.* (1991) is a 30–50% reduction in salinity.

Sprich *et al.* (1990) tested 12 makes of sprayable film as possible alternatives to polyethylene. None was found to be as effective as polyethylene, but eight of them caused significant reduction in viability of

O. crenata and further developments of this principle could well prove to be of economic interest.

It is interesting to note that levels of carbon dioxide and of ethylene are much increased under polyethylene sheeting while oxygen levels are generally unchanged. Rubin and Benjamin (1984) have pointed out that these changes in gas balance, in conjunction with raised temperature, may be responsible for stimulating germination of weed seeds, hence contributing to greater thermal sensitivity. However, there is no evidence that the control of *Orobanche* is associated with any suicidal germination by ethylene and it is assumed that most of the damage is associated with direct thermal killing of the dormant seeds.

Time of planting

It has long been recognized that the infestation of faba beans and other crops such as lentil and chickpea by *O. crenata* in the Mediterranean basin is profoundly affected by date of planting. Early plantings in October, November or early December, depending on the region, tend to be most heavily infected while those planted later, in November, December or January (according to region), are much less heavily parasitized (see Fig. 3.6). This date of planting effect has been well documented for Spain (Mesa-García and García-Torres, 1982), for Egypt (Nassib *et al.*, 1984; Raaimakers *et al.*, 1988), and for Syria (van Hezewijk *et al.*, 1987). Van Hezewijk *et al.* (1991) have subsequently confirmed that this seasonality

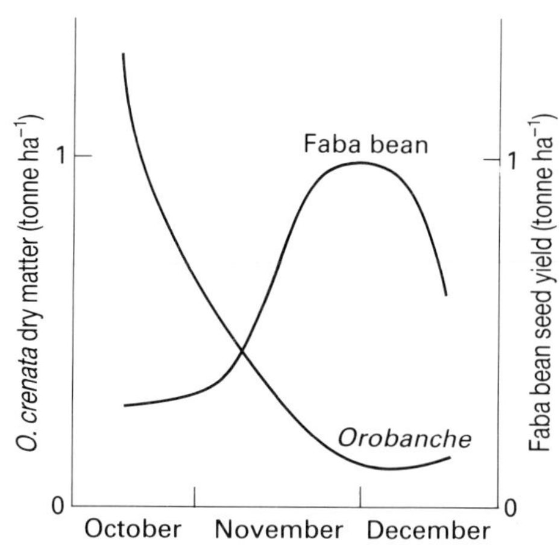

Fig. 3.6. Effects of planting date on *Orobanche crenata* weight and yield of faba bean in Syria (from van Hezewijk *et al.*, 1987).

of behaviour is dictated by the germination response of *O. crenata* to temperature. Soil temperatures during December and January in Syria were shown to be continuously below 10°C, well below the optimum 15–20°C for this species. Unfortunately, delayed sowing seriously reduces the yield potential and a compromise may be needed depending on the strategy being followed. If maximum yield is the objective, November planting of faba bean may still give the highest yield in Syria and December planting in Spain, but if it is felt essential to minimize *O. crenata* emergence as part of an integrated control system, then later planting could reduce parasite emergence by some 50%.

For other *Orobanche* species there appears to be less scope for manipulating infestation by changing planting date, but Castejón-Muñoz *et al.* (1993) report that in Spain, infection of sunflower may be reduced by planting early, not later than 1 March.

Method of planting

There are few reports of planting arrangement being critical for *Orobanche* development but Pieters and Aalders (1986) reported that density of *O. crenata* in faba bean was negatively correlated with crop sowing density. Hence dense planting may be beneficial but it can also be suggested that planting in rows, and wider row-spacing rather than narrow row-spacing, will prove easiest for hand-pulling and/or directed herbicide application.

Fertilization

As noted above, there tends to be an association of *Orobanche* with less fertile soil conditions, and some experience of suppression of the parasite by high levels of, for instance, chicken manure (Abu-Irmaileh, 1979b) and by fertilizer, especially nitrogen.

Pieterse (1991) has shown that both the germination and the vigour of *O. crenata* are seriously reduced by exposure to ammonium or urea nitrogen at the relatively high concentration of 4 mM when present during both conditioning and germination phases, while nitrate N has no effect. Jain and Foy (1987) had similar results when *O. aegyptiaca* seeds were exposed only during the conditioning phase.

There have been some reports of successful control of *Orobanche* species in pot experiments, using high levels of nitrogen with or without high P. Abu-Irmaileh (1979b, 1981) reported good suppression of *O. ramosa* in tomato and tobacco by nitrogen in the form of ammonium sulphate at about 4 g kg^{-1} of soil (0.8 kg N kg^{-1}), but tomato yields were reduced unless phosphorus and potassium were also increased. Lower levels of N were less effective and P and K alone had no significant effect. Jain and Foy (1987, 1992) came to very similar conclusions with *O. aegyptiaca* on tomato, 1 g N kg^{-1} of soil (as ammonium nitrate) prevented emergence of *Orobanche* but reduced tomato growth, while the further addition of potassium phosphate at 1 g PO_4 kg^{-1} soil resulted in excellent tomato growth, three times that of the infested control. In their test, potassium phosphate alone again had no significant influence. The levels of N in these reports are quite massive (equivalent to field doses approaching 1000 kg N ha^{-1}) and there have been few published results of equivalent field experiments, yet van Hezewijk *et al.* (1991) were able to demonstrate significant reductions of *O. crenata* in faba bean by ammonium sulphate equivalent to only 14 and 28 kg N ha^{-1}. Two applications of 28 kg N ha^{-1} reduced emergence by 34–46%. The intention of this study was to confirm the hypothesis that reduced (ammonium or urea) N was more effective than nitrate, and show that nitrification inhibitors might delay conversion to nitrate and so improve the effectiveness of the applied ammonium or urea. The results were inconclusive in this respect, because rainfall was inadequate to move either the fertilizer or the nitrification inhibitor into the main zone of *Orobanche* germination. Dicyandiamide ('Didin') gave more promising indications of delayed nitrification than 'N-Serve' but still failed to enhance the control of *O. crenata*.

The fact that high levels of N are needed and that nitrate is apparently less effective both *in vitro* and in soil, tend to suggest that the mechanism of the N effect is mainly a direct one on the germinating parasite seedling, rather than an indirect effect via the host, as seems more likely in the case of *Striga*. But further work is still needed to confirm this.

Infestation by *O. minor* in subterranean clover (*Trifolium subterraneum*) in Australia has been effectively controlled by application of superphosphate at normal field rates, but it was thought that this was mainly an indirect effect through suppression of the favoured alternate weed host, *Chondrilla juncea* (Southwood, 1971).

While the evidence for practical benefit from fertilizer application is still very limited, it is clearly a worthwhile option to consider, wherever such a treatment is likely to be economic. But the results so far suggest that high levels of N are necessary and that they need to be balanced by high P.

Fire

Burning the stubbles of infested crops can greatly reduce the carry-over of new seed in the soil.

Resistant and Tolerant Varieties

Sunflower

The potential for using resistant varieties to combat the *Orobanche* problem was apparently recognized in about 1912 when scientists in Russia first started looking for sunflower varieties which would resist attack by *O. cernua* (Pustovoit, 1967, 1976). Cubero (1986, 1991) in valuable reviews describes how, by 1916, a range of material had been developed with complete resistance and by 1925, 95% of the sunflower crop grown in Pustovoit's region was based on these resistant lines. However, when they were distributed more widely in the following three years, the resistance failed. This failure was soon explained by the existence of different races of the parasite, the original being designated race A and the more virulent, race B. Unfortunately, race B proved to be a complex of different subraces, making it difficult for the breeders, and by 1957 there were still no cultivars resistant to all populations of race B. The latest situation in Russia is believed to be that in each region sunflower varieties are available with virtual immunity at least on a local basis though it is not clear whether any have universal resistance against all races. Elsewhere, there appear to be further races not corresponding to A and B. In Bulgaria, five races are recognized (Bchvarova, 1978, 1979), all or some of which differ from A and B. One of the more troublesome of these originated by transfer from the wild host *Artemisia maritima* (Petrov, 1970). In Romania likewise, five races with increasing virulence have been recognized, while sunflower varieties have also been selected with five corresponding levels of resistance, gene Or5, present in inbred line P-1380-2 conferring resistance to all races (Vranceanu *et al.*, 1981). Other material with near immunity includes the inbreds R-10519 and P-13181, also the Romanian hybrids Romsun 53 and Sorem 80. The last two also resisted a sample of *O. cernua* from Turkey. The Vranceanu paper is recommended for its information on a wide range of sunflower materials from a number of countries. In Spain too, a different range of races has been recognized, some of which overcome the resistance of known 'resistant' lines such as Record, but broad-spectrum immunity is still demonstrated by lines P1380-2A and Zhanovski-8281 (Melero-Vara *et al.*, 1989). In Yugoslavia, Acimovic (1980) has established that there are at least seven physiological races. Unfortunately, there does not so far appear to be any agency determining the relationship, if any, between the races occurring in different regions. Petzoldt and Sneyd (1986) working with *O. cernua* from Turkey and Hungary found one of the older Russian lines VNIIMK 8931 to be susceptible, but identified lines 81 1 and 81 14 (of unstated origin) to be much less susceptible.

The sources of resistance were originally from within sunflower landraces, but other sources have been exploited including Jerusalem artichoke (*Helianthus tuberosus*) and *H. petiolaris*.

There have been differing opinions on the **genetics** of the resistance. Some workers have concluded that a single dominant gene is involved but Cubero (1991) concludes that there is a quantitative genetic action. These differences may well reflect differences according to the race of parasite involved.

The **mechanism** of resistance in sunflower was proposed by Antonova (1978) to be based on a lignification response in the host root, which blocked further penetration of the haustorium. In Israel, Shalom *et al.* (1990) also noted thickened cell walls and the deposit of phenolic materials in the roots of resistant sunflower varieties in response to haustorial penetration. Wegmann *et al.* (1991) have proposed that phytoalexins are almost certainly involved in the resistant response, reporting that a line resistant to *O. cernua*, cv. 81-14, produced twice as much scopoletin in its roots as did a susceptible line 'Giganta'. It is possible, however, that the mechanism may differ according to the race of parasite, as Pustovoit (1967) noted that whereas resistance to race A involved swellings in the root, resistance to race B did not. There is no reported evidence for the existence of resistance based on low stimulant exudation.

Resistance of sunflower to other *Orobanche* species has not been so carefully studied and there is no certainty of cross-resistance from one species to another.

Faba bean and vetch

Faba bean In spite of intensive screening and breeding work, there are still no faba bean varieties with complete immunity to *O. crenata*, even on a local basis. Cubero (1986, 1991) has summarized the history of work on this crop and describes how workers in Italy, Spain and Egypt have shown varying levels of susceptibility in different ranges of material, but until recently the only line sufficiently resistant to be brought into breeding programmes has been the line F-402, identified in Egypt (Nassib *et al.*, 1979, 1982, 1984). Results with this line have not been consistent, and it appears that its resistance can be influenced by environmental conditions, showing up much more impressively under relatively dry conditions (minimum 25% field capacity or two-monthly irrigation) than under wetter conditions (minimum 75% field capacity or monthly irrigation) (ter Borg and van Ast, 1991).

Now there are breeding lines based on F-402 and its derivatives Giza-402 and VF-1071, which have much improved agronomic characteristics, and are showing promising resistance in both Syria and in Morocco. The best of these are 18009, 18035 and 18105 (ICARDA, 1992).

The resistance of F-402 and other sources of partial resistance has proved not to be a simple matter and Cubero concludes that there is a quantitative genetic basis in most cases (Cubero, 1991).

The **mechanism** of resistance in F-402 is still not altogether clear, but probably results from a combination of characters. Nassib *et al.* (1979, 1984) suspected that it could involve both its relatively small root system, and some mechanical hindrance in the host root tissue. Ter Borg and van Ast (1991) confirmed that under moist conditions the lower number of *O. crenata* on F-402 could be attributed to a smaller root system, but under drier conditions, the numbers per unit length of root were much smaller in F-402 than in a susceptible variety Giza-2, suggesting some other mechanism. Zaitoun *et al.* (1991b) have confirmed that there is localized formation of corky tissue in the roots of F-402 which inhibits full development of the parasite. This was not observed in susceptible varieties Giza-3 or Reina Blanca. Wegmann *et al.* (1991) could not confirm that phytoalexins played any role in the resistance of faba bean, though they did demonstrate enormously greater production of the phytoalexins maackiain and medicarpin in the roots of a 'resistant' chickpea variety ILC-280 than in a susceptible line FLIP-81/35W.

At present it seems unlikely that faba bean will be the subject of genetic engineering for herbicide resistance, but attempts have been made to achieve resistance through induced mutations. Salem *et al.* (1991) report success with fast neutron bombardment but it is not clear that the resistance was achieved without serious side-effects.

Vetch In vetch (*Vicia sativa*) a much higher degree of resistance has been found than in faba bean amounting almost to immunity in the line 473-A (Gil *et al.*, 1982, 1984). Much lower numbers of *O. crenata* succeed in establishing on the root, and very few, if any, emerge under field conditions. The genetics of the resistance are complex, a number of genes being involved which show 'no dominance or, if any, a very low level of partial dominance for resistance'.

Tolerance, the ability of a variety to be attacked but to show little damage (see Box 1.4, p. 57), has been observed in a number of faba bean studies, especially in those of Radwan *et al.* (1988a,b). Wegman *et al.* (1991) suggest how tolerance might be a function of osmotic pressure, protecting the host against water loss to the parasite, but the degree of tolerance and its potential usefulness in practice have yet to be clearly quantified.

Tomato

Most attempts to select *Orobanche*-resistant tomato varieties have yielded only a range of varying susceptibility. Dalela and Mathur (1971a) in India found only one line ('Early crop') 'moderately resistant' to *O. cernua* out of 41 studied over a four year period. Abu-Gharbieh *et al.* (1978) in Jordan found only 'slight resistance' to *O. ramosa* in eight out of 100 lines studied, though, also in Jordan, Kaswari and Abu-Irmaileh (1989) reported high resistance in one line (LA372) of *Lycopersicon peruvianum*. Kabulov and Tashpulatova (1974) named seven lines 'relatively resistant' to *O. aegyptiaca* out of 30 tested in Samarkand, and a few years later two of these lines , Gibrid-88 and Bolgarskii, are again mentioned by Mukumov and Faizieva (1977) as being the 'most tolerant'. In Astrakhan, greater success was achieved by Avdeev and Shcherbinin (1976, 1977, 1978) when they tested 206 lines over a five year period and identified UZ-1 as almost immune to *O. aegyptiaca*. By further selection within UZ-1, a homozygous line PZU-11 was isolated, in which it was concluded that resistance was controlled by a single dominant gene (Avdeev, 1986). Unfortunately, the resistance of this line has not yet been confirmed elsewhere. Foy *et al.* (1988) included it in a major study with *O. aegyptiaca* in Israel and failed to demonstrate useful resistance in this or any other of the 1361 lines tested, which included a number of *Lycopersicon* species other than *L. esculentum*. It is just possible that the different population of *O. aegyptiaca* in Israel made an important difference but the authors comment that their technique, which involved an extremely heavy inoculum of 30 mg *Orobanche* seed per pot (ca. 10,000 seeds per pot), may have resulted in a masking of any partial resistance that might have been present. As Boorsma (1980) suggests that the optimum infestation rate for screening faba bean varieties against *O. crenata* is 900 seeds per pot, it seems highly likely that seed numbers were too high in the tomato study. Cubero (1991) comments that PZU-11 'has not shown resistance when sown in other countries' but does not indicate his evidence for this. Further evaluation would clearly be justified.

Tomato varieties have been screened for resistance to the herbicide glyphosate, but none was found sufficiently tolerant for practical purposes (Foy *et al.*, 1988). It is highly probable, however, that new genetically engineered varieties will in due course be available with target resistance to glyphosate or other herbicides suitable for control of *Orobanche*.

Tobacco

In tobacco, as with tomato, most studies aimed at identifying resistance have proved largely fruitless. Cubero (1986, 1991) has reviewed most of the published results from the former USSR, Romania and India which have involved many hundred varieties of tobacco and samples of other *Nicotiana* species, exposed to several different *Orobanche* species. Some promising, but still not fully acceptable, resistance to *O. ramosa* and *O. aegyptiaca* was identified by Vinogradov *et al.* (1981) among some 700 cultivars, hybrids and chemically induced mutants. Tolerant and resistant lines were found in Trebizond-285, Corojo, Florida-513, NS-2512, Hicks Resistant, Khatana, R X T and Jamaica K-444. The partial resistance was believed to be controlled by one or two pairs of recessives and some modifier genes. Racovitza (1973) found quite wide differences in susceptibility of tobacco cultivars and a range of wild *Nicotiana* species to *O. ramosa* in Romania, at least in some years. There was a wide range of variation in stimulant production among wild *Nicotiana* species, and some degree of correlation with susceptibility in the field. The most resistant cultivar in the field, Joiner, also showed relatively low exudation of germination stimulant, but another low-stimulant variety, Djebel-264-V, was quite susceptible, so this possible mechanism of resistance is far from certain. Dalela and Mathur (1971a), working in India with 21 cultivars found only one, Vasttakpal (Type G), with 'moderate resistance' to *O. cernua*.

There are clearly some differences in the susceptibility of tobacco genotypes to *Orobanche* species, but it seems unlikely that a high level of resistance will be found. The emphasis of tobacco breeding is now likely to be focused on genetic engineering for herbicide resistance (see below).

Other crops

Eggplant The one reported study of varietal resistance in eggplant (Dalela and Mathur, 1971a) compared 128 cultivars against *O. cernua* over a period of years and found seven to be 'highly resistant'. These included Pusa Purple Long × Manjri Gota, Pusa Purple Long × Nurki, Running King, Verma's Giant, D-12-2-66, DC-4-1-67 and E-147.

Cucurbits Cubero (1986, 1991) has summarized the several studies which have been published involving *O. aegyptiaca* and melon, cucumber, water melon, pumpkin and squash. No certain resistance was detected in most of these, though Kabulov and Kabulova (1977) made anatomical studies of the roots of susceptible and 'resistant' lines of pumpkin, melon and cucumber, showing that the differences were associated with difficulties in penetration of the host stele.

***Brassica* species** Among 48 varieties of rapeseed (*B. campestris*) and mustard (*B. juncea*) Dalela and Mathur (1971b) were able to name three rapeseed and four mustard as 'resistant' to *O. aegyptiaca*.

Sobrino-Vesperinas (1985) also found a range of response to *O. ramosa* in 33 rapeseed varieties, one of which, CE-TO-32-2/82, from Spain, showed only 3% plants attacked. In Nepal, Vasudeva Rao *et al.* (1988) observed that while 'toria' and 'sarson' (both *B. campestris*) were seriously affected by *O. aegyptiaca*, 'rai' (*B. juncea*) was much less affected. In Israel, likewise, brown cultivars (*B. juncea*) from India were much less susceptible than 'toria' (*B. campestris*) (Shchory, 1983).

Hemp Cubero (1991) describes how varieties of hemp with resistance to *O. ramosa* were widely grown in France during the last century but that this resistance later failed, presumably as a result of the introduction or build-up of more virulent populations of the parasite.

Carrot In Egypt, Zahran (1982) observed that local red-rooted 'scarlet' carrots were much less affected by *O. crenata* than the improved variety Chantenay.

Lentils, chickpea, pea Cubero (1991) observes that preliminary work is in progress on each of these with more promising indications for chickpea and pea than for lentil.

Chemical Control Methods

Fumigants

Fumigants are highly volatile compounds which have the ability to permeate and kill ungerminated seeds, unlike normal herbicides which can only damage plants after they have at least started germinating. The seeds must, however, be imbibed and hence physiologically active.

Fumigants have been more widely tested and used for control of *Orobanche* species than for other parasitic weeds. **Methyl bromide** is the most reliable treatment and this has been used routinely in the USA in both California and Texas for control of localized infestations of *O. ramosa*. It has also been used on large-scale farms in Israel. The cost of the chemical and the plastic covers that are needed to enclose the gas are exceptionally high and it is mainly used where the treatment is part of a government-funded quarantine practice or in high value situations, especially glasshouses, where there is a combined problem of *Orobanche* and nematodes or other soil-borne pathogens. Owing to the toxicity of the gas, its use is often restricted to trained operatives. Conditions required for success include a well-tilled soil which has been kept moist (70% field capacity) for 1–2 weeks beforehand, and a temperature above 10°C. Wilhelm (1962) describes how the normally recommended dose of 500 kg ha^{-1} can be substantially reduced to about 250 kg ha^{-1} by using a formulation in which the methyl bromide is dissolved in kerosene and injected through tines 30 cm apart into the soil at 15–20 cm depth. The soil is immediately covered closely with polyethylene sheeting. This contrasts with the standard method in which the gas is released into a space between the soil surface and the sheeting supported above it.

Equipment is now available which lays the sheeting mechanically in strips. One edge of the strip is glued to the previous strip while the other is buried in the soil to ensure a seal. The sheeting may be removed after 48 hours but the soil must be ventilated for a further few days before planting crops. Residues of bromine can occur in some crops and although not generally damaging, it is important to ensure that any residue is acceptable under local regulations. Another possible side-effect is increased ammonium N as a result of selective toxicity to soil bacteria, which can be undesirable for tobacco. Although the soil microflora usually recovers in due course from fumigation there can be a serious temporary suppression of *Rhizobium*. Hence for a legume crop, inoculation of the seed, or nitrogen fertilization, may be necessary following fumigation.

Chloropicrin is often included at low concentration in mixture with methyl bromide as a lachrymatory warning agent. It is also used in its own right as a soil fumigant for control of pathogens. It has some activity on *Orobanche* species but is generally inferior to methyl bromide and does not appear to offer any advantage over it.

The main alternative low-boiling point fumigants, **ethylene dibromide** (EDB) and **dibromochloropropane** (DBCP) have now been banned for agricultural use. Of those that are still available, **1,3-dichloropropene/1,3-dichloropropane** mixtures ('DD' or 'Telone') or pure **1,3-dichloropropene** ('Telone II') have relatively low activity on most weed seeds but Jacobsohn *et al.* (1991b) report successful control of *O. crenata* at 150–300 l of 'Telone II' ha^{-1} and of *O. ramosa* and *O. aegyptiaca* at 300–450 l ha^{-1}. The liquid formulation was injected to 20 cm soil depth at 30 cm centres. The soil had been well irrigated two weeks previously and received sprinkler irrigation immediately following application, to provide a seal. A range of other annual weeds were controlled at least partially at these doses. **Methyl isothiocyanate** at 176 kg a.i. ha^{-1} has given partial control of *O. minor* in New Zealand (James and Frater, 1977).

The remaining alternatives for fumigation are the two relatively non-volatile compounds, dazomet and metham-sodium, both of which break down in wet soil to release the volatile methyl isothiocyanate. **Dazomet**, which is available as a dry granular formulation containing 98% active ingredient, was shown effective against *O. ramosa* in Turkey at 600 kg ha^{-1} (Nemli *et al.*, 1991) and in Bulgaria at 300–600 kg ha^{-1}, the higher dose being needed on heavier soils (Fetvadzhiev, 1970). In Ethiopia, doses of 250–500 kg ha^{-1} were effective against *O. cernua*, the higher dose being needed where the soil was not covered by plastic sheeting (Sherif *et al.*, 1989). In New Zealand, 400 kg ha^{-1} gave only partial control of *O. minor* (James and Frater, 1977). Dazomet is usually spread on the soil surface and then ideally incorporated into the soil by rotary cultivation before application of irrigation to seal the soil surface. Where irrigation is not available the soil surface may be rolled to achieve the same effect. Plastic sheeting is not normally used with dazomet, but is beneficial where it is feasible. Residues of the chemical persist much longer than those of methyl bromide and the soil needs to be cultivated and ventilated for at least four weeks before crops can be safely sown. In the Ethiopian trial only two weeks were allowed and tomatoes were seriously damaged. Fetvadzhiev (1970) allowed 4–6 weeks before sowing tobacco.

Metham-sodium is formulated as a liquid containing 30–50% of sodium N-methyldithiocarbamate and has performed well in New Zealand, where 163 kg a.i. ha^{-1} gave excellent control of *O. minor* (James and Frater, 1977) when sprayed onto the soil surface and immediately watered in thoroughly to 22 cm soil depth. Yields of tobacco were increased from 1775 to 2488 kg ha^{-1}, providing an economic return on investment. Quality was unaffected when the crop was planted four weeks after treatment, but there was damage when the interval was only two weeks (James, 1976).

In Israel the use of metham-sodium has not always given reliable results, perhaps because of the heavy soils in which control was attempted or because of the technique used, of application via drip irrigation ('chemigation') (Jacobsohn, 1986). Later work has suggested that it is best to apply the chemigation to dry soil rather than wet, but for complete control of *O. aegyptiaca* it is still necessary to cover the soil with plastic during and for one week after the chemical application so that the fumigant gas is retained while the *Orobanche* seeds become conditioned and hence susceptible (Kleifeld *et al.*, 1991).

Germination stimulants

As an obligate parasite, it is vital that *Orobanche* germinates close to a host root and attaches to it within a few days if it is to survive. Hence any germination in the absence of a suitable host root is effectively suicidal, and any chemical which can trigger germination has potential as a means of control.

As already noted, the response of *Orobanche* species to **ethylene** is inconsistent and there have been no reports of its successful use in the field. Among the first strigol analogues to be developed, however, **GR5** and **GR7** were immediately shown to have activity on *O. ramosa* as well as on *Striga* (Johnson *et al.*, 1976). These are synthetic compounds, related to, but simpler than, strigol, the natural *Striga*-stimulating substance exuded by the roots of cotton (see Chapter 1). In Egypt, GR7 was shown to reduce *O. crenata* by 90% at only 0.3 kg ha^{-1} in acid soil (pH 4.5) while a dose of 3 kg ha^{-1} was needed to achieve the same result in alkaline soil (pH 8) (Zahran, 1982). In the field the best result from 2 kg ha^{-1} applied in the irrigation water to alkaline soil, was only a 30% reduction. Further analogues, especially **GR24** and **GR28** were also highly active *in vitro* at concentrations well below 1 mg l^{-1} and Saghir (1986) has usefully summarized a series of studies in Lebanon. These confirmed that a soil drench with a 1 mg l^{-1} solution of GR24 applied six weeks before sowing tomato gave well over 90% control of *O. ramosa*. Unfortunately, there is at present no prospect of these compounds being made available comercially and they remain only a theoretical possibility.

An alternative range of germination stimulants has been the subject of study in Bulgaria, and field tests with compounds coded **D55** and **V53** have apparently given effective control of *O. ramosa* in tobacco (Chalakov *et al.*, 1986; Tchalakov *et al.*, 1991). In related studies, Zhelev (1987) refers to the activity of isothiocyanates and suggests that the natural root exudate substances may have this type of structure. He also refers to the stimulant activity of derivatives from mustard oil, including **phenylisothiocyanate** and **allylisothiocyanate** but it is not clear whether the coded materials D55 or V53 also have isothiocyanate structure. Further information on these chemicals and their means of application is awaited with interest.

Herbicides

Soil-applied herbicides No really promising soil-applied herbicides emerged from *Orobanche* research until the late 1980s. A number had been reported earlier to show selectivity, including **terbucarb** in faba bean, and **trifluralin** and **diphenamid** in tomato, but as reviewed in detail by Foy *et al.* (1989), none proved effective or reliable enough to become accepted field treatments. Then García-Torres *et al.* (1989, 1991) showed that a number of new compounds had more promising potential, especially imazethapyr and chlorsulfuron. Latest results (García-Torres and López-Granados, 1991a,b; Jacobsohn and Eldar, 1992) confirm that **imazethapyr** has good selectivity against *O. crenata* in faba bean, lentil, chickpea and pea at 50–100 g a.i. ha^{-1}, and against *O. cernua* in sunflower at 20–80 g a.i. ha^{-1}. The herbicide can be applied either as a pre-emergence incorporated treatment or to the soil surface after planting. The related **imazaquin** and **imazapyr** appear somewhat less selective in the legumes, but imazapyr looks promising at about 25 g a.i. ha^{-1} in sunflower. **Chlorsulfuron** at 2–6 g a.i. ha^{-1} has controlled *Orobanche* in both faba bean and sunflower, giving better results on the soil surface than when incorporated, and some selectivity has been shown against *O. ramosa* in tomato (Kotoula-Syka and Eleftherohorinos, 1991). Factors influencing the effectiveness and safety of these herbicides have yet to be fully resolved, and residues on following crops may prove problematic in some situations; but imazethapyr in particular promises to provide an effective new means of *Orobanche* control in several crops. There is also now the very real possibility of herbicides such as chlorsulfuron and imazethapyr being used in conjunction with genetically engineered crop varieties. This has already been demonstrated experimentally for chlorsulfuron on tobacco (Joel, 1992) and is discussed after the following section.

The fungicide **benomyl** has been claimed to decrease *Orobanche* infestation when used as a seed dressing (Petzoldt, 1979) or as a post-emergence foliar spray over the crop at 3000 mg l^{-1} (Khalaf, 1982). Petzoldt believed that seed dressing resulted in a decrease in *Rhizobium* and hence a reduction of potential sites of entry for the parasite, but there has been no further report to confirm this. Petzoldt included benomyl as a component of a relatively successful integrated control package, but more work is needed to confirm the success of this easy and inexpensive treatment in the field. A number of other fungicides were evaluated by Krishnamurthy *et al.* (1982) and, among others, **tridemorph** greatly decreased *O. cernua* in tobacco. However, other attempts to confirm this activity, including pot experiments at the Long Ashton Research Station (unpublished), have not been promising.

Post-emergence herbicides The potential for a herbicide, applied to host shoots, being translocated through the host and damaging the parasite at an early stage on the roots was proposed by Whitney (1973) when he had demonstrated that *2,4-D* applied to faba bean was accumulated in the *O. crenata* tubercle to a level 14 times that in the host root. Even with this differential concentration, damage to the host was worse than that on the parasite and no selective control was achieved. However, following the observation of some tolerance of faba bean to **glyphosate** in routine herbicide evaluation at the Weed Research Organization in 1972, Kasasian (1973b,c) was the first to test this herbicide for selective control of *O. crenata* and immediately showed that rates of 0.2 to 0.3 kg a.i. ha^{-1} could give selective control when sprayed over the beans at six weeks after sowing. The first fully successful field trials were reported by Schmitt *et al.* (1979) and a recommendation made for field use in Morocco, at 60 g a.i. ha^{-1}, to be applied twice, once when *O. crenata* tubercles were first visible on the roots and again two weeks later. Many other workers have reported corresponding results in countries around the Mediterranean, including Italy, Cyprus, Israel, Egypt and Spain (see Plate 10C). Results have not always been satisfactory and the reasons for this were analysed by Mesa-García and Vasquez-Cobo (1985) in Spain.

They concluded that the initial application was often applied a little too late, after the *Orobanche* shoot buds had begun elongating. And on early-sown faba bean only a single dose had been applied, where two were needed to control the parasite over the more extended period of development. A single application can be adequate where there is a lighter infestation and/or shorter season, while three applications may be needed after very early sowing. Crop damage was recorded in 20% of cases. Mesa-García *et al.* (1984) looked critically at the risks of damage to faba bean by treating uninfested crops with doses ranging from 60 to 360 g a.i. ha^{-1}. Single doses of 60 or 120 g ha^{-1} caused less than 5% yield reduction when applied after the beginning of flowering. Double doses of 120 g ha^{-1} and applications during the vegetative stage were more damaging.

It has been shown that the action of glyphosate on *O. crenata* is attributable to its selective accumulation in the young parasite up to a level four times as high as that in the faba bean host root three days after spraying (Arjona-Berral *et al.*, 1990). It has also been shown by Müller and Distler (1991) that when there are large numbers of parasite nodules, there is competition for the available glyphosate and concentrations may not reach a toxic level. This may explain some of the poor results reported from severely infested sites, but it is not certain how this can be overcome and whether under these conditions it is safe to use a higher dose of glyphosate.

The potential for use of glyphosate has been explored in many crops other than faba bean. Among other legumes, lentil has sometimes been safely treated in Spain and Morocco (Kelili *et al.*, 1983; Arjona-Berral *et al.*, 1988) but slightly lower doses of about 40 g a.i. ha^{-1} are commonly suggested and it is not considered selective enough in Syria. Vetch (*Vicia sativa*) may be sufficiently tolerant (Jacobsohn and Levy, 1986) but pea has generally proved too sensitive to the herbicide for reliable selectivity. Carrot and celery have been treated very successfully in Israel and in Cyprus for control of *O. crenata* and *O. aegyptiaca* respectively, using doses of 50 g a.i. ha^{-1} (Jacobsohn and Levy, 1986; Americanos, 1991). Early work on tobacco by Kasasian (1973c) suggested the possibility of selectivity against *O. aegyptiaca* at low doses but, in spite of promising results reported by Lolas (1986) against *O. ramosa*, it is understood that further work in Greece has failed to confirm its reliability. In tomato, likewise, some selectivity was indicated by Kasasian (1973c). Foy *et al.* (1988) evaluated a wide range of tomato cultivars and some wild *Lycopersicon* species for possible variation in glyphosate tolerance but concluded that there was none adequate to ensure reliable selectivity against *O. aegyptiaca* in this crop. In sunflower too, there have been some encouraging results against *O. cernua* (e.g. Petzoldt and Sneyd, 1986) but latest conclusions are that selectivity is inadequate (Castejón-Muños *et al.*, 1987, 1990). Cabbage has been selectively treated with glyphosate in Cyprus (Americanos and Vouzounis, 1992).

Hence, in summary, the crops in which the use of glyphosate is well confirmed include faba bean, carrot and celery. Those in which it has failed to show sufficient selectivity include lentil, pea, tomato, eggplant, tobacco and sunflower. Those in which there is still inadequate information include cabbage, and other brassica crops, potato and lettuce. The range of crops in which this herbicide is useful is likely to be gradually widened with the further exploitation of the glyphosate-resistance gene from *Escherichia coli* by genetic engineering (see next section).

The imidazolinone herbicides have proved of increasing interest as post-emergence as well as pre-emergence herbicides. **Imazaquin** applied to faba bean twice at 40 g a.i. ha^{-1} has proved even more effective than glyphosate in Syria (Linke, 1992) and it was promising also in pea at the lower dose of 20 g ha^{-1}. In Cyprus, imazaquin at 5–10 g a.i. ha^{-1} has been selective post-emergence against *Orobanche* in cabbage when applied twice starting as the heads began to form (Americanos and Vouzounis, 1992). Vasilakakis *et al.* (1988) have also reported promising results with imazaquin post-emergence in tobacco. **Imazethapyr** has proved selective as a post-emergence treatment for faba bean in Spain (García-Torres and López-Granados, 1991a) and for pea in Israel (Jacobsohn and Eldar, 1992) and it seems likely that this class of herbicide will find increasing use as post-emergence treatments, if only as a supplement to pre-emergence treatments where prolonged control is needed.

Among other post-emergence translocated herbicides, **maleic hydrazide** (MH) has controlled *O. aegyptiaca* and *O. ramosa* on tobacco and *O. aegyptiaca* on melon. These results are reviewed and discussed by Foy *et al.* (1989) but this herbicide is now rarely recommended. Foy *et al.* also refer to some promising selectivity of **2,4-DB** against *O. minor* in groundnut but there has been no confirmation that this is successful in the field.

Spraying the host with growth regulators has been reported to reduce *Orobanche* infestation, and/or protect the crop from damage. Hassan *et al.* (1991) showed benefits from **cycocel** (CCC) and **gibberellic acid** (GA_3) against *O. ramosa* on tomato, while El-Ghamrawy and Neumann (1991) reduced growth of *O. crenata* and improved growth of faba bean using kinetin.

Some other chemicals have been used as directed applications for the control of the emerged parasite shoots. **Allyl alcohol** was recommended and used for some time in Cuba and in India but that use is assumed to have been discontinued on safety grounds. Oils have also been proposed. Krishnamurthy *et al.* (1976) found swabbing with **kerosene** gave better results than allyl alcohol on *O. cernua*, and Linke and Saxena (1991b) comment that 95% of *O. crenata* and *O. aegyptiaca* were controlled by 0.1 ml kerosene per parasite shoot. Kerosene has to be used carefully to avoid damage to the crop but recent reports from India (Krishnamurthy and Nagarajan, 1991; Krishnamurthy and Chari, 1991) show that a range of **vegetable oils** can be equally effective, at 'one or two drops' per parasite shoot, and are safer on tobacco than kerosene. The oils tested successfully include castor, coconut, cotton, gingelly, groundnut, linseed, mustard, neem, palm, safflower, sunflower, pongamia, ricebran, soyabean and tobacco. The young flower bud of *O. cernua* is normally killed within a few days and the whole plant within a week. One or other of these oils should be available locally to most farmers and should be valuable in preventing seeding as well as having some benefit in crop yield though this has not yet been quantified. Furthermore it is suggested that a PVC tube with a cotton stove-wick can be used for application – an admirably simple device for the small farmer.

Herbicide resistance

The possibilities for genetic engineering for herbicide resistance, and its application to control of *Orobanche*, were first discussed in some detail by Foy *et al*. (1989) and it was pointed out that there was scope especially for glyphosate (in crops such as tobacco and tomato) and also for some of the other herbicides which are the commercial targets for the companies involved, such as chlorsulfuron, imidazolinones and glufosinate. Fortunately, most of these involve a resistance mechanism at the target site of action rather than an enzyme system which degrades the herbicide. Hence there is no interference with the free movement of the herbicide through the host and its accumulation in the parasite sink. The first confirmation of the effectiveness of this approach has now been reported by Joel (1992). Transgenic tobacco with resistance to sulfonylurea herbicides was grown in *O. aegyptiaca*-infested soil, then sprayed with chlorsulfuron 15 g a.i. ha^{-1} at 10, 20 and 30 days after planting. The parasite was completely controlled and the crop yielded about 95% of the uninfested controls. The practical exploitation of this technique will depend on the clearance of such genetically engineered material for general use, but this is not now expected to cause substantial delay. Some varieties with resistance which has been selected rather than transferred may meanwhile also become available and provide exactly the same opportunities. Crops other than tobacco in which herbicide resistance is likely to be developed in the near future include tomato and sunflower while the herbicides other than sulfonylureas include glyphosate and the imidazolinones such as imazethapyr.

Box 3.2. Factors affecting the action of herbicides against parasitic weeds.

Soil-applied herbicides may act directly on the parasite seed as, or shortly after, it germinates. They can be highly effective and selective for control of *Cuscuta* as this germinates independently, from a relatively shallow depth in the soil. For the root parasites, however, there is a problem of the depth at which the parasite seed germinates and attaches to the host root. This may be much deeper than for a non-parasitic weed, often 10–15 cm in both *Striga* and *Orobanche*. Any herbicide depending on direct contact with the seed or seedling is likely to require soil-incorporation, irrigation or heavy and timely rainfall, or it will only affect the seedling as it reaches up to the treated soil layer. The latter is the way in which trifluralin and oxyfluorfen control *Striga*. They prevent emergence but are not killing the seedling at a sufficiently early stage to prevent damage. Against the root-parasites, success should be more certain with herbicides that are taken up and freely translocated by the host. This may occur with the sulphonyl ureas and imidazolinone herbicides which are showing promise against *Orobanche*.

Translocated herbicides have been especially successful in the control of both *Orobanche* and *Cuscuta*. These are applied over the foliage of the host (and parasite in the case of *Cuscuta*) and are translocated into the parasite via the host phloem and the haustoria. There must be reasonable selectivity in favour of the host but some crop damage is usually accepted when faba bean and lucerne are treated with glyphosate for control of *O. crenata* and *C. campestris* respectively. It has also been shown that the damage to the host is reduced by the presence of the parasite.

Translocated herbicides have not yet been so successful for control of *Striga* on cereals, and Samb and Chamel (1992) have pointed out that this may be due to the different nature of the host–parasite bridge. In *Orobanche*, *Cuscuta* and in *Striga gesnerioides* there is much closer linkage of phloem systems in host and parasite, allowing better transfer of the systemic herbicides which are mainly phloem-transported. For successful systemic control of *Striga* on cereals and of mistletoes it may be important that the herbicide is moved predominantly in the xylem and apoplast, rather than the symplast. With such herbicides the rate of transpiration in the parasite may be critical in the transfer of adequate herbicide.

Perhaps the simplest way of applying a herbicide is to direct it onto the parasite itself – **directed** application. In the case of root parasites emerging some way from their host, it should be possible to do this without wetting the host, though specialized application equipment may be desirable, such as shielded low-pressure sprayers or wipers. Water pistols have been used for directing 2,4-D and other herbicides onto *Striga* in cereals (Ogborn, 1987) and special wiping sticks have been devised for applying vegetable oils to *Orobanche* in tobacco (Krishnamurthy and Chari, 1991). Even if the host is not contacted there must be concern whether the herbicide could be translocated from the parasite into the host. Fortunately, although tracer studies have sometimes shown traces of radioactivity moving from parasite to host, the risks of herbicide damage in this way are small.

Where there is no clear separation of host and parasite, some contact with the host will be unavoidable and the herbicide will need to have some selectivity. 2,4-D is selective enough for application to *Striga* in sole crop cereals but not when it is mixed with more sensitive legume crops. Against mistletoes, 2,4-D is sometimes sufficiently selective even when trees are in leaf but there is greater safety when it can be applied to an evergreen parasite in a deciduous host tree during the leafless season.

Biological Control

By insects

Biological control of *Orobanche* species by the insect *Phytomyza orobanchia* has been successfully applied in the former USSR and a number of other countries in Eastern Europe. As the insect is already very widely distributed in *Orobanche*-infested areas, it has not been a matter of 'classical' bio-control by introduction to new areas, but one of 'enhancement' of the natural population to increase its effectiveness at the local level. Useful reviews of the literature on the subject include Kovalev (1973) and Girling *et al.* (1979), and Mihajlovic (1986) has more recently summarized work in Yugoslavia. The technique is described in detail by Klyueva and Pamuchki (1982). *Orobanche* plants are collected at the end of the season, complete with pupae of the fly in the capsules and stem. These are stored over the winter in a room or chamber with temperature and humidity controlled at 6–8°C and 75–85% RH. Of the 2–3 generations per season, the first is least affected by parasites so the earliest emerged *Orobanche* should be collected. In the following season, at the time of first *Orobanche* emergence, bundles of the stored shoots are hung at intervals in the field, one bag per hectare, containing sufficient to yield about 1000 pupae. The bags are suspended 60–70 cm above the ground and a hole cut in the side of the bag. Sugar solution or honey is smeared on the bag as food for the adults as they emerge. The adults hatch and infest the young *Orobanche* shoots, resulting not only in greatly reduced seed production but in more serious damage or death of many shoots due to the burrowing of larvae in the stems, often aggravated by fungal infection (see Fig. 3.5). Care is clearly needed to avoid spillage of seeds from the stored plants. Trenchev (1981) has described procedures used in Bulgaria and recommends one pair of flies per *Orobanche* stalk per m^2, or double this if there is much parasitization of the pupae. He also comments that the beneficial effect of the fly is reduced by autumn ploughing, presumably because the effects of the collected and released pupae can be enhanced to the greatest extent by those remaining in the field if they are not buried too deeply. Klyueva and Pamuchki (1983) have devised procedures and equations for assessing the effectiveness of releases and although there are few critical reports of evaluation, success is claimed to be 85% or more in terms of reduced seed production, and yields are also claimed to be substantially increased.

Results are affected in most cases by parasites of the *Phytomyza* which include, in Yugoslavia, the chalcids *Crataepiella carlinarum* and *Sphegigaster orobanchiae*.

A further problem is that of insecticide use which can largely eliminate *Phytomyza*. Some attention to this has been given in eastern

Europe, but there is no clear information on the type or timing of insecticides by which this problem can be avoided.

By pathogens

It is common to see *Orobanche* plants collapse from fungal attack, and a number of causal organisms have been identified and some attempts made to exploit them for biocontrol. The fungi most commonly identified are *Fusarium* species, and Panchenko (1981) lists four species from *O. aegyptiaca* of which he considered *F. oxysporum* var. *orthoceras* the most suitable for biocontrol purposes. It is possibly this form which had already been formulated and distributed as 'Product F' (Kott, 1969). In this earlier work it was prepared on a chopped straw/maize meal medium and reduced *Orobanche* on tobacco by 86–100% and on melons by 84–88%. Yield of melons was increased from 6 to 9 t ha^{-1}. Inoculum was placed in the planting hole before transplanting. Effectiveness could last for 80 days and was best at high soil moisture. Panchenko describes the preparation of cultures on river sand and maize meal, and the addition of this to soil at 100 g m^{-2}, resulting in 90–95% reduction in *O. aegyptiaca* on water melon in a sandy soil low in organic matter, while 150 g m^{-2} in a heavier, more organic soil gave reduction of only 38–50%. In field-scale development work, the culture on a substrate of rice husk/maize meal (95:5) was incorporated into the soil to 15 cm depth along tomato rows and gave highly profitable results. *Orobanche* was infected at an early stage and usually failed to emerge. Various cucurbit crop plants and tomato exposed to the culture were found to be infected by the fungus but to show no symptoms of damage. This raises the very interesting question whether the fungus can be distributed systemically through the host and infect *Orobanche* nodules at a distance from the point of application. Kabulov and Khalimov (1974) obtained excellent kill of *O. aegyptiaca* using *Fusarium 'orobanches'* cultured on potato agar, on which viability was retained for 2–3 years. For field application a 1:1 mixture of cattle concentrate feed and chopped straw was used, first soaked and sterilized by steam, and placed at 15–25 cm in tobacco planting holes. Optimum conditions were 18–25°C and good soil moisture.

Outside the former USSR, there has been less attempt to exploit fungal pathogens, but Hodosy (1981) in Hungary used a complex of fungi, including *F. oxysporum*, to achieve over 90% control of *O. ramosa* in tomato, and Bedi and Donchev (1991) describe successful trials with *F. oxysporum* var. *orthoceras* for control of *O. cernua* on sunflower in Bulgaria, using various culture media involving maize meal, barley, wheat straw and soil. The crop was unaffected by the fungus. Wan and Wang (1991) also report successful use of *Fusarium 'orobanche'* against *O. aegyptiaca* in China; and Mazaheri *et al.* (1991) have used *F. oxysporum* successfully to control *O. aegyptiaca* in Iran, achieving 75% reduction of the parasite and an 80% increase in yield of tobacco in the field. In the USA, a strain of *Rhizoctonia solani* was isolated from soils in which *O. ramosa* was apparently unhealthy (Gold *et al.*, 1979). This strain proved non-pathogenic to tomato but when added to sterilized soil resulted in an 80% reduction of *O. ramosa*. There has not apparently been any attempt to exploit this organism as a mycoherbicide, but it remains an interesting possibility.

Most recently, Linke *et al.* (1992) have isolated *Ulocladium atrum* from *O. crenata* and found in preliminary tests that it has promising potential for biocontrol at least under humid conditions. It did not infect 11 crops on which it was inoculated but a requirement for high humidity may limit its usefulness.

In the light of the successful development of mycoherbicides based on native organisms for a number of non-parasitic weed problems, it would seem that the *Fusarium* and other fungal pathogens attacking *Orobanche* should be receiving more attention.

Prevention

The means of spread of *Orobanche* are the same as those for *Striga* and the precautions are generally similar. Spread with crop seed has been shown to occur even with sunflower, as a result of seed blown onto the seedheads by wind (Castejón-Muñoz *et al.*, 1988). The difficulty with *Orobanche*, however, may be that seeding is not so readily prevented by hand-pulling when the parasite plants are hidden by crop foliage. There is also the additional risk of movement with irrigation water which cannot be easily prevented other than by ensuring no parasites are allowed to grow and mature on the sides of irrigation channels.

One of the most serious risks is that of transfer on soil with the seedlings of tomato or tobacco when they are transplanted. Hence seedbeds should always be on uninfested land or treated with fumigation by methyl bromide if necessary.

Transfer can occur in animal manure; Jacobsohn *et al.* (1987) showed that seeds of *O. aegyptiaca, O. cernua* and *O. crenata* could all survive passage through the digestive tract of sheep.

Integrated Control of *Orobanche*

Summary of general methods

Methods of control of *Orobanche* not specific to a particular crop include:

Rotation Rotation into any non-host crop will prevent further build-up of the problem, provided of course that any host weed species are controlled. Selection of trap-crops such as sweet pepper, which are known to stimulate parasite seed germination should be preferable, contributing to a more rapid decline in the soil seed bank, but the economic viability of the alternative crop is of overriding importance.

Flooding More work is needed to establish the conditions under which flooding is beneficial on each of the major *Orobanche* species – the length of flood, suitable soil types, temperature, etc. – but flooded rice is generally observed to be beneficial.

Fumigation Use of methyl bromide is the 'ultimate' *Orobanche* control method, providing near-complete control when correctly applied. Costs are high and, even with the benefits from control of nematodes and other pathogens and weeds, are justified only in certain high value cropping situations. Dazomet and metham-sodium are less reliable unless plastic covers are used and then prove almost as costly as methyl bromide. They are, however, less hazardous to use, and combinations of irrigation, incorporation and rolling can be developed to ensure reliable results on specific soil types.

Solarization Solarization may be thought of as a (relatively) poor-man's fumigation. It is unlikely to work where air temperatures do not naturally approach 40°C, and is virtually dependent on irrigation facilities to moisten soil in an otherwise dry, cloud-free season. But the timing, during the hottest part of the dry season, is least likely to interfere with normal cropping patterns.

Prevention of *Orobanche* seeding Hand-pulling may be feasible in many situations, depending on the intensity of infestation, ease of access and visibility in the crop, and labour cost and availability. The usefulness of hand-pulling may be increased where the pulled plants are kept as a source of *Phytomyza* for the following season (see below). Spot treatment with vegetable oils may offer a significant advantage in terms of speed and of avoiding damage to crop roots.

Phytomyza Where emergence and seeding of *Orobanche* cannot be completely prevented, it should be worth encouraging *Phytomyza orobanchia* as much as possible. This may require some revision of the use of insecticides where they are being used for control of other insect pests. Otherwise it will involve the collection and suitable storage of hand-pulled *Orobanche* plants and their distribution the following season and/or reduced tillage to avoid deep burial of the *Orobanche* residues in which the insect overwinters. In areas where *Phytomyza* is not already present, its introduction should be considered, but only in conjunction with a bio-control laboratory which will ensure there is minimum risk of introducing parasites with the *Phytomyza*.

Mycoherbicide Whilst the use of fungal cultures such as *Fusarium* has not yet been fully developed outside Eastern Europe, it is an approach which deserves much wider attention, and relevant research establishments should be pressed to conduct the necessary development work in infested areas, especially where transplanted crops are involved.

Crop-specific methods of control

In most crops it is necessary to think in terms of integrated control, using combinations of relevant control methods. A recent review by Pieterse *et al.* (1992) summarizes this approach for a number of crops, and has been helpful in the preparation of the following sections.

Sunflower In most major *O. cernua*-infested sunflower-growing regions there is likely to be some varietal selection and breeding work already in progress, from which the latest information on resistant varieties can be obtained. In most such areas, immune varieties are available which may virtually solve the problem, though care is still needed to look out for and destroy occasional plants of *O. cernua* which may represent potentially dangerous new races of the parasite. Where other *Orobanche* species occur on sunflower, the *O. cernua*-resistant varieties should be obtained and tested, though at present it is far from certain they will prove resistant. In this and other situations where resistant varieties are not suitable for whatever reason, there is scope for the use of herbicides, especially imazethapyr. In due course, genetically engineered herbicide-resistant varieties will become available in which the use of glyphosate or other herbicide should be completely safe and effective. Of the general methods, rotation, hand-pulling and *Phytomyza* are fully relevant, and flooding where irrigation is available. Fumigation and solarization may not be economic, and mycoherbicide is less easily applied in a non-transplanted crop.

Faba bean No single method is likely to provide complete control of *O. crenata* in faba bean, but there is a choice of methods which, when combined, are likely to give a high degree of relief. Partially resistant lines, originating from Giza-402 or other sources should become available in the near future from ICARDA and other collaborating institutions in the Mediterranean region. Delayed planting can greatly reduce emergence of *O. crenata* and correspondingly improve the chances of good control by glyphosate (Linke and Saxena, 1991b). Other herbicides, especially imazethapyr, and sulfonyl ureas such as chlorsulfuron, are also likely to be recommended as alternatives to glyphosate or as additional treatments. The use of nitrogen and phosphorus fertilizer at 33.7 kg N and 71.4 kg P_2O_4 was recommended as part of an integrated package in Egypt (Nassib and Hussein, 1989). An improved package has been tested more recently in which the variety Giza-402 at a high seedrate is later treated with glyphosate at 46 g a.i. ha^{-1} and a foliar application of liquid fertilizer (Nassib *et al.*, 1991a,b). In Syria, Linke (1992) proposes a combination of slightly delayed planting date with low applications of imazethapyr pre-emergence, or glyphosate or imazaquin post-emergence. A similar combination is suggested in Spain (García-Torres and López-Granados, 1991a). Linke and Saxena (1991b) emphasize the potential for solarization and hand-pulling as further components of an integrated system. The importance of hand-pulling, wherever it is feasible as a supplement to other methods (or where infestations are low) to ensure complete prevention of seed production, can hardly be overemphasized as it is likely to make all the difference to the longer-term prospects of complete eradication of the problem. Of the other general methods, rotation and *Phytomyza* are relevant but flooding, fumigation, solarization and mycoherbicide are less so for the same reasons as for sunflower.

Lentil The range of options are similar to those for faba bean, except that there is no recognized 'resistant' material. Variety may, however, be important in the success of delayed planting, which requires an early-maturing variety such as ILL 8 (Linke and Saxena, 1991b). Imazaquin and imazethapyr are both suitable but the former, pre-emergence, has been best. Hand-pulling is all the more feasible in lentil as the *Orobanche* is likely to be visible above the crop. The kerosene or vegetable oil treatment may thus be useful as an alternative.

Chickpea The problem in chickpea is often avoided completely by late planting, but where it is planted earlier, Linke and Saxena (1991b) suggest a combination of solarization, herbicide (reduced doses of glyphosate at 2×30 g a.i. ha^{-1}), and hand-pulling. Other general methods apply as for faba bean.

Pea No special recommendations have been made for pea but delayed planting should help where the problem is from *O. crenata*, and further herbicide evaluation may well lead to reliable use of imidazolinone or sulfonyl ureas. Other general methods as for faba bean.

Carrot and celery There has been little or no screening of either of these crops for resistance to *Orobanche*, but there is evidence for variation in susceptibility of carrot, which deserves more systematic evaluation. Otherwise the use of glyphosate should be especially helpful, together with other general methods as for faba bean.

Tobacco While *Orobanche*-resistant varieties of tobacco are not yet available, there is the prospect in due course of herbicide-resistant varieties allowing the use of chlorsulfuron, etc. Meanwhile, maleic hydrazide is apparently successful in some areas. Among the general methods, hand-pulling and the use of oils for direct kill of emerged parasite are the main options for the small farmer, while flooding, solarization, fumigation, *Phytomyza* and mycoherbicide could all be useful and relevant in many situations. The cost of fumigation may be especially justified where there are nematode problems.

Tomato and eggplant With more work, it is possible that varieties of eggplant with good resistance to *Orobanche* could be selected or bred, but it is likely to take many years to develop such varieties with the other agronomic and quality characteristics needed for the market, and in the case of tomato it is probable that herbicide-resistant varieties will become available sooner. These should then largely solve the *Orobanche* problem in tomato which is otherwise currently one of the most difficult of all. Until any new varieties are available, the most relevant of the general methods for both crops are fumigation or solarization (control of nematodes and other pathogens may help to make these economic), and flooding. Hand-pulling or direct application of oil may be feasible in some small-scale growing systems but not on a larger scale. Mycoherbicides are potentially useful, applied in the planting holes, while *Phytomyza* is likely to be precluded by routine insecticide use.

Cucurbits There has been very little attempt to look for *Orobanche*-resistant cucurbits and some more effort would be justified. Maleic hydrazide has been successful in some areas. Hand-pulling and direct application of oils should be feasible in the low-growing species and the use of *Phytomyza* has been especially successful in cucurbits in the former USSR. Fumigation may be economic on the larger scale, and, as cucurbits are often associated with very hot growing conditions, solarization may be appropriate, while the availability of irrrigation would also allow flooding treatments. Mycoherbicides could readily be applied to planting holes but their success in this far-spreading type of crop may perhaps depend on systemic spread of the fungus through the host.

***Brassica* crops** More work is justified, to look for resistant or at least less susceptible varieties in the seriously affected areas. Some relief may be gained by switching from one *Brassica* species or type to another where the alternative is economically acceptable, but the relative susceptibility of the different types is still not well documented. No selective herbicide treatments are known. Hand-weeding and spot treatment with oils are not likely to be easy in this close-planted crop but at present there is little alternative to recommend. Flooding, or rotation with flooded rice, should be helpful where it is feasible. Where it is not, the use of catch- or trap-cropping may need to be tried.

Conclusion

As for *Striga* in cereals, there are very few simple, complete solutions to *Orobanche* problems. There are prospects for the development of herbicide-resistance in a few of the major crop species which should become an important control option, especially for tomato and tobacco. Genetic engineering might also, in due course, provide fully parasite-resistant cultivars, but for many of the less important host crops this solution will be remote indeed. Meanwhile, as for *Striga*, priority in field research must be placed on the development of well integrated long-term systems using the available partial control methods relevant to each crop and locality. Given the relatively high value of many of the horticultural crops involved, the range of potential methods is somewhat wider than in the subsistence cereal crops affected by *Striga*. Hence fumigation, solarization and herbicides may be relevant as well as the more basic agronomic approaches. The potential for mycoherbicides is one of the most deserving research priorities.

4 Other Root Parasites

A number of parasitic genera of the Scrophulariaceae, other than *Alectra* and *Striga*, include species of some agricultural significance. There are also the less well known genera of Orobanchaceae in the tropics and some representatives of other parasitic families, none of which is widespread or of major importance, but which can cause problems locally.

Scrophulariaceae

Of the genera covered here, the more temperate, such as *Odontites* and *Rhinanthus*, have tended to decrease in importance with intensification of agriculture in Europe. Others are more tropical in distribution and could possibly become more important. Much less is known about the biology of these tropical species but the following sections summarize the most readily available information.

Buchnera hispida Buch.- Ham. ex Don

Buchnera hispida in the vegetative stage is comparable to *S. hermonthica* in stature and leaf shape and it dries black like *Striga* species, but the leaves are slightly broader and less rough. The flowers are quite different in their blue colour and in being almost radially symmetrical, without the enlarged lower lobes of all *Striga* species and without a distinct bend in the corolla tube (see Plate 11A).

Distribution Widespread in tropical Africa, also in subtropical southern and southwestern Africa, Madagascar and India.

Host range This is not well known but Nwoke and Okonkwo (1974) confirmed that *B. hispida* is a facultative parasite capable of maturing without a host. By analogy with other facultative parasites in the Scrophulariaceae it is likely to be able to form haustoria on the roots of many associated plants, but direct evidence is lacking. Pot experiments by Nwoke and Okonkwo (1974) confirmed that it could parasitize the cereals sorghum and pearl millet but, even when it occurs among cereal crops in the field, it is far from certain that it is attached to the crop plants; wild hosts may be more favoured. In Ethiopia it has been seen in dense populations on *Eleusine indica* and *Cynodon dactylon* on field edges adjacent to maize fields but not within the crop.

Importance Nwoke and Okonkwo (1974) recorded severe damage to both sorghum and pearl millet in pot experiments to the extent that sorghum was killed by even a single plant of *B. hispida*. Field observation of damage to cereals is not common but Parkinson (1989) records that sorghum, millet and maize were all attacked, at least mildly, in the Kano and Kaduna districts of northern Nigeria and the pot experiments reported by Nwoke and Oknokwo clearly demonstrate that it does have quite serious potential and deserves closer study.

Biology and ecology The seeds of *B. hispida* are significantly larger than those of most *Striga* species, about 0.55 mm long, and this modest extra size gives them just enough additional resources to establish independently of a host root. As in the slightly smaller-seeded facultative parasite *S. angustifolia*, it is important that seedling reserves are not wasted trying to reach the soil surface from excessive depth. Thus they have a requirement for light which is under the control of a phytochrome system (Okonkwo and Nwoke, 1974; Nwoke and Okonkwo, 1980). Light is also beneficial during the conditioning period of 6–8 days that is required, as in other *Striga* species, before they will germinate. Host root exudates and ethephon have little or no influence.

The distribution of *B. hispida* often overlaps that of *S. hermonthica* and it is apparently favoured by some of the same factors of infertile soils and intermittent rainfall. It is most often seen on lighter, sandy soils but is not thought to be seriously restricted by soil texture.

Control No specific control methods have been attempted on *B. hispida* and it can only be suggested that isolated patches are dealt with by hand-pulling or spot spraying with herbicide such as 2,4-D. In the case of more general infestation, it may be necessary to rotate out of the susceptible cereal and at the same time control weeds which could act as alternative hosts.

Ramphicarpa fistulosa (Hochst.) Benth. (= *R. longiflora* Benth.)

The genus *Ramphicarpa* is distinguished from *Striga* by its long, straight corolla tube and its asymmetrical fruit which is obliquely ovoid (see Plate 11B). *R. fistulosa* also differs from most *Striga* species in having pure white flowers and very finely divided leaves. The white flowers, opening at dusk, and the long corolla tube are associated with pollination by moths at night.

Distribution Widespread in tropical Africa, also in South Africa and in Madagascar.

Host range This is not well known but includes mainly grasses and perhaps some sedges. Rice is the crop most often attacked but maize, sorghum and cowpea are also listed by Kuijt (1969).

Importance Rice is damaged locally in West Africa and Madagascar but the areas affected are usually quite restricted by the apparently limited conditions under which it thrives, i.e. when soil is saturated but barely flooded. Infected rice plants are often severely damaged.

Biology and ecology The seeds are about 0.55 mm long, similar in size to those of the facultative *B. hispida*, but there is no evidence for *R. fistulosa* being able to establish without a host and it is assumed to be an obligate parasite, dependent on host root exudates for its germination.

Most typically, *R. fistulosa* occurs in a ring at the water level around the edge of a rice paddy or around the edge of a depression in a partially flooded paddy. The reason for this localization is not known.

Control No specific control measures have been developed and hand-weeding should normally be enough. From its observed localization at the water level, it may be assumed that control of water level, or drainage, may also be effective in suppressing it.

Ramphicarpa veronicaefolia Vatke (= *Cycnium veronicifolium* (Vatke) Engl.)

The flowers of this species have a striking resemblance to those of *Striga hermonthica* in their size and pink colour. The lobes, however, are generally broader and the two united upper lobes are almost as long as the others. The foliage is also different in being distinctly toothed or shallowly lobed.

Distribution The species is apparently restricted to East Africa, especially Kenya and Tanzania, mainly in the lowlands, but extending up to 2000 m elevation.

Host range This includes a wide range of wild hosts, but sorghum and cowpea can occasionally be attacked as confirmed in Tanzania by Fuggles-Couchman (1935) and there have been recent reports of damage to maize in eastern Kenya.

Importance This is rarely of significance but it is apparently capable of causing acute damage on a local basis.

Biology and ecology This species is presumed to be a facultative parasite but no direct study has been made. It is associated especially with *Commiphora* bush vegetation.

Control No specific methods are known; hand-pulling or spot-spraying should normally suffice.

Odontites verna (Bell.) Dum. (= *Bartsia odontites* Huds. and including *O. serotinus*)

This species, now sometimes known as *O. vernus*, is generally a small, branched plant 15–25 cm high with lanceolate, serrate leaves up to 5 cm long (see Plate 11C). The dull pink flowers are rather like those of Labiatae in structure with a straight corolla tube extended by elongated upper lobes forming a hood. Two subspecies are distinguished, the more northern ssp. *verna* with chromosome number (2n) = 40, and the more southerly ssp. *serotinus* with 2n = 20.

Distribution Mainly southern Europe but extending north into the British Isles and east into Turkey; also now locally in North America, e.g. in Manitoba and in Wisconsin.

Host range *O. verna* probably parasitizes a very wide range of wild hosts but is of significance where it attacks pasture grasses and occasionally wheat and lucerne.

Importance This is regarded as a weed of pastures in Manitoba, Canada (Forbes, 1964), and of lucerne in Wisconsin (Donald and Fawcett, 1975). In Europe it has been recorded occasionally as a minor weed in wheat (e.g. Brenchley, 1920) but it is now rarely seen except under unusually low fertility conditions such as the long-term trial on Broadbalk Field at Rothamsted Research Station, UK, where wheat has been grown continuously without fertilizer for well over 100 years (Last, 1960a).

Biology and ecology The seeds are relatively large, 1.5 mm long, and have a dormancy that is broken by chilling (Chancellor, 1973), in keeping with its mainly temperate distribution. There was some exposure to light in Chancellor's study but it is not clear whether this was essential. Germination under natural conditions in the UK was found to be almost all in April. It failed to respond to ethylene or root exudates from a number of potential host species. In detailed studies by Govier and Harper (1964) and Govier *et al.* (1967, 1968) it was confirmed that seedlings can establish without a host and eventually mature, flower and produce capsules. Young roots have root hairs and the uptake and translocation of phosphorus was demonstrated. However, growth was very much more vigorous and final size and seed production was greatly increased when attached to a host such as *Lolium perenne*. Xylem to xylem connections were demonstrated in the haustorium. Using ^{14}C it was shown that translocation of carbon was almost all from host to parasite though traces did move from parasite to host. The relative quantities of labelled sugars and of nitrogenous materials transferred and appearing in the parasite tissues varied strikingly depending on the host species. Grown on barley or *Stellaria media*, 80–90% of the soluble radiocarbon was in the form of monosaccharide sugars while on white clover (*Trifolium repens* L.) it was only 30–60% and the balance mainly in amino acids. Sucrose, as in other parasites, was not detected, and is presumably converted to monosaccharide in the haustorium. In the young parasite, before stomata were fully developed, it was shown that the flow of material from host to parasite may be enhanced by exudation from three types of gland or hair. It is suggested that this exudation may be a means of disposing of excess nitrogenous materials when growing on a leguminous host.

The damaging effect of *O. verna* on lucerne was assumed by Donald and Fawcett (1975) to result from an allelopathic effect. It is not clear whether they were aware of the parasitic connection or not. However, Schewe and Staniforth (1981) pursuing this approach failed to confirm any damaging effect on lucerne from leachates of living or dead foliage.

O. verna is a plant of infertile grassland and field borders, often on wet ground adjacent to waterways. Although it thrives on leguminous hosts such as white clover and is tolerant of high nitrogen in its host, it appears to be favoured by low fertility and conversely discouraged by high soil fertility. The reason for this is not clear but the shading effect of rank vegetation is likely to be one reason.

Control *Odontites* is known to be susceptible to 2,4-D and MCPA at about 0.5 kg a.i. ha^{-1} (Forbes, 1964). It is also likely to be suppressed by manuring of grasssland or cereals and by cultivations timed to destroy the main flush of germination in spring.

Rhinanthus Species

Two main species can occur as weeds. They were once included together in *R. crista-galli* L. but are now distinguished as ***R. minor*** L. (often split into several subspecies) and ***R. angustifolius*** **C. Gmelin** (= *R. major* Ehrh. = *R. serotinus* (Schoenheit) Oborny). Both are branched annual herbs up to about 50 cm with bright green, sharply toothed leaves about 5 cm long and 1 cm wide. *R. minor* has flowers about 1.5 cm long, bright yellow, the two upper lobes joined to form a narrow lip and the three lower spreading to about 5 mm across. The calyx characteristically enlarges in fruit to form a bladder about 1 cm across in which the seeds, with their characteristic wings, are loose and rattle, hence the common English name of 'yellow rattle'. *R. angustifolius* (see Plate 11D) is distinguished by being slightly larger in all its parts, the flowers up to 2.5 cm long and the purple teeth on the upper lip of the corolla being about 2 mm long, much longer than broad, whilst those of *R. minor* are rounded and shorter than broad. The chromosome number (2n) = 14 in both species.

Distribution Throughout Europe, western Asia and North America.

Host range This is known to be extremely wide but some host preference has sometimes been noted (Gibson and Watkinson, 1989).

Importance *Rhinanthus* species occur quite densely in pastures especially those on infertile soils, and although it is now more usual for them to be of interest as objects for conservation than for control they are known to cause some damage to their hosts. Sowing 500 seeds of *R. angustifolius* per m^2 reduced growth of meadow fescue (*Festuca pratensis*) by nearly half (Rabotnov, 1959). The cereal crops wheat and barley can both be attacked. Damage is rarely serious, but in Bulgaria yields of winter wheat have been increased two to three times by use of herbicide to control *R. angustifolius* (Benkov, 1970).

Biology and ecology Seeds are even larger than those of *Odontites*, 3.5 mm long excluding the 'wings'. Like *Odontites, Rhinanthus 'crista-galli'* has been shown to require stratification to overcome dormancy (Vallance, 1952). Otherwise there is no more specialized mechanism involved. Like *Odontites, Rhinanthus* can complete its life cycle without a host but rate of growth is vastly increased after attachment. In studies on the photosynthesis and acquisition of resources from the host, differences from *Odontites* included the fact that sucrose is found in *R. angustifolius* and that the amino-acid complex in the parasite appears to be more independent of that in its host (Hofstra and Klaren, 1973; Klaren *et al.*, 1976; Klaren and Janssen, 1978). Rate of photosynthesis in the parasite is much lower than that of barley before attachment but increases two- to three-fold after attachment, perhaps because utilization of the products of photosynthesis is no longer limited by shortage of organic nitrogen. Carbon:nitrogen ratios show a striking fall in the attached parasite. Transfer of materials from host to parasite is helped by high transpiration rates, as in many other hemi-parasites (Press *et al.*, 1988). Photosynthesis in the host is not changed as a result of parasitism. Ter Borg and Bastiaans (1973) studied why *R. angustifolius* is associated with shorter, thinner patches in grassland and concluded that it was due to a combination of better growth of the parasite under less shaded conditions and some damage to host plants from the parasite. Growth reduction in the host was more severe under good growing conditions than under poor conditions. There was some evidence of a preference for leguminous hosts, but very fast-growing annuals such as *Senecio vulgaris* were not so suitable, nor were more highly fertile conditions.

Pollination by various types of bee has been studied by Kwak (1979), who showed that hybridization can result between *R. minor* and *R. angustifolius*.

Control In cereals, Hakansson (1956) failed to control *R. angustifolius* well with MCPA in Sweden but it has been controlled by 2,4-D in Bulgaria (Benkov, 1970) while in grassland, both compounds have been effective; also 2,4-DB which is safer to many legumes (Uzunov *et al.*, 1974). Cultural methods such as nitrogen fertilizer and cutting or grazing before seed-formation are also likely to bring about a gradual reduction.

Melampyrum arvense L.

Known as 'cow-wheat' in English, *Melampyrum* species are morphologically similar to *Rhinanthus*, but in *M. arvense* the lower leaves are without teeth, the flowers are more elongated, yellow below but purplish towards the tip and the calyx is small and inconspicuous (see Fig. 4.1). The upper leaves and bracts are pectinate (comb-like) at least in their lower half, and the upper bracts are pale purplish, matching the flowers. Chromosome number (2n) = 18.

Fig. 4.1. *Melampyrum arvense* and fruit (from *Drawings of British Plants*, 23, Stella Ross-Craig (1966) G. Bell and Sons, London).

Distribution Throughout Europe and western Asia.

Host range This is not precisely known but is presumably very wide and includes the temperate cereals, wheat and barley.

Importance This is normally a minor weed of cereals in southern and eastern Europe, but is apparently capable of causing serious harm. Benkov (1978) reported 29% reduction in wheat spikelets and 79% reduction in grains per ear from 24 *M. arvense* per m^2 in Bulgaria.

Biology and ecology In these respects *Melampyrum* species are not thought to differ significantly from *Odontites* and *Rhinanthus*. Seeds are relatively large, about 5 mm across.

Control *M. arvense* is reported to be controlled in wheat by a mixture of bromoxynil and MCPA in Turkey (Cetinsoy, 1980) and by 2,4-D in Bulgaria (Benkov, 1978).

Seymeria cassioides (Walt.) Blake

S. cassioides (known as 'senna seymeria') is a much-branched annual up to 1 m high, covered in sticky glandular hairs (see Fig. 4.2). The leaves are finely divided with segments 1–3 cm long. The flowers, 1 cm across, are yellow with a purple centre.

Distribution Southeastern USA; also introduced to the Bahamas.

Host range *S. cassioides* has a narrow host range, thriving only on slash, longleaf and loblolly pines (*Pinus elliottii, P. palustria* and *P. taeda* respectively) (Fitzgerald, 1975; Musselman and Mann, 1978). In pot experiments haustoria also attached to firs (*Abies*), spruce (*Picea*) and cedar (*Cedrus*) species but they were less numerous than on pine (Mann and Musselman, 1979) and it has not been observed on any of these trees in the wild.

Importance Mann *et al.* (1969) described the damage caused to slash pine (*Pinus elliottii*) while Musselman and Mann (1978) refer to *Seymeria* as 'the most important genus of native root parasites in the south' (of the USA).

Biology and ecology Seeds are about 1 mm long and germination apparently occurs without any specialized mechanism other than light (Musselman and Mann, 1978; Gwynn *et al.*, 1978). Early root growth is normal and it can grow independently provided plenty of nutrients are

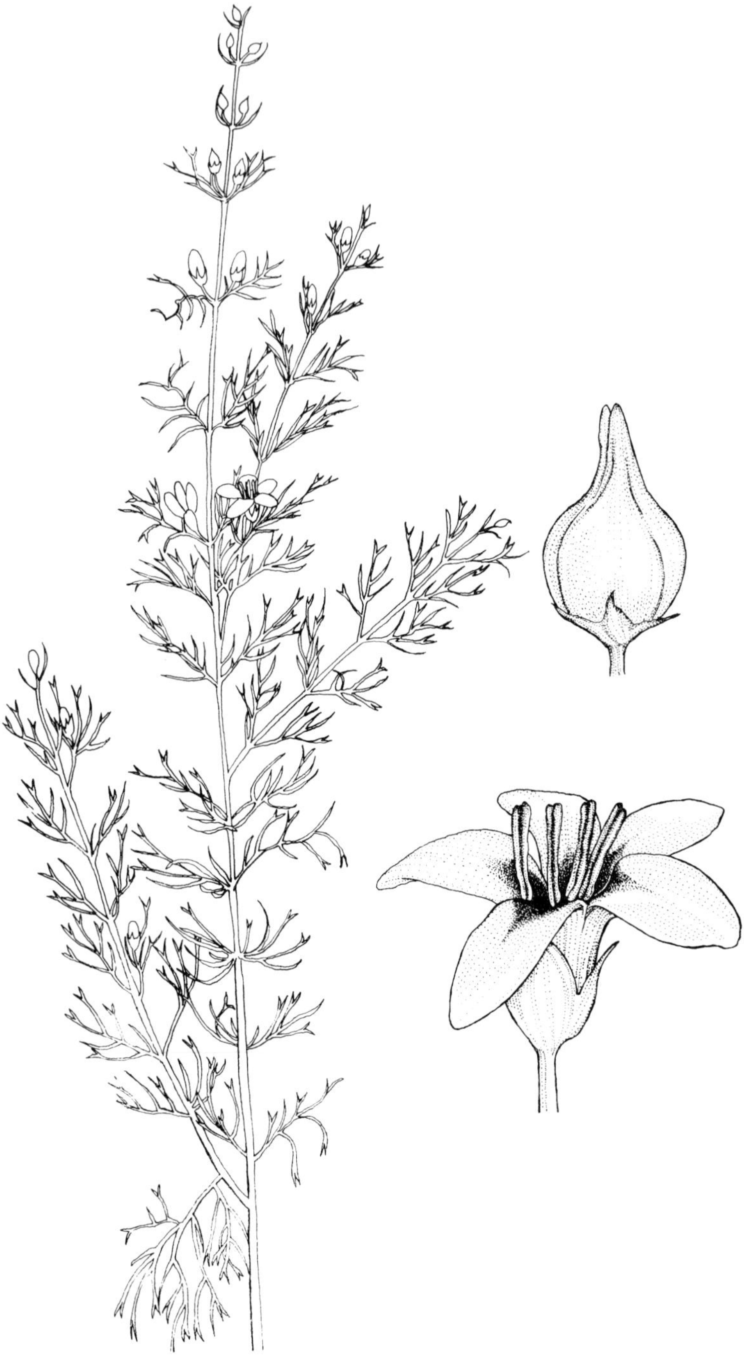

Fig. 4.2. *Seymeria cassioides.*

available, but shoot growth is very limited until it attaches to a suitable host. Grelen and Mann (1973) describe it as an obligate parasite, which is perhaps correct on the basis that it is not known to be able to complete its life cycle without a host. It is unusual, however, in that it has a very narrow host range combined with apparently no specialized germination mechanism. Perhaps in the forest context there is little danger of suitable host roots being unavailable whenever the seeds germinate. Processes of germination, early growth and establishment of the haustorial connections are described by Fitzgerald (1975). Stangle and Musselman (1981) report that shading results in death, even when the host is still illuminated.

Control Fitzgerald and Terrell (1975) have shown that 2,4,5-T and fenoprop at about 1 kg ha^{-1} killed most of a young population of *S. cassioides* but there were some symptoms in the host trees, and Gwynn *et al.* (1978) confirmed that 2,4-D and dicamba could be translocated from parasite to host, so such herbicides would need to be used with care. No other methods have been reported.

OROBANCHACEAE

Aeginetia species

Most *Aeginetia* species differ from *Orobanche* in having little or no stem above ground, the flowers being solitary on long individual stalks up to 50 cm high. The most important species, ***A. indica*** **Roxb.**, has large flowers about 5 cm long, purplish with yellow or white markings (see Fig. 4.3).

Distribution *A. indica* ranges from India, China and Japan to the Philippines and Indonesia.

Host range *A. indica* affects sugarcane in India, Taiwan, Japan and the Philippines. Although sugarcane is the main crop affected, rice and maize are listed by Kuijt (1969) as economic hosts, and pot experiments have confirmed that sorghum can also act as a host. Kuijt comments that '*Aeginetia* is restricted mostly to grasses'.

Importance No distinctive symptoms occur on sugarcane but it can cause reduced yield and also affect the quality of the sugar juice. Stools already weakened by other problems may be killed. Two other species are also recorded attacking sugarcane, namely ***A. pedunculata*** **(Roxb.) Wall.** in India and ***A. saccharicola*** **Bakh.** in Indonesia (Hughes, 1961), but there have been no recent reports of their occurrence.

Fig. 4.3. *Aeginetia indica* growing on sorghum in pots.

Biology and ecology The seeds of *A. indica* are about the same size as those of *Orobanche* species (about 0.3 mm across) and are reported to number 70,000 per capsule (Kuijt, 1969). Germination generally requires host root exudate but Hughes (1961) reports that while many species stimulate germination, relatively few grass species can be parasitized.

French and Sherman (1976) found that germination was inhibited by light but could be stimulated by sodium hypochlorite. Chennaveeraiah *et al.* (1971) obtained germination with a combination of 2,4-D and cytokinin. In addition to reproducing by seed, Kuijt (1969) points out that *Aeginetia* spp. are capable of vegetative reproduction, uniquely for Orobanchaceae. 'Root-like' organs spread to make new haustorial contacts from which further inflorescences can emerge. The system is presumably perennial. Flowering is known to occur independently of the daylength and of the flowering of the host (Tashima *et al.*, 1972).

Control *A. indica* has been controlled in the Philippines by spraying with 2,4-D (Rivera, 1953; López and Barile, 1964,1965). The cane variety NCo-310 has been reported highly resistant in Taiwan (Lo, 1955).

Christisonia wightii Elmer

Christisonia is a genus closely related to *Aeginetia* and there have been reports in the rather distant past of its occurrence on sugarcane in the Philippines, but there is no recent confirmation of this.

SANTALACEAE

Santalaceae is a large family of about 30 genera, all of which are believed to be facultative hemi-parasites. The family includes the well known sandalwood (*Santalum album* L.) and many other large tree species which do not superficially differ from any other forest tree, but normally supplement their nutrition by forming haustoria on contact with the roots of other trees. Others are shrubby, again behaving as root parasites (e.g. *Osyris* and *Thesium* species), or as shoot parasites like mistletoes (e.g. *Dendrotrophe*, *Phacellaria*).

Thesium Species

Out of about 250 species of *Thesium*, only a few are known to parasitize crops. ***T. humile*** **Vahl** is an erect or more often prostrate annual to 50 cm long (see Fig. 4.4). Leaves are fleshy, linear, up to 2 cm long. Flowers are yellowish, 3 mm across. ***T. resedoides*** **A. W. Hill** and ***T. australe*** **R.Br.** are apparently similar but the latter is perennial.

Distribution *T. humile* is a plant of the western Mediterranean, occurring in Spain and North Africa. *T. resedoides* occurs in Southern Africa, *T. australe* in Australia.

Fig. 4.4. *Thesium humile* and fruit (from *Students' Flora of Egypt*, V. Tackholm (1974) Cairo University).

Host range Abou-Raya *et al.* (1976) concluded that *T. humile* could parasitize a wide range of hosts including barley and a range of wild grasses, also *Artemisia* and *Plantago* species. Less is known of the other two species but both occur on sugarcane and on a range of wild species.

Importance *T. humile* is known to attack barley in Libya and to cause yield reductions of 10–15% (Abou-Raya and El-Sharkway, 1979). It can also be a problem on cereals in Spain (King, 1966) and on onions in Jordan. *T. resedoides* is known to cause serious damage to sugarcane in South Africa and Swaziland (Visser, 1981); infested cane was only 30 cm high while surrounding unaffected plants were 2 m high. *T. australe* R.Br. is also reported damaging sugarcane in Australia (Myatt and Rehbein, 1954; Hughes, 1961).

Biology and ecology Fruits of *T. humile* are 2 × 3.5 mm with a pericarp which becomes highly mucilaginous on wetting. This mucilage provides a substrate for microorganisms which may have a role in germination, a view supported by observations of Chabrolin (1965), who showed that certain fungi were essential for germination. There is no need for host root exudate for germination. After germination root growth is normal and there is good development of chlorophyll, allowing independent growth for some time. Abou-Raya and El-Sharkway (1979) believe that only limited development is feasible without a host, but there is no certainty that *T. humile* is effectively an obligate parasite. Abou-Raya *et al.* (1976) describe the haustoria as being typical for Santalaceae. They also note that autoparasitism is common.

Control Hughes (1961) reports that young shoots of *T. australe* can be killed by 2,4-D and it would seem likely that this would prove effective on most *Thesium* species.

Osyris alba L.

O. alba is a perennial shrub up to 2 m high with narrow, linear leaves about 2 cm long and small creamy-white flowers, 2–3 mm across.

Distribution Mainly Mediterranean.

Host range Joel *et al.* (1991b) comment that a wide range of trees and shrubs are parasitized in the eastern Mediterranean, including pine and oak; also *Cupressus sempervirens* L. around the edge of fruit orchards from which it had spread to almond and pear trees. Grapevines are also affected.

Importance King (1966) records that *O. alba* has occurred on grape vines, fruit and forest trees in Yugoslavia and on vines in southern France. Joel *et al.* (1991b) describe infestations in almond and pear orchards that caused chlorotic symptoms and reduced yield and quality of fruit.

Biology and ecology The seeds develop in small fleshy fruits and are apparently eaten and spread by birds (Joel *et al.*, 1991b). No specialized germination requirement is known. Once established on a host a strong rhizome system is developed at depths down to 50 cm, from which shoots emerge. In contact with host roots very large haustoria, up to 10 cm diameter, are formed.

Control Joel *et al.* (1991b) suggest repeated mowing but this did not completely kill the parasite. Herbicides glyphosate and trichlopyr applied as 10 and 15% a.i. solutions by rope-whip applicator gave more lasting control.

BALANOPHORACEAE

This family has about 18 genera, mainly tropical in distribution but including the species *Cynomorium coccineum* in southern Europe. The flowers in this species are so extremely reduced as to be barely recognizable. Combined with the fleshy nature of the inflorescence it is often mistaken for a fungus and is known commonly as Maltese fungus. *Thonningia sanguinea* is the only species so far reported to have any agricultural significance.

Thonningia sanguinea Vahl.

T. sanguinea consists mainly of an extensive, robust underground system, occasionally producing a fleshy red inflorescence which emerges only a few cm above the soil (see Fig. 4.5). The rhizome-like system spreads extensively and establishes haustoria on contact with host roots. It also swells to form quite large tuberous haustoria (Hepper and Gasson, 1986). Although referred to as rhizomes, there is evidence that these organs are modified roots (Kuijt, 1969). There are separate male and female inflorescences but whether the plant is monoecious or dioecious is still in some doubt. The inflorescence consists of a dense mass of minute flowers in a Compositae-like head, surrounded by rings of pointed red scales which also clothe the peduncle.

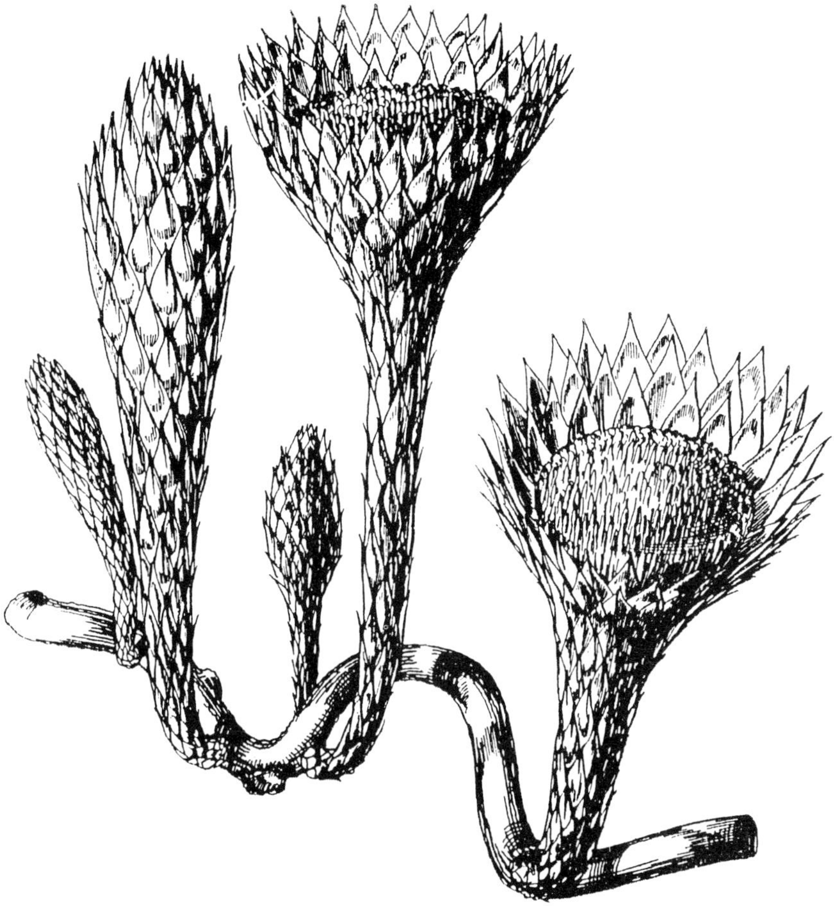

Fig. 4.5. *Thonningia sanguinea* (from *Flora of West Tropical Africa*, Volume 1, J. Hutchinson and J.M. Dalziel (eds) 1958) British Crown Copyright. Reproduced with permission of the Controller, Her Majesty's Stationery Office, and the Trustees, Royal Botanic Gardens, Kew).

Distribution Throughout most of tropical Africa.

Host range This is not clearly known, but is believed to include a wide range of forest species including rubber.

Importance *T. sanguinea* is reported to cause serious damage to rubber trees, though the exact symptoms have not been described. It is believed a single plant of *Thonningia* can spread to attack 20 or more trees over an area of 0.5 ha.

Biology and ecology Little is known about the biology of this interesting plant. Animals are thought to be involved in seed dispersal (Kuijt, 1969). Kuijt also interprets the published information on 'rhizome' and haustorial structure. In ecological terms, *Thonningia* is very much a plant of the forest and is not seen in the savannah.

Control No control method is known.

5 *CUSCUTA* SPECIES, THE DODDERS; AND *CASSYTHA FILIFORMIS*

Any connection between the name 'dodder' for *Cuscuta* species and the English word meaning to tremble with frailty is doubtful, but this group of parasites can indeed induce a severe frailty in its hosts, comparable to that caused by *Striga*. This results from a powerful diversion of sugars from host to parasite, often causing a total failure of the host crop to set fruit. Seed production is less profuse than that of the root parasites but serious problems can build up, especially where carry-over of seed in the soil is supplemented by the sowing of contaminated crop seed, as often occurs with lucerne and niger seed.

TAXONOMY AND IDENTIFICATION

The genus *Cuscuta* is sometimes placed in its own family the Cuscutaceae but is here treated as part of the otherwise non-parasitic bindweed family, Convolvulaceae. Apart from being parasitic, *Cuscuta* species are distinguished from the rest of the family by their fused calyx and the presence of infra-staminal scales (see Figs 5.13, 5.14, pp. 196, 197). The *Cuscuta* species, known as dodders in English, are shoot parasites with a very simplified structure consisting of mainly yellow twining stems with scale leaves and clusters of small flowers, each a few millimetres across with calyx and perianth each of 4–5 lobes. Although not usually showing any green colour, they do contain some chlorophyll and are therefore not strictly holo-parasites. However, photosynthesis in most species is negligible and they are all obligate parasites, almost totally dependent on their hosts from a few days after germination. The taxonomy of *Cuscuta* is difficult and the number of species varies between 100 and 200 according to the authority, but the most useful monograph, by Yuncker (1932), classifies 121 full species. Identification depends almost totally on detailed examination of the flowers, the number and shape of perianth segments, the shape and arrangement of the stigma and styles, the shape and dehiscence of the capsule, and the shape of the infra-staminal scales attached to the filaments. As the last two characters are often difficult to determine, even in fresh material, the following key relies mainly on other features. But even these can be difficult to see in the dried specimen and professional help is usually needed to confirm identifications. Severova (1991) illustrates differences in the scale and calyx characters of many European species. Useful keys to European species are also given by Feinbrun (1970) and in *Flora Europaea* (Tutin *et al.*, 1972).

A single species, usually known as *C. campestris*, is responsible for the great majority of *Cuscuta* problems but a number of other species can be troublesome locally. The following key shows how the main species are separated but it almost certainly omits some others which might occur. Hence it should not be relied upon if there is any doubt about the range of species occurring in a particular area. There have been many misidentifications of *C. campestris* because this North American species has only relatively recently been introduced to many countries of the Old World. Using available keys to the local flora, many specimens of this weed have been wrongly named *C. australis* or *C. chinensis* because they share the relatively easily seen character of capitate stigmas. As a result of this difficulty it is still far from certain just how important these two species are.

The genus *Cassytha*, although superficially similar, is quite unrelated to *Cuscuta*. It is sometimes placed in its own family the Cassythaceae, but more often in the laurel family Lauraceae, in which it is the only parasitic genus. This is dealt with separately at the end of this chapter.

Key to the Most Important *Cuscuta* Species

A One style, supporting two stigmas (section Monogyna)

Style shorter than the elongated stigmas, flowers 6–8 mm long, white with purplish rim. Calyx very short. Capsule conical 5–8 mm long, seeds 3–3.5 mm. Mainly Central to East Asia. . . . . ***C. reflexa***

Style about as long as stigmas, all extremely short, flowers 3–4 mm, **calyx with broad fleshy lobes**, almost equalling corolla tube. Capsule elongated, cone-shaped, 6 mm long. Seeds 3–3.5 mm. Mainly in the Middle East. . . . . ***C. monogyna***

Style about twice as long as stigmas, flowers 3–4 mm long, **in elongated clusters**, sometimes red-spotted, calyx much shorter than corolla tube, the lobes narrower than above. Seeds 2–3 mm long. Mainly in Europe. . . . . ***C. lupuliformis***

Style much longer than the short stigmas, flowers 3–4 mm long **in elongated clusters**. Seeds about 3 mm long. Mainly in East Asia. . . . . ***C. japonica***

B Two styles, stigmas linear, without knobs (section Cuscuta)

Perianth mostly 4-parted

Flowers 2–3 mm, pedicelled, **in loose heads of 3–8 flowers**. Stigmas subsessile. Capsule round, closely enclosed by corolla. Seeds about 1.25 mm. Mainly West and Central Asia. . . . . ***C. pedicellata***

Flowers 1.5–2 mm, sessile **in very small, dense heads** 4–6 mm across; corolla lobes with erect hooded tips. Capsule round. Seeds about 1 mm. Mainly eastern Mediterranean. . . . . ***C. palaestina***

Perianth mostly 5-parted

Calyx lobes fleshy at least at the tip, flowers 1.5–2.5 mm, sessile in heads 5–6 mm across. Capsule round, enveloped in corolla. Seeds about 1 mm. Widespread. . . . . ***C. planiflora***

Calyx lobes membraneous:

Flowers 3 mm long in heads 10–15 mm across; styles plus stigmas shorter than the ovary. Capsule roughly round. Seeds about 1.2 mm. Only in flax and linseed fields. Widespread. ***C. epilinum***

Stems slender, reddish. Flowers 3–4 mm in dense heads 7–10 mm across, styles plus stigmas slightly longer than ovary. Seeds about 1 mm. Mainly Europe. . . . . ***C. epithymum***

C Two styles, capitate, with knobs (section Grammica)

Flowers granulate, covered with minute protuberances, 2–2.5 mm long *on distinct pedicels*. Seeds about 1.5 mm. Mainly North and Central America and Caribbean. . . . . ***C. indecora***

Flowers not granulate

Capsule enclosed in corolla:

Flowers 2–4 mm long, **pedicelled, in a loose head**, somewhat glandular, **corolla lobes deflexed**. Corolla persisting as a cap on the capsule. Seeds about 1.5 mm. Mainly North America. . . . . ***C. gronovii***

Flowers 2–3.5 mm **in dense heads. Corolla lobes fleshy at the tip.** Capsule 3–4 mm across, enclosed tightly by corolla, circumscissile. Seeds about 1.2 mm. Mainly in East Asia. . . . . *C. chinensis*

Capsule exposed:

Flowers about 2 mm, in compact heads. **Corolla lobes obtuse**. Seeds about 1.5 mm. Sometimes reddish-glandular on capsule. Distinct crater between styles. **Infra-staminal scales bifid**. Widespread through Europe and Asia. . . . . *C. australis*

Flowers 2–3 mm, in compact heads 10–12 mm across. **Corolla lobes acute, often flexed upwards**. Capsule round, 2–3 mm across, not concealed by corolla. **Infra-staminal scales exserted, fimbriate, not bifid**. Seeds 1–1.5 mm. Almost worldwide. . . . . *C. campestris*

THE MAIN *CUSCUTA* SPECIES

Cuscuta reflexa Roxb.

C. reflexa is one of the more robust species with a vine 1–2 mm thick when fresh, reddish or yellow, rather than orange and with a tinge of green sometimes, as a result of a significantly higher level of chlorophyll than in many other species (see Plate 12A, Fig. 5.1). This can cause confusion with *Cassytha* in the vegetative stage but the latter can be distinguished by the presence of hairs, at least on the scales; *Cuscuta* species are all quite glabrous. The flowers are large, up to 10 mm long, white, with a very

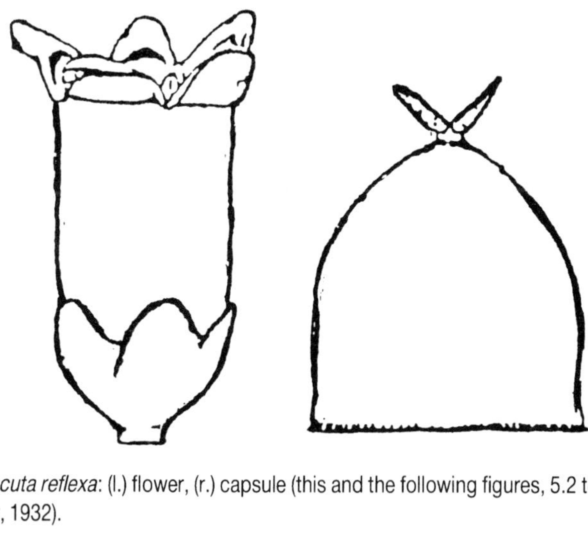

Fig. 5.1. *Cuscuta reflexa*: (l.) flower, (r.) capsule (this and the following figures, 5.2 to 5.14, are drawn from Yuncker, 1932).

short calyx, and an elongated conical capsule. The style is so short as to appear almost non-existent. The seeds are large, 3–3.5 mm long.

Distribution This is an Asian species, common in India and in the Himalayas, but extending west through Pakistan to Afghanistan, south to Sri Lanka and east to China and Indonesia. It may have been introduced and become established occasionally elsewhere.

Host range *C. reflexa* commonly occurs on woody species including tree crops such as citrus, coffee, peach and litchi. Bhattarai *et al.* (1989) recorded 39 host species in 28 plant families in Nepal. These were mainly perennial, woody species but included a few annuals. These authors distinguish between 'primary' and 'secondary' hosts, the latter only being attacked where the parasite is already established on the former. Two grasses were recorded as secondary hosts but it was noted that there was a lack of haustorial development, making the use of the term 'host' questionable. See Box 5.1, p. 198.

Importance In Holm *et al.* (1979), *C. reflexa* is listed as a 'principal' or 'serious' weed in Afghanistan, Nepal, India and Pakistan. The main crops affected are not recorded in this publication but are likely to include a wide range of woody and herbaceous fruit and vegetables and perhaps some forest species. The exact degree of damage has not often been measured, but this is an extremely vigorous species sometimes covering whole canopies of trees and is assumed to cause serious harm, in the same way as *C. campestris*.

Cuscuta monogyna Vahl.

This is the most robust of the species occurring as weeds, with a vine 2–3 mm thick, but flowers are smaller than in *C. reflexa*, 3–4 mm long, and at least half this length is enclosed by calyx (see Fig. 5.2). The very short style and stigmas distinguish it from the other species in this group. Capsule and seeds are as in *C. reflexa*.

Distribution *C. monogyna* is a species mainly of the Mediterranean and Middle East, but extending from France through southern Europe, east to Afghanistan.

Host range The hosts of *C. monogyna* are predominantly woody shrub and tree species, e.g. *Rosa*, *Salix*, *Zizyphus*, *Vitis*, *Tamarix* and *Citrus*, but can also include herbaceous hosts especially in Leguminosae.

Fig. 5.2. *Cuscuta monogyna*: (l.) flower, (r.) capsule.

Importance *C. monogyna* is listed by Holm *et al.* (1979) as a 'principal' weed in Iran, Pakistan and the former USSR and 'common' in Italy, Morocco, Spain and Turkey. It is known also to be troublesome locally in Afghanistan, Iraq, Jordan, Israel and Cyprus. The crops most commonly attacked include citrus, olive and other fruit crops, and vegetables including onion. The damage caused has not been defined in detail but is presumed to be equivalent to that of other species such as *C. campestris*.

Cuscuta lupuliformis Krock.

This is a robust species with quite thick reddish vines, comparable to *C. monogyna* and *C. japonica* in most characters, but distinguished from the first by its shorter calyx, while the style length is somewhat longer than in *C. monogyna* but much shorter than in *C. japonica* (see Fig. 5.3).

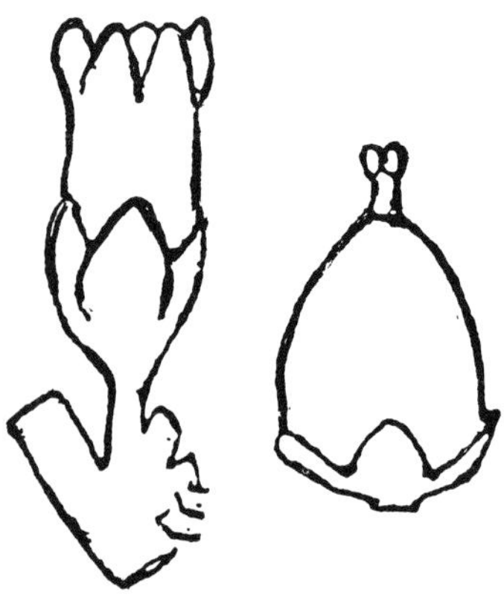

Fig. 5.3. *Cuscuta lupuliformis*: (l.) flower, (r.) capsule.

Distribution Mainly Europe, but introduced sporadically elsewhere, including India, Afghanistan and Brazil.

Host range Almost exclusively on trees and shrubs.

Importance Listed by Holm *et al.* (1979) as a 'principal' weed in the former USSR. *C. lupuliformis* is one of the *Cuscuta* species used by Wolswinkel (1973) to confirm their devastating effects on faba bean hosts.

Cuscuta japonica Choisy

This is closely similar in most respects to the last species, but differing in the distinctly longer style (see Fig. 5.4).

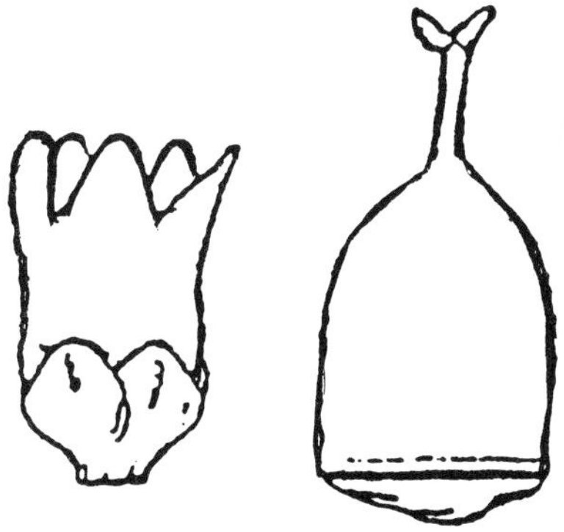

Fig. 5.4. *Cuscuta japonica*: (l.) flower, (r.) capsule.

Distribution Eastern Asia.

Host range Both woody and herbaceous hosts are attacked. Zaroug and Ito (1988) record a range of wild hosts in Japan, also potato and eggplant but not tomato, which showed a hypersensitive response, nor pea, cucumber, clover or *Brassica campestris*.

Importance Locally important in China and Japan, but not well documented.

Cuscuta pedicellata Ledeb.

A fairly distinct species among those with two styles and elongated stigmas, having 4-parted flowers in very loose heads (see Fig. 5.5).

Distribution Eastern Mediterranean and Middle East to Central Asia.

Host range A wide range of wild hosts, including many Leguminosae and occasionally on forage legumes of economic importance.

Importance A relatively minor species, recorded as a common weed of clovers and lucerne in Egypt and of lentil and cumin in Israel.

Fig. 5.5. *Cuscuta pedicellata*: (l.) flower, (r.) capsule.

Cuscuta palaestina Boiss.

This is the smallest and most fine-stemmed of the weedy species, with 4-part flowers only 1.5–2 mm across in dense heads only 4–6 mm in diameter (see Fig. 5.6).

Fig. 5.6. *Cuscuta palaestina*: (l.) flower, (r.) capsule.

Distribution Mainly in the eastern Mediterranean.

Host range Similar to that of *C. pedicellata*.

Importance A localized problem in clovers, reported to be common in Lebanon.

Cuscuta planiflora Ten. (including *C. approximata* Bab.)

Known as the small-seed dodder in the USA, this is a very widespread species with many varieties. The form sometimes treated as a separate species, *C. approximata*, is distinguished mainly on the basis of a yellow shiny base to the calyx (see Fig. 5.7).

Dawson (1984a) has pointed out how *C. planiflora* differs from some other species in that the main stems twine around the stems of the host. He suggests that this habit may limit the rate of spread compared with *C. campestris* and *C. indecora* in which only the lateral tendrils do so.

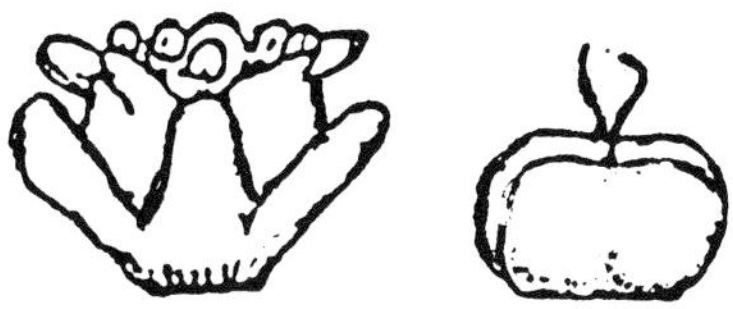

Fig. 5.7. *Cuscuta planiflora*: (l.) flower, (r.) capsule.

Distribution Mainly Mediterranean and central Asia but introduced to many other areas, especially North America.

Host range A wide range of species is attacked, including the Leguminosae.

Importance Listed by Holm *et al.* (1979) as a 'serious' or 'principal' weed in Pakistan and Iran and as common in Italy, Morocco and the USA. It is also the commonest problem species in Egypt (Al-Menoufi and Hassan, 1976) on lucerne and clovers, including berseem (*Trifolium alexandrinum*).

Cuscuta epilinum Weike

This is one of the most distinctive species, in that it has a very restricted host range, and is seen only in flax or linseed fields. The flowers have no very distinct characters other than the two elongated stigmas, but the erect corolla lobes give it a more elongated appearance than that of *C. campestris*, and the capsule continues to be enclosed by the corolla at maturity (see Plate 12B, Fig. 5.8). The stems are a similar thickness to those of *C. campestris* but are usually paler, yellowish in colour rather than orange. The seeds are small, little over 1 mm long, and should be readily separated from all but the smallest seeded crop samples. Yet spread in seed samples apparently continues to occur, at least on a local basis.

Distribution This species is believed to originate from southwest Asia but is now widespread in Europe and occurs sporadically in other parts of the world as a result of introduction, e.g. to Ethiopia, Egypt, Morocco, Cyprus, Canada, the USA and South Africa.

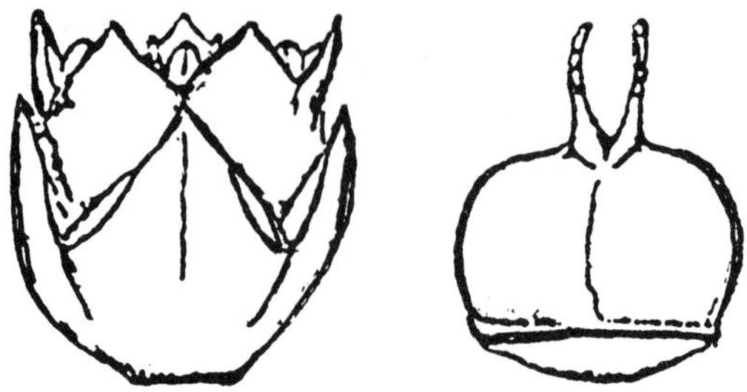

Fig. 5.8. *Cuscuta epilinum*: (l.) flower, (r.) capsule.

Host range Very much restricted to flax and linseed, though some weed species in these crops are also sometimes attacked. These are presumably acting as secondary hosts. Gaertner (1950) records a few further primary hosts.

Importance Listed by Holm *et al.* (1979) as a 'principal' weed in the former USSR and common in Morocco, *C. epilinum* is rarely an important problem, but presumably causes significant yield loss where it does occur, reducing flax fibre length as well as yields of seed or fibre.

Cuscuta epithymum (L.) Murr. (including *C. trifolii* Bab.)

This is a thin-vined dodder with reddish stems and very small pinkish, scented flowers (see Fig. 5.9).

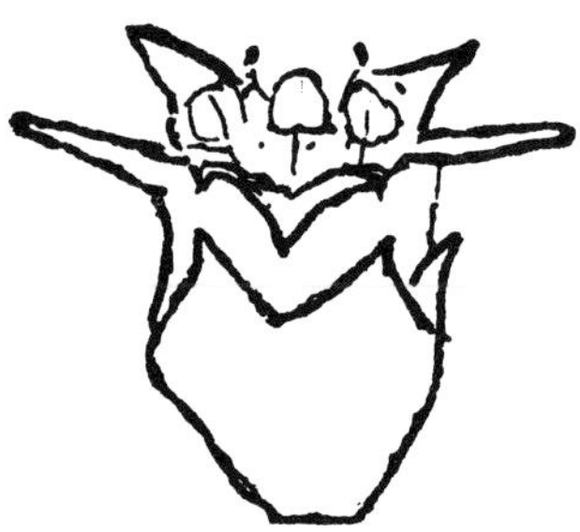

Fig. 5.9. *Cuscuta epithymum*: flower.

Distribution Europe and West Asia, but occasionally occurring elsewhere as a result of introduction, e.g. North America, Japan, Australia, South Africa, South America.

Host range Mainly wild hosts, herbaceous and shrubby, but often also on clovers and other forage legumes.

Importance Listed as a 'serious' or 'principal' weed in Afghanistan, Hungary, Czechoslovakia, Poland, Turkey, Italy, Portugal and the former USSR, and as common in Canada, Iran, South Africa, Spain and Venezuela (Holm *et al.*, 1979). Perhaps some of these are based on misidentification, but it can apparently cause acute problems at least locally, especially in forage legumes and in carrot.

Cuscuta indecora Choisy

Known as large-seed dodder in the USA, this species is distinguished by its white, fleshy flowers with a granular texture (see Fig. 5.10).

Distribution Mainly restricted to North and South America, but recorded sporadically elsewhere, as in Morocco and Greece.

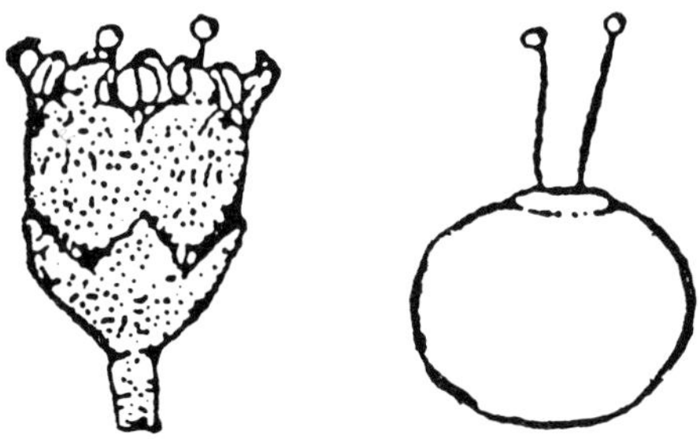

Fig. 5.10. *Cuscuta indecora*: (l.) flower, (r.) capsule.

Host range Mainly Leguminosae.

Importance Locally important on lucerne in the USA, Argentina and perhaps in Morocco.

Cuscuta gronovii Willd. *ex* Roem. and Schult.

Known as the swamp dodder in the USA, this is mainly a species of wet places, distinguished from the other main species with capitate stigmas by its loose, pedicelled flower head (see Fig. 5.11). The flowers are also usually somewhat glandular.

Distribution Common and widespread in North America but present also in Europe, especially along water courses.

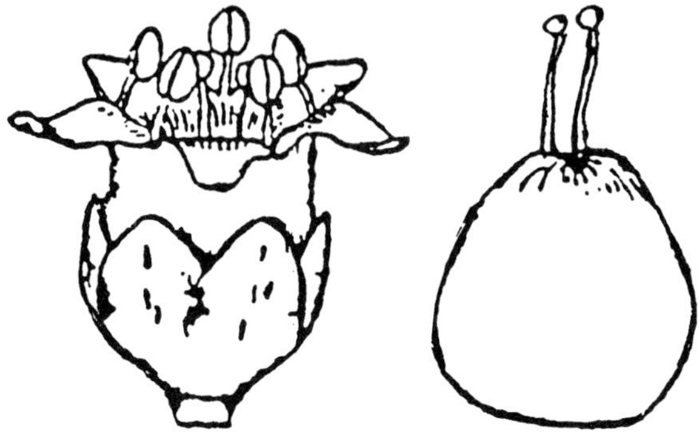

Fig. 5.11. *Cuscuta gronovii*: (l.) flower, (r.) capsule.

Host range A wide range of woody, shrubby and herbaceous species of wet places, including *Salix* and *Populus* species, also the crop cranberry (*Vaccinium macrocarpon*).

Importance *C. gronovii* is of importance on cranberry in North America, capable of causing at least 50% yield reduction (Bewick *et al.*, 1988a). It is also serious on carrot and occurs less seriously on mint, celery, potato and sweet pepper.

Cuscuta chinensis Lamarck

The main key character in Yuncker (1932) is the circumscissile capsule (splitting around the circumference) but this is not always easy to detect (see Fig. 5.12). The flowers do however have a distinct appearance with the thickened corolla lobes and, as in *C. gronovii*, the capsule tends to remain enclosed in the corolla rather than being much exposed as in *C. campestris*. Knepper *et al.* (1990) indicate how the seeds differ from *C. campestris* (see below) but it is not certain how readily this character is seen. Otherwise it has similar vegetative characters.

Distribution Mainly in South and East Asia, especially China and Japan, but occurring sporadically elsewhere, as in Europe, the former USSR and Egypt.

Host range Probably very wide, but not well documented. Recorded on lucerne in Egypt, soyabean in China and from a number of host crops in India.

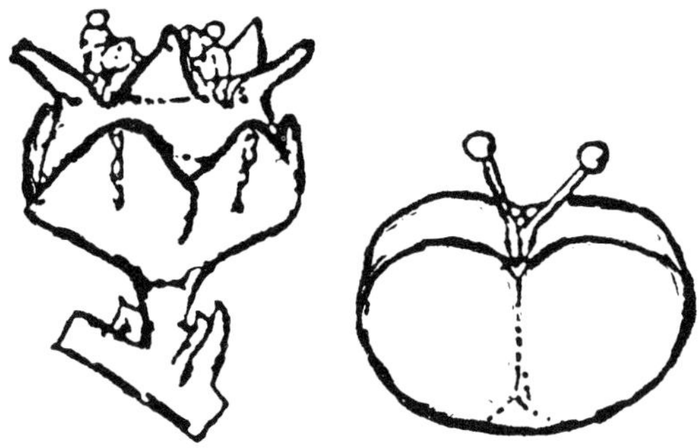

Fig. 5.12. *Cuscuta chinensis*: (l.) flower, (r.) capsule).

Importance This species, like *C. australis*, is of uncertain status as a weed because of confusion with *C. campestris*. It is reported as a problem on a number of crops including niger seed and mung bean in India, but there have to be some doubts about these records. Assuming their identification was correct, this was the species involved in studies by Kumar *et al.* (1989). By use of fumigation they created differential infestations in mung bean from which they deduced a yield loss of 94%. Reports from China are probably more reliable, where it is regarded as a serious parasite of soyabean.

Cuscuta australis R.Br.

Among the species with capitate stigmas, *C. australis* is closest in appearance to *C. campestris*, from which it is distinguished by its obtuse, erect corolla lobes and bifid scales (see Fig. 5.13). There is also sometimes a reddish glandular surface on the capsule.

Distribution In spite of its name this species is widely distributed from southern Europe through to East and Southeast Asia. In Japan, it is apparently well recognized but Takabayashi *et al.* (1981) comment how it has been largely replaced by the introduced *C. campestris* and is not now regarded as an agricultural problem.

Host range It is apparently restricted to herbaceous hosts in a wide range of families but especially on *Polygonum* species. It is also reported as one of the species attacking soyabean in China.

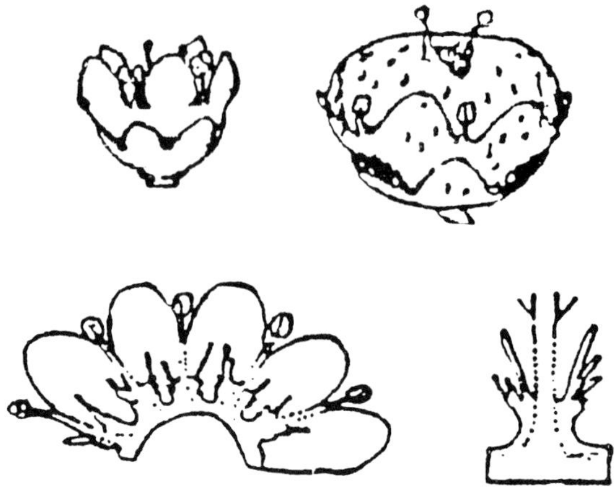

Fig. 5.13. *Cuscuta australis*: (l.) flower, (r.) capsule, and (below) dissected flower and enlargement of infra-staminal scale.

Importance It is far from certain how important this species is, owing to the frequent confusion with *C. campestris*. It is listed by Holm *et al.* (1979) as a 'principal' weed in Italy and as common in Australia, Japan and Morocco, and is reported to be spreading in Malaysia, but these records are possibly not reliable.

Cuscuta campestris Yuncker (including *C. arvensis* Engelm.; *C. pentagona* Engelm.)

Known as field dodder in the USA, this is by far the most important single *Cuscuta* species, native to North America, but now occurring at least sporadically through all the other continents and causing acute local problems. It is characterized by orange-yellow stems with virtually no trace of green. Corolla lobes are spreading rather than erect but the acute tips are usually flexed upwards (see Plates 12C, 12D and Fig. 5.14). The capsule is round, very much exposed and with an indentation between the two styles. In common with *C. indecora* and perhaps many other species, tendrils are produced opposite to the scale leaves, unlike the branches, which arise in the axils of the scales (Dawson, 1984a). These tendrils are responsible for contacting and twining around the host while the main and branch stems are relatively insensitive to contact with the host and grow more-or-less straight and spread a great deal more rapidly – up to 8 cm per day – than those of *C. planiflora* where the main stems are responsible for twining.

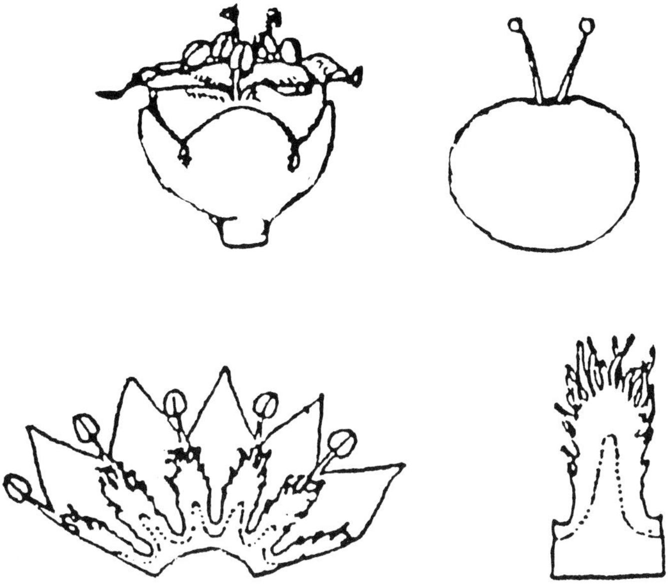

Fig. 5.14. *Cuscuta campestris*: (l.) flower, (r.) capsule, and (below) dissected flower and enlargement of infra-staminal scale.

Distribution Although originating in North America, this species now causes problems in many countries of South America, Europe, the Middle East, East and Southern Africa, South and East Asia and Australia. It is still very localized in West Africa and in parts of other regions but is continuing to spread.

Host range The host range of *C. campestris* is extremely broad, but there are a number of especially favoured hosts, on which it grows most vigorously. Among these are the crops lucerne and niger seed with whose seed the parasite is spread. Many weeds are also favoured hosts, including *Convolvulus arvensis*, many Compositae such as *Artemisia* and certain Acanthaceae. Although most hosts are herbaceous, woody species can be attacked, including coffee. Grasses often appear to be attacked at least as secondary hosts but the haustorial connections are generally found to be incomplete.

Importance Holm *et al*. (1979) list *C. campestris* as a 'principal' or 'serious' weed in Afghanistan, Botswana, Hungary, Arabia, Pakistan and the Soviet Union, and common in Canada, Morocco, Zimbabwe, South Africa, the USA and Argentina. It is perhaps most important as a weed of

Box 5.1. Host range, host preference and host specificity.

Host range is a general term signifying the total range of host species which can be attacked by a particular parasite species or, more precisely, a particular population, race, or strain of a parasite species. It can be of special importance in considering rotational approaches to control of parasites, so that susceptible crops, or even susceptible weeds within those crops, can be avoided.

Within the host range of a species or strain, it can often be recognized that some are much more favoured hosts, being attacked more heavily or more frequently than others. We can then speak of **host preference** and **favoured hosts**. *Cuscuta campestris* has an extremely wide host range but clearly shows such host preference in respect to crops such as lucerne and niger seed, and weeds such as *Convolvulus arvensis* and *Artemisia* species.

In the case of the *Cuscuta* species we have to recognize a refinement of host preference in the form of primary and secondary hosts. **Primary hosts** are those on which *Cuscuta* can establish and develop from the seedling stage. **Secondary hosts** are those on which the parasite is apparently unable to establish from seedlings but on which it is able to make attachments and sink effective haustoria after it is well established on a primary host. *C. epilinum* appears unable to establish on any host crop other than flax/linseed, but once established it is able to attack and damage adjacent weed species. The case of *Cuscuta* on grass species is still not quite clarified. Although it is almost certain that they cannot act as primary hosts, grasses very often appear to be acting as secondary hosts. Rath and Mohanty (1988) recorded 14 grasses and one *Cyperus* species which were 'naturally' infested by *C. chinensis* though none could be infected artificially. In some cases workers have looked closely and shown that there has been no effective penetration of haustoria, but we believe it could be found that at least some *Cuscuta* species are able to attack some grasses. *Cassytha* has no problem in doing so.

In the case of *Striga* species on cereals, host range is usually quite broad, but there is clear evidence for variation in host range between different populations or strains of *S. asiatica*. Broadleaved species are rarely attacked by the cereal *Striga* species, but they can occasionally occur as primary hosts, while there is also some suggestion that they can behave as secondary hosts only after establishment on a cereal (Kumar and Solomon, 1941). This could certainly occur where the 'secondary host' does not stimulate germination but is susceptible to attachment by secondary haustoria. Whether this is a common occurrence and whether significant damage can result from such secondary attachment remains to be clarified.

In mistletoes, the term 'secondary host' is sometimes used in the sense of 'less favoured', as for certain hosts of *Arceuthobium* species which appear only to be attacked when there is extreme infection pressure from neighbouring primary hosts.

Host specificity is a term sometimes used loosely in place of host range, but should be used only for the extreme of host preference when host range is limited to a single host species, or a very narrow range of specified hosts. Such specificity is rare in Scrophulariaceae but can be almost applied to certain strains of *Striga gesnerioides* (e.g. that attacking cowpea) and to *Cuscuta epilinum*. It is especially appropriate to some *Arceuthobium* species, which have only one or a few closely related hosts. In Orobanchaceae, host range varies from very broad in *O. ramosa* to very narrow in some non-weedy species and in the two main forms of *O. cernua* which are restricted to Solanaceae and Compositae (especially sunflower) respectively.

In general the facultative parasites, such as *Rhinanthus* and *Odontites* species, have extremely broad host range, attacking a large proportion of neighbouring species, though some host preference has been noted. A possible exception to this generalization is *Seymeria cassioides* which is apparently a facultative parasite, yet has a very limited host range. This may be associated with its occurrence in relatively pure stands of forest species.

The mechanism of host preference is discussed in the individual chapters.

---

lucerne, both in North America and elsewhere, but vegetable crops and some tree crops such as coffee can be damaged locally, as in Ethiopia. Over 80% of lucerne and red clover fields were infested in Yugoslavia (Stojanovic and Mijatovic, 1973) and 20% of these crops were abandoned. Sugarbeet is seriously affected in Eastern Europe, and Belyaeva *et al.* (1978) reported 40% of crops affected in Kirgizia. The association with niger seed is important in India and in Ethiopia (see Plate 13A). Other crop hosts listed in the review by Cooke and Black (1987) include asparagus, carrot, chickpea, grapevine, honeydew melon, lespedeza, onion, potato, red beet, red clover and tomato. Eggplant can also be attacked. Ornamentals can be seriously affected and chrysanthemums are especially susceptible in Australia, the Canary Islands and Ethiopia. The severe damaging effect of *C. campestris* and other species on their faba bean host has been described by Wolswinkel in a series of publications (see the section 'Effects on host' below).

Relatively few crop-loss studies have been published, but Dawson (1989a) arrived at a figure of 57% reduction in forage yield of lucerne over a two year period from artificial infestations of *C. campestris*. Belyaeva *et al.* (1978) estimated that infestation of sugarbeet by *C. campestris* reduced sugar yields by 3.5–4 t ha^{-1}. Also in sugarbeet, comparing uninfested and infested plants, Stojsin *et al.* (1991) measured decreases of 23–41% in root weight and 1.3–2.6% in sugar content.

There is little doubt that yield loss can be almost total, and furthermore phenolic content of the crop can be increased, greatly reducing the palatability of forage. The *Cuscuta* itself can also prove mildly toxic, and infested forage may need to be destroyed altogether (Movesian and Azarian, 1971).

Other Species

Other *Cuscuta* species which have occasionally been recorded as weeds (Holm *et al.*, 1979) include ***C. americana* L.**in Mexico and South America; ***C. babylonica* Aucher and Choisy** in Israel; ***C. europaea* L.** in Europe, Afghanistan, Australia, the USA and the Middle East; ***C. hyalina* Roth** in India, Pakistan, Mali and Sudan; and ***C. suaveolens* Ser.** in Argentina, Southern Africa and Spain.

BIOLOGY AND PHYSIOLOGY

Most of the information in this section will be based on *C. campestris* with additional information from other species as available.

The Seeds

The seeds of *Cuscuta* species vary in size from about 1 to 3 mm across. Most are approximately round in outline, but irregular, with one or more flattened areas, depending on how many seeds of the maximum four developed together in a capsule, and with a surface varying from smooth to coarsely rugose. Those of *C. campestris* are generally 1–1.5 mm across, but quite irregular in size and in shape, with a rough surface which helps to distinguish them from the smooth, regular seeds of lucerne and provides the basis for seed cleaning equipment (see later). Knepper *et al.* (1990) provide a key, based on Gaertner (1950), for separating a number of species occurring in seed shipments to the USA, based on seed size, seed surface, hilum and scar characters. Section Monogyna (see key) is distinguished by seeds over 2 mm, testa cells not collapsed and scar area rectangular. Within this section, two species are separated as follows:

1. Hilum hidden, scar area narrow-rectangular and sunken . . . . *C. monogyna*
2. Hilum exposed, scar area square and flush with seed surface . . . . *C. japonica*

The other two sections both have seeds less than 2 mm, the testa cells alveolate and the scar area circular. In the section Grammica the seeds are 1–2 mm long and the hilum a short line 0.13–0.27 mm long, while in section Cuscuta seeds are smaller – about 1 mm – and the hilum less than 0.1 mm long. Within section Grammica some species are separated as follows:

1. Seeds 1.2–1.8 mm, wider than long, hilum vertical, scar area not distinct . . . . *C. indecora*/*C. gronovii*
2. Seeds 1.2–1.8 mm long, longer than wide, hilum horizontal, flush or only slightly raised . . . . *C. campestris*/*C. australis*
3. Seeds 1–1.2 mm, hilum horizontal, distinctly raised . . . *C. chinensis*

Germination

The seeds of *Cuscuta* have no specialized germination requirement to ensure that they germinate only in the presence of a host, as do *Striga* and *Orobanche* species. A proportion of the seeds do, however, have a dormancy, based on an impervious, hard seed coat which ensures that they germinate over a long period and so run little risk of all germinating simultaneously when there might happen to be no host available. The proportion of hard seeds in *C. campestris* varies considerably from e.g. 77% in Hungary (Czimber, 1969) to 95% in the USA (Hutchison and Ashton, 1980).

Breakdown of the seed coat depends on mechanical abrasion in the soil, on the action of soil microflora and on chilling during the winter months. Different studies showed widely differing proportions of seeds germinating in the season after burial. Allred and Tingey (1964) in Utah, USA, showed 64% emergence in the first season and only 3% in the second. Dawson (1965) in Washington State, USA, and Hutchison and Ashton (1980) in California, USA, obtained much more gradual germination to only 10% over two years. These differing results can almost certainly be linked to the colder winter conditions in Utah than in the other two states, but they can also be modified by factors such as depth of burial. Seeds buried 5 cm or more apparently lose their dormancy much less rapidly. The overall result is a typical flush of germination each spring followed by a more or less prolonged period as further seeds lose their dormancy. Optimum temperature for germination of *C. campestris* is 30–33°C but it can germinate over a range from 15 to 39°C (Stojanovic and Mijatovic, 1973; Hutchison and Ashton, 1980).

Seeds of *Cuscuta* are capable of long survival, up to 50 years or more in dry storage depending on the species (Gaertner, 1950) and at least 10 years in the field (Menke, 1954).

The larger-seeded species can emerge from several centimetres depth in the soil but *C. campestris* has difficulty from any depth greater than 3 cm.

Contact, Attachment and Penetration of the Host

Regardless of depth of germination, the root makes no further growth once the shoot has emerged, and serves only to support the shoot until it locates a host stem (see Fig. 5.15). This is usually achieved by circumnutation, the whole shoot swinging round in a circular anti-clockwise motion, about once every hour. Light and a temperature of about 25°C are required for the circumnutation of *C. campestris*. The length of the shoot varies according to species but is about 5–10 cm in *C. campestris*, and a host must be within about 3 cm for successful contact to be made (Sitkin, 1976). *C. campestris* has a relatively short independent life and its seedling loses its ability to attach to a host and dies after about eight days (Sitkin, 1976). In the larger-seeded species *C. lupuliformis* and *C. gronovii* the shoot can grow up to 15 and 35 cm long respectively and survive for several weeks. Such differences no doubt depend partly on

Fig. 5.15. Typical seedling of *Cuscuta* species, showing short, swollen root (shaded).

seed size but also on the chlorophyll content and efficiency of photosynthesis.

On contact with a stem the *Cuscuta* shoot winds round tightly making up to three complete coils. This coiling action is a response to any physical contact and can occur round inanimate objects too, but the coiling action requires high-energy blue or far-red light and is suppressed under red light or darkness (Zimmerman, 1962; Lane and Kasperbauer, 1965). It is also much reduced under deep shade (Dawson, 1966). A first reaction then is the development of pre-haustorial initials on the inside of the coil. Tsivion (1978a) has shown these to result from the coiling itself and do not require contact with the host. Cutin is then exuded to form a bond (Weinert and Barckhaus, 1975). The development of the haustorial initials and a penetrative organ can occur even into inert material such as a cotton string, but the full development of the haustorium and its vascular tissue requires cytokinin (Tsivion, 1978a). The *Cuscuta* tissue can generate low concentrations of cytokinin but full development may require additional cytokinin from the host tissue.

Intrusive organs develop at the haustorial surface and searching hyphae penetrate the host tissues by a combination of separation and penetration of cells (Lee and Lee, 1989). Once they have penetrated to the host stele, connections with the xylem are formed by a bridge of tracheids while connection with the phloem has been shown by Dörr (1972, 1987) via specialized transfer cells. There are no direct connections between the sieve tubes of host and parasite but plasmodesmata are shown by Dörr (1987) to create a symplastic bridge between searching hyphae and host parenchyma cells.

Once a coil of *Cuscuta* stem has made a successful haustorial connection, new shoot buds are formed in this zone, and in *C. campestris*, up to 20 shoots can arise from a single point of attachment (Dawson *et al.*, 1984). Truscott (1958) made a detailed study of the haustorium of *C. gronovii* and described how it was able to regenerate shoots after complete removal of the external tissues. He compared this process with the generation of adventitious shoot buds in the roots of certain non-parasitic plants.

Detached fragments of *Cuscuta* stem and/or tendril are capable of establishing themselves on the host. This characteristic has been used in northeastern India as a means of spreading *C. campestris* deliberately for control of the twining weed *Mikania micrantha* in tea.

While *Cuscuta* species normally behave as annuals, it is possible for several, including *C. epithymum* and *C. planiflora*, to overwinter as haustoria in host stems and regrow in a second year (Stojanovic and Mijatovic, 1973).

Dependence on the Host

Most if not all *Cuscuta* species have functioning chloroplasts in their stems and can photosynthesize both before and after attachment to the host. The green colour is usually masked by yellow pigments but is sometimes apparent in embryos and young seedlings and in the tips of shoots which are relatively starved (Kuijt, 1969). Chlorophyll contents are low but the ratio of forms a and b is usually normal. Chloroplasts in *Cuscuta 'chinensis'* (almost certainly a misnaming of *C. campestris*) are less than 10% as numerous as in a green plant but their structure is similar, albeit less well differentiated, with grana in poorly defined stacks (Tuohy *et al.*, 1986). Machado and Zetsche (1990) further show that the structure and function of the chloroplasts can differ greatly from one species to another. In *C. reflexa* chlorophyll content was about 2.5% of that in an *Ipomoea* species and Rubisco activity and photosynthesis about 4% of normal, while all these parameters were zero in *C. europaea*. Corresponding to these activities, plastid thylakoids were 'strongly reduced' in *C. reflexa* and zero in *C. europaea*. The plastids in *C. lupuliformis* are also somewhat abnormal (Fer, 1979) and it is only under conditions of high light intensity and very high carbon dioxide concentration that photosynthesis can balance respiration in this species (Fer, 1976a). Hence the bulk of the parasite's carbon is obtained from the host, mainly in the form of sucrose (Tsivion, 1978b). The transfer of sugars from host to parasite is enhanced by a phloem-unloading mechanism, possibly depending on relatively high levels of abscisic acid in the haustoria (see below under 'Effects on Host'). Transpiration is sometimes said to be high in *Cuscuta* as in other parasites, but Fer (1981a) comments on low transpiration due to a thick cuticle in *C. lupuliformis*, and stomata were also found to be relatively few in some Indian species (Pant and Banerji, 1965). Fer (1981b) shows that a very large proportion of the transfer of resources from host to parasite is via the phloem link and little from the xylem.

The root of *C. gronovii* is able to absorb phosphorus (Fer, 1976b) and nitrogen (Srivastava and Chauhan, 1977) from the soil and they are transferred to and utilized by the parasite shoot, but once the parasite is attached to the host and the root has withered, *Cuscuta* must be totally dependent on the host for all its minerals. Mineral balance in *Cuscuta* tends to differ from that in its host. Wolswinkel has shown how K and Mg are selectively absorbed by the parasite via the phloem connection (Wolswinkel, 1975, 1979). In *C. lupuliformis* levels of K were high while those of Ca and Na were normal (Fer *et al.*, 1990). Some amino-acids are transferred from host to parasite (Wolswinkel, 1974a; Fer, 1979), but enzymes necessary for amino acid synthesis are present in the parasite and it is not believed to be dependent on these ready-elaborated N sources.

There is evidence for a strong connection between the flowering of *Cuscuta* and the flowering of its host (e.g. Fratianne, 1965), and there is usually no flowering of the parasite until host flowering has at least been initiated, whether by short or long day conditions. However, the parasite is sometimes observed to flower on a non-flowering host, and *C. reflexa* will flower in the absence of a host, on a culture medium, under short day conditions (Baldev, 1959). From experiments in which defoliation of the (non-flowering) host allowed the parasite to flower, Fratianne concluded that the lack of flowering is due to a flowering hormone inhibitor, rather than the lack of flowering hormone. This could also explain how *Cuscuta* may begin to flower when the host has been severely damaged. Wolswinkel (1989) confirms the general observation of a link between the flowering of host and *Cuscuta*, but cautions that the explanation of a flowering inhibitor could be mistaken and that the flowering of the parasite may be controlled by changed nutritional factors resulting from the flower initiation in the host. He also points out that *C. reflexa* is taxonomically somewhat separate from most other common species and its short day character is not necessarily a general feature of the genus.

Effects on the Host

Cuscuta species are not known to have any substantial effect on either the water balance or the photosynthesis of the host but they do have a profound physiological effect on the host, the haustorial connections acting as exceptionally powerful sinks, competing to divert sugars from host sinks such as developing fruit. Wolswinkel (1973, 1974a,b, 1979) has studied the effects of a complex of three species (*C. campestris, C. europaea* and *C. lupuliformis*) on faba bean and demonstrated how there can be a complete inhibition of pod-setting and fruit growth, mediated by an enhanced phloem-unloading mechanism at the point of haustorial attachment. Abscisic acid (ABA) has been shown to be high in the haustorial tissue of *C. reflexa* and it has been suggested that this is an important factor in the unloading process (Ihl *et al.*, 1984, 1987; de Bock and Fer, 1992). The unloading mechanism is apparently very similar to that involved in the unloading of assimilate from the seed tissues to the developing embryo in the legume but when both host seeds and parasite are competing the parasite wins. Inevitably, such a drain can result in drastically reduced yield in infected crops. Infested crops almost always suffer yellowing and reduced vegetative growth as well as reduced flowering and fruiting.

Ecology

Climate

Cuscuta is a mainly temperate and subtropical genus which is reflected in the response of species such as *C. campestris* to overwinter chilling treatment which releases many of the hard seeds from dormancy. On the other hand, this same species has a relatively high optimum temperature requirement for germination which helps to ensure that germination occurs rather later than other plants in the same ecology, when potential hosts are already established.

Moisture is not generally a critical factor for *Cuscuta* species, though drought stress has been thought to contribute to increased attack in conjunction with low soil fertility (see the next section). In Ethiopia, *C. campestris* is often seen to grow poorly and sometimes die back during the rainy season. This dieback may be due to enhanced pathogenic attack but this has not been confirmed.

Light/Daylength

As discussed above, *Cuscuta* species may require relatively high light for the twining process and establishment on the host. They are also sensitive to daylength for flowering.

Soil Type/Fertility

There have been few reports of any association of *Cuscuta* species with particular soil types but a recent paper by Stojsin *et al.* (1991) suggests a tendency for *C. campestris* on sugarbeet in Yugoslavia to be favoured by low soil fertility combined with drought stress. In an experiment comparing fertilizer treatments, attack was greatest in plots receiving no fertilizer, or potassium only.

Insects

The floral biology of *Cuscuta* has not been well studied but Gaertner (1950) states that *C. campestris* and *C. gronovii* are autogamous and hence require no insect cross-pollination, while others need pollination by insects. *C. epithymum* may commonly be cross-pollinated but is also self-fertile when insects are not present.

Cuscuta species are commonly damaged by a number of insects, especially those in the genera *Smicronyx* (Curculonidae), *Melanagromyza* (Agromyzidae) and *Herpystis* (Tortricidae) (Baloch *et al.*, 1967a; Girling *et al.*, 1979). The potential for some of these for biological control has been studied in Pakistan and is discussed under bio-control below.

Pathogens

A number of fungi are known to cause serious damage to *Cuscuta* species, including *Alternaria*, *Curvularia* and *Colletotrichum* species. The potential for some of these for bio-control is discussed below.

CONTROL OF *CUSCUTA* SPECIES

Cultural and Mechanical Control

Hand-pulling

Hand-pulling is the simplest and most effective means of controlling all light, scattered infestations of *Cuscuta* but it may be necessary to pull the infested host plant together with the parasite, or prune the affected branch of woody species if regrowth is not to occur from the haustoria embedded in the host tissue. If flowering and seed set has already occurred, the pulled material must be removed from the field and eventually burnt.

Rotation

Given the very wide host range of most *Cuscuta* species, rotational crops may need to be selected with care. In the case of *C. epilinum* it should be sufficient to avoid flax or linseed, though the number of years these crops have to be avoided is not well established. At least five years may be necessary. With most other *Cuscuta* species, the safest will be grass or cereals. Even cereals, however, must be well weeded if development and seeding is not to occur on broadleaved weed species. While some development is likely to occur on most broadleaved crops, there are considerable differences in their relative susceptibility to each *Cuscuta* species and perhaps occasionally to the local population of a species. Soyabeans, for instance, are attacked by *C. chinensis* and *C. australis* but are not normally attacked by *C. campestris*. Other crops reported to be resistant to *C. campestris* include bean (*P. vulgaris*), squash (*C. pepo*), cucumber, cotton and tomato. Changing to an alternative crop, even if it is susceptible, may be useful if it allows easier control by herbicide.

Land preparation

Deep ploughing of *Cuscuta*-infested land should greatly reduce the chances of the parasite emerging and establishing from the most recently shed seed but older seed in the soil may be brought to the surface by this practice. It will also be necessary to follow deep ploughing by shallow or minimum tillage for some years to avoid bringing seed back to the surface.

Time of planting

In the USA it is known that *C. planiflora* will germinate at lower temperatures and establish significantly earlier than *C. campestris*. Hence delayed planting will help to avoid the worst of this species. In most regions there is also clearly a flush of germination of *Cuscuta* early in the season from an accumulation of seeds which have lost their dormancy over a cold winter. Where possible, it is worth waiting for this flush and planting a little later after they have been destroyed by tillage (Dawson, 1971a). Planting in autumn, instead of spring, achieves a similar or even greater advantage as the crop will be well established and creating dense shade by the time of maximum *Cuscuta* germination.

Method of planting

There are arguments both for dense uniform planting and for well spaced rows. Dense planting can result in a shading which suppresses the twining and attachment of *Cuscuta,* but this will be ineffective if the main flush of germination occurs before the crop is well established. It is most relevant in a perennial forage crop such as lucerne, where a delay in the timing of cutting, early in the second and subsequent years, may help to reduce new infection. In annual crops there can be some advantage from row-planting as scattered infections will spread only along the row, reducing the number of infested crop plants which have to be removed or destroyed with the parasite (Dawit Mulugeta, 1992).

Post-planting cultivations

Shallow harrowing may be safe and effective in some crops, when carried out shortly before and after emergence of the crop and weed, e.g. in lucerne (Dawson *et al.*, 1984) and in sugarbeet (Lukovin and Kitenko, 1974).

Fertilization

There is one report (Stojsin *et al.*, 1991) of a reduction of *Cuscuta* in sugarbeet by mineral fertilization. This could perhaps arise from greater crop vigour and shading. Care is needed, however, when using farmyard manure, to be certain it is not contaminated with *Cuscuta* seed. Its use may otherwise be counterproductive.

Mixed cropping

An interesting possibility for control by mixed cropping has been identified by Reddy and Rao (1987). They showed that cluster bean or guar allowed attack by '*C. chinensis*' (possibly *C. campestris*) but the parasite was grossly deformed and did not flower normally. This observation was followed up by a field experiment (Rao and Reddy, 1987) in which they grew mung bean and guar bean separately or in 1:1 mixture and infested them artificially with the *Cuscuta*. The mung bean was very susceptible to the *Cuscuta* and allowed rampant development throughout the crop. Guar, however, caused stunted, abnormal, pale green growth ('pseudodichotamous gracillarioid habit') and when mixed with the mung bean greatly suppressed the spread of *Cuscuta* from mung bean to mung bean. Starting with 10% of mung bean infected, a pure crop yielded only 7 g m^{-2} while that interplanted with guar yielded 87 g m^{-2}. They also comment that seeding of the *Cuscuta* was much reduced in the mixed crop, suggesting that guar could be the source of a flowering inhibitor. It is not clear whether the guar flowered in this experiment. Apart from growing a fully mixed crop, the authors also suggest using guar as a border around susceptible crops to prevent ingress of *Cuscuta* from field edges.

Solarization

We have found no evidence for the effectiveness of solarization in control of *Cuscuta*. To judge from the reported failure of this technique on certain hard-seeded weed species (Horowitz *et al.*, 1983; Sauerborn and Saxena, 1987) it would seem unlikely to be effective.

Mowing

A mechanical equivalent of hand-pulling, which is feasible to some degree in lucerne, is the use of extra close mowing – to about 3 cm rather than the usual 10 cm – resulting in a much larger proportion of the infested crop tissue being removed and reducing the chances of regrowth of the parasite (Dawson, 1987). Flail-mowing is recommended in lucerne as a follow-up after pre-emergence herbicide begins to lose its effectiveness (Cudney *et al.*, 1992). It proved substantially cheaper than fire. These treatments need to be carried out before *Cuscuta* seeding.

Fire

Fire has been one of the standard methods for controlling *Cuscuta* spp. in lucerne in the USA. The crop is damaged but generally recovers much more successfully than the parasite. Propane/butane burners are used either for spot treatment or for whole-field or 'solid' burning, the latter involving tractor-mounted equipment. It is common practice for the burning to be done twice, one day or a few days apart – the first, using about 200 l propane per ha, to kill and dry the foliage and the second, using half that quantity, to achieve a more complete burn. Orloff (1987) reported 22% of California lucerne farmers still relying on this method in 1985 but also emphasized the serious damage done to first-year lucerne by the practice, yields and crop stand both being significantly reduced. Spot burning reduces the extent of crop damage and cost of fuel and, although labour costs may then be higher, it is recommended as part of an integrated system for complete prevention of seeding of *C. indecora*, following herbicide and flail-mowing (Cudney *et al.*, 1992). Flaming has also been reported effective in sugarbeet (Lukovin and Rudenko, 1975) and pea and cucurbit crops (Kaculas and Tarziev, 1965). When propane or butane burners are not available, alternative ways of using fire on patches of *Cuscuta* include: (i) vegetation can first be sprayed by kerosene or other fuel and then burned; (ii) vegetation can first be sprayed by contact herbicide and then burned; (iii) dry material can be heaped onto the patch and then burned.

Resistant and Tolerant Varieties

There have been relatively few attempts to assess varietal differences in susceptibility to *Cuscuta*. Nemli (1987) exposed five varieties of tomato, three of sweet pepper and two of eggplant to attack by *C. campestris* and found all tomato varieties resistant and all eggplant and pepper susceptible. Rao and Rao (1991a) found some variation in susceptibility among 15 varieties each of black gram (*Phaseolus mungo*) and mung bean to *Cuscuta* (probably *C. campestris*) in India, but no more than partial resistance. Few other such comparisons seem to have been made and there are no reports of striking differences within any crop species.

It might be suspected that tomato could show such differences, as it is usually found to be resistant to *C. campestris*, yet is seriously attacked by this species in California, USA (Ashton and Santana, 1976). The wild relatives, *Lycopersicon chmielewskii* and *L. hirsutum* show much greater resistance than a standard tomato variety in California (Al-Menoufi and Ashton, 1991). It is also known that young plants of variety New Yorker are susceptible while plants more than four weeks old are resistant (Sitkin, 1976). It is likely therefore that varieties would differ in the length of the susceptible phase. So far, however, there is no evidence for such differences, and anomalies in tomato response are perhaps more often due to variation within the *Cuscuta* and/or differences in tomato-growing methods. In particular, direct-seeded tomato is likely to be damaged in the early stages while a transplanted crop is usually immune.

Where resistance has been studied in resistant crop species, there have been differing observations on the resistance mechanism. Nemli (1987) observed a hypersensitive response in tomato, cotton and bean, while this reaction was not seen in cucumber or squash. Capderon *et al.* (1985) noted the development of a suberized layer in cotton in response to invasion by *C. lupuliformis*. The resistance of potato and sweet potato to *C. reflexa* was apparently associated with higher levels of calcium in these hosts, compared with those in susceptible hosts such as lucerne, bean, mustard and sunflower (Nagar and Sanwal, 1984). Higher levels of calcium were shown to be inhibitory to the enzymes involved in haustorial penetration.

Chemical Control Methods

Fumigants

Fumigation is unlikely to be practical for control of *Cuscuta* on a field scale, and effectiveness on hard seeds would be expected to be a problem, as for solarization. However, Ashton and Santana (1976) suggest that **metham-sodium** and **dazomet** are suitable for control of *Cuscuta* in tobacco seedbeds and in Australia, Milvain (1990) has recommended control of *C. campestris* by 'Di-trapex' (**dichloropropane plus methyl isothiocyanate**). Hence it may be of value for seedbeds of transplanted crops.

Herbicides to prevent establishment

Considering the rudimentary nature of the *Cuscuta* root and the short time that it is in contact with the soil, it is surprising how many soil-acting herbicides are effective in preventing *Cuscuta* establishment. A number have proved highly selective for control of *Cuscuta* spp. in established lucerne. One of the first to be used was **chlorpropham** (known as **CIPC** in North America), applied at 4–6 kg a.i. ha^{-1}, prior to *Cuscuta* germination in the spring (Lee and Timmons, 1956). This and related volatile herbicides control *Cuscuta* through the action of their vapours (Slater *et al.*, 1969), not normally preventing germination, but reducing growth and twining of the seedlings. Results are best when the crop is at least 15 cm high, ensuring a microenvironment in which the vapour concentration can reach the necessary levels. Because of its volatility and susceptibility to breakdown by soil microorganisms, chlorpropham persists in the field for only a few weeks before it loses its effectiveness. Kaufman and Kearney (1965) identified the bacteria that degrade chlorpropham in soil and Kearney (1965) characterized an enzyme, isolated from one of these bacteria, that catalyses the hydrolysis of chlorpropham. It was then shown that certain N-methylcarbamates competitively inhibited the activity of this enzyme (Kaufman *et al.*, 1970). Dawson (1972) later showed that by combining the herbicide with one of these compounds, PCMC (*p*-chlorophenyl *N*-methylcarbamate), the period of control could be extended from three to five weeks. PCMC was not cleared for general commercial use but Dawson (1984b) later showed that the freely available carbamate insecticide carbaryl at 0.8 kg a.i. ha^{-1} could also be used in the same way though with slightly less effectiveness. Chlorpropham with or without carbaryl remains an option for use in lucerne and also in a number of other crops including clovers, lespedeza, blueberries, cranberries, carrots and onion (Dawson, 1987) and in niger seed (Tosh and Patro, 1975). It may be used in tomatoes but only as a directed spray or granules after the crop is well established, also in safflower and in established sugarbeet (Dawson *et al.*, 1984). If the carbaryl mixture is not used, repeat applications may be needed to give season-long contol.

Chlorpropham works best when applied to the surface of moist soils and is much less effective when incorporated into the soil. It is used in both spray and granular forms. The latter are especially suitable when crop or weed foliage might otherwise intercept an excessive proportion of a spray. Another carbamate herbicide, **propyzamide** (known as *pronamide* in North America), has lower volatility than chlorpropham but is active at lower doses of about 2 kg a.i. ha^{-1} and has longer persistence. Apart from lucerne, it has proved selective in chickpea (Graph *et al.*, 1982), sugarbeet and onion (Foschi and Rapparini, 1977; Rapparini, 1989a), cranberry (Bewick *et al.*, 1989) and niger seed (Tosh *et al.*, 1978; Misra *et al.*, 1981). In niger seed it was most effective when applied 20–25 days after sowing.

Among other pre-emergence herbicides, the most widely used in lucerne has been **chlorthal-dimethyl** (known as **DCPA** in North America). Although registered only for use in seed crops it shared about half the *Cuscuta*-control market in USA with chlorpropham in 1984 (Dawson, 1984c). It is used at the even higher dose of 10–11 kg a.i. ha^{-1} and persists rather longer than chlorpropham. It is non-volatile and presumably acts by direct penetration into the *Cuscuta* whether from the soil or from deposits on the host (Bayer *et al.*, 1965; McNeely *et al.*, 1966). Its use in lucerne has been largely replaced by pendimethalin but it continues to be of value in carrot (Torell, 1966; Shlevin and Golan, 1982). Other crops in which it has proved selective include chickpea (Graph *et al.*, 1982), tomato, onions and ornamentals (Dawson *et al.*, 1984; Dawson, 1987). In potatoes, two applications of 10–11 kg a.i. ha^{-1} are necessary for a good result (Schauer, 1969; Selleck *et al.*, 1979). Danielson (1976) described a novel method of using this herbicide, dipping ornamental transplants such as *Chrysanthemum* in a suspension of the herbicide (15–30 g a.i. l^{-1}) before transplanting. The *Cuscuta* was then unable to coil and attach normally on the foliage. A petroleum wax emulsion was included to reduce loss by rain or overhead irrigation but further post-emergence sprays might still be needed to maintain full control. Chlorthal-dimethyl can also act as a post-emergence treatment on young established *Cuscuta* (Dawson and Saghir, 1983).

Dichlobenil has been tested in lucerne and recommended for use in seed crops only (Dawson *et al.*, 1984). It has comparable vapour action to that of chlorpropham at somewhat lower doses of about 2 kg a.i. ha^{-1}. Its use in lucerne is limited by its higher cost and lower margin of safety in that crop but it may still have a place in higher value ornamental shrubs and in blueberry and cranberry (Dawson, 1987).

Another major class of volatile herbicides with potential for *Cuscuta*-control have been the dinitroanilines. The best known member of the group, **trifluralin**, gives incomplete results as a liquid formulation, even when incorporated, but a granular form gives improved results comparing favourably with chlorpropham (Orloff *et al.*, 1989; Dawson, 1990a) and its cheapness makes it a feasible option at about 4 kg a.i. ha^{-1}. The related **prodiamine** at 2–4 kg a.i. ha^{-1} was found to have greater reliability, giving a longer persistence of control than either chlorpropham or chlorthal-dimethyl (Orloff, 1987; Dawson, 1987, 1990a). Current use of prodiamine is not certain and may be limited, due to its excessively long persistence. **Pendimethalin**, however, with similar high activity, has now largely replaced chlorpropham as a standard herbicide for control of *Cuscuta* in lucerne grown for seed in northwestern USA though its relatively high cost may limit its use in crops for forage. As it is less volatile than trifluralin, incorporation is not necessary, but results are best when application is followed by overhead irrigation or rain within ten days. It is not sufficiently selective for use in seedling lucerne. A mixture of pendimethalin with **linuron** and **diuron** has been marketed in Hungary as 'Cuscutex' and at a total dose of 5 kg a.i. ha^{-1} proved superior to chlorthal-dimethyl at 15 kg ha^{-1} (Gimesi, 1979). Pendimethalin has been shown to have potential also in potato (Rapparini, 1989b) and in black gram (*Phaseolus radiata*) (Rao and Rao, 1991b). Of the other compounds in this group, **oryzalin** was shown to have less persistent effect than pendimethalin in lucerne (Dawson, 1990a); **fluchloralin** at 1.25 kg a.i. ha^{-1} was superior to chlorpropham at 2 kg ha^{-1} in several legume crops, including mung bean, black gram, moth bean and kenaf (Rao and Gupta, 1981; Rao *et al.*, 1985b); **isopropalin** at 2.5.kg a.i. ha^{-1} was effective in seedling chilli (Mikhailova and Genov, 1978) and **butralin** has shown selectivity in soyabean (Lang *et al.*, 1989).

Among the thiocarbamate group, which also has significant volatility, **CDEC** (also known as **sulfallate**) and **pebulate** have both been recommended for control of *Cuscuta* in both direct-seeded and transplanted tomatoes, CDEC as a pre-emergence or pre-planting incorporated treatment at 5–7 kg a.i. ha^{-1} (Hutchison, 1977) and pebulate as a pre-planting soil-incorporated treatment at similar doses (Dawson *et al.*, 1984). The pebulate has little direct effect on *Cuscuta* but reduces many of the broadleaved weeds on which it may otherwise get established. Commercial availability of CDEC is now in some doubt. **Ethofumesate** has been used successfully both pre- and post-planting in sugarbeet at doses of 0.6 to 1.6 kg a.i. ha^{-1} (Foschi and Rapparini, 1977) but it is damaging to lucerne (Dawson, 1984c).

One further group of soil-applied herbicides with vapour action are the dinitro-cresols among which **dinoseb** was tested successfully as a pre-emergence treatment by Dawson (1971b). Dinoseb and **DNOC** have been more often used as non-selective contact herbicides but their use is now generally discontinued on grounds of mammalian toxicity.

Various natural products have been tested on *Cuscuta* and many found active. Oganyan (1976) showed that the fungus *Tricothecium roseum* inhibited early germination and seedling growth of *C. gronovii* and found that the antibiotic, **tricothecin**, isolated from the fungus, could completely prevent germination at 0.1%. Muromtsev and Agnistikova (1969) confirmed the activity of tricothecin and found **cycloheximide** ('Actidone'), and **Blasticidin S** were also active, together with a range of fungicides including maneb, dinocap and dichlone. There is a surprising lack of any more recent work on these interesting possibilities.

Extracts of the neem tree (*Azadirachta indica*) were found to reduce germination and seedling growth of *C. reflexa* and the active principle was shown to be the flavonoid quercetin (Chauhan *et al.*, 1989). Habib and Rahman (1988) detected activity against *C. campestris* in extracts from *Cynodon dactylon* and *Chenopodium murale* and found the active principles to be phenolic compounds including chlorogenic, iso-chlorogenic and p-coumaric acids and scopoletin.

An alternative to the killing of the young *Cuscuta* by treating the soil, is to treat the surface of the host plant to make it toxic to the parasite, or to interfere with the normal attachment process. Apart from the use of chlorthal-dimethyl in this way, mentioned above, Nagar and Sanwal (1984) have proposed the repeated use of **calcium salts** and shown that they can control *C. reflexa* on a host such as *Brassica campestris* by spraying 0.1M calcium chloride once a week for five weeks. The idea for this treatment was based on the observed differences in calcium content, 0.02 to 0.04% in *Cuscuta*-susceptible species, and 0.15 to 0.20% in resistant potato and sweet potato. They present evidence for the importance of calcium in inhibiting a range of cell-wall-degrading enzymes involved in the penetration of the *Cuscuta* haustoria into host tissue. There have not as yet been any further confirmatory reports on this technique.

Herbicides to control established Cuscuta

Several of the herbicides mentioned in the above section have been used successfully as post-emergence herbicides applied over crop foliage. In most cases, however, they can only control germinating or very young seedlings of *Cuscuta* and are ineffective on established parasites.

Two main classes of post-emergence treatment can be distinguished for control of established *Cuscuta* – contact and translocated. Among the contact herbicides used for *Cuscuta* control, **diquat** has been successful in lucerne and red clover in eastern Europe at doses of about 0.8 kg a.i. ha^{-1}

(Gimesi, 1966; Sarpe *et al.*, 1973). Selectivity is best soon after cutting lucerne when there is relatively little crop foliage to be damaged. Both crop and parasite are largely desiccated by the treatment but the crop normally recovers rapidly while the parasite does not. Damage to lucerne is sometimes unacceptable, however, and it has not been widely used. The related **paraquat** is generally less well tolerated by lucerne but is more generally available and may still be useful for the non-selective destruction of small patches of *Cuscuta*, in place of burning or hand-pulling. *Diesel oil* has been used in the same way, also **DNOC**, but the latter is no longer generally available.

As with *Orobanche* there is scope for exploiting the physiological sinks created at the point of attachment of the parasite and it was confirmed by Fer (1983, 1984) and by Dawson and Saghir (1983) that the translocated herbicide **glyphosate** would preferentially accumulate in *Cuscuta* tissue and cause severe damage to the attached parasite. Field trials soon showed that at suitable doses of 75–150 g a.i. ha^{-1} selective control could be achieved in established lucerne with minimum damage to the crop (Hacquet *et al.*, 1983; Dawson, 1989a,b). In Russia, Moshalenko and Dement'ev (1991) report that a local formulation ('Utal') at 0.4 to 0.8 l product ha^{-1} 7–10 days after the first mowing provided near season-long control. Seedling lucerne can also be treated safely with the same doses once the seedlings are 10–20 cm tall (Dawson, 1990b,c). Dawson also showed that the closely related herbicide **SC-0224** ('**sulfosate**') can be used in exactly the same way. It is interesting to note that the risk of damage to lucerne is **reduced** by the presence of *Cuscuta* because so much of the herbicide is diverted into the parasite, away from the host meristems. Application is best made when the *Cuscuta* is still vegetative with shoots not more than 60 cm long. Kill of *Cuscuta* is not complete and repeat applications may be needed to suppress regrowth from the haustoria as well as from new infections, especially in seedling lucerne where there is insufficient shade to suppress new attachment.

Further uses of glyphosate seem likely to be developed in due course but the only reported successes in other crops so far have been those by Abu-Irmaileh and Fucik (1989) showing that *C. monogyna* could be controlled on citrus using glyphosate solutions of 50 mg a.i. l^{-1} at 2–3 l per tree, and by Bewick *et al.* (1988a) who used it in cranberry at repeated doses of 0.6 kg a.i. ha^{-1} and in carrot (not too young) at 0.4 kg a.i. ha^{-1}. As glyphosate has been successfully used against *Orobanche* in some other vegetable crops, including carrot and celery, there is clearly some scope in such crops but doses have to be somewhat higher than for *Orobanche* and use may only be justified where infestation would otherwise cause severe loss.

Maleic hydrazide has been reported to control *C. campestris* selectively on sugarbeet in Kirgizia (Evtushenko *et al.*, 1973; Belyaeva *et al.*, 1978). A dose of about 7 kg a.i. ha^{-1} controlled flowering parasite and was tolerated by well established sugarbeet, resulting in increased sugar and root yields. Young sugarbeet was insufficiently tolerant.

Most recently it has been shown that **imazaquin**, already of interest for control of *Orobanche*, may also have potential for *Cuscuta*. Liu *et al.* (1991) treated foliage of mung bean and soyabean with ^{14}C-labelled imazaquin and demonstrated accumulation and damage in attached *C. australis*. Pot experiments suggested that selective control should be possible in soyabean.

Biological Control

By insects

There have been detailed studies of natural enemies of *Cuscuta* spp. in Pakistan, the former USSR and Yugoslavia. In Pakistan, 30 insect species were identified on *Cuscuta* of which five were thought to have no other host. Three of these were studied in detail – *Melanagromyza cuscutae, Smicronyx cuscutae* and *Acroclita* sp. (Baloch, 1968). In Kazakhstan, nine insects were identified and three considered to have potential for biocontrol (Marikovskii and Ivannikov, 1966; Shinkarenko, 1982). In Yugoslavia, several species were observed, of which *Smicronyx jungermanniae* was of greatest interest (Lekic, 1970).

The larva of the agromyzid fly *M. cuscutae* bores into the stems and fruits of *Cuscuta* species, including *C. lupuliformis* and *C. europaea* in Europe and *C. reflexa* in India and Pakistan. It is believed to be very strictly specific to *Cuscuta* spp. but there are a number of parasites which attack it (Baloch *et al.*, 1967b). Attempts to introduce and establish this insect on *C. indecora* and *C. americana* in Barbados apparently failed (Julien, 1987) but it is still thought to have potential especially for the larger-vined species (Girling *et al.*, 1979). In Kazakhstan, where it already occurs naturally, its effectiveness has been augmented by collecting and storing infested vines over the winter, to reduce natural mortality of the pupae (Ivannikov, 1972).

Smicronyx species are gall-forming weevils in the Curculionidae. *S. roridus* was studied in Pakistan and was found to be specific to *Cuscuta* spp. Four generations per season develop in the stems of the thick-vined *C. reflexa*. Other species occur in the thinner-vined species, e.g. *S. rufovittatus* Anderson on *C. campestris*. Both these species were introduced to Barbados in 1967 but neither apparently became established. *S. roridus* was also released in Bangladesh but is not thought to have established (Julien, 1987). Tyurebaev (1977) studied *S. jungermanniae* and introduced it from southeast Kazakhstan to the Irtysh River delta. Release of 200–300 parasite-free beetles per m^2 was achieved and significant damage caused to *C. campestris*. Horvath and Bujaki (1990) also report effective control of *C. campestris* by *S. jungermanniae* in Hungary. The same species has been studied in Yugoslavia (Lekic, 1970) where it affects *C. campestris, C. europaea* and *C. trifolii* (= *C. epithymum*) but only builds up heavy populations in uncultivated areas. In crops it does not have the opportunity to build up to the same degree, but there is perhaps scope for it to be collected from waste areas and introduced to crop areas (with care not to introduce *Cuscuta* seeds). *S. seriepilosus* is also recorded from *C. campestris* in Yugoslavia, and *S. tartaricus* from *C. lupuliformis* in Kazakhstan. Shinkarenko (1982) reports the useful transfer of *S. tartaricus* from south to north Kazakhstan for control of *C. campestris*.

The tortricid moth *Herpystis cuscutae* was studied in Pakistan. The larvae feed on fruits of *C. reflexa* and it has many features suitable to a bio-control agent (Baloch *et al.*, 1969). Unfortunately, it may not be quite strictly host specific and no attempts have been made at its exploitation.

By pathogens

The fungus *Alternaria cuscutacidae* is reported to have been used successfully on *C. campestris* in the former USSR (Miusov and Bashaeva, 1968), also on *C. cupulata* in lucerne, but less successfully in sugarbeet (Rudakov, 1961, 1963). Effectiveness may depend on the additional action of secondary pathogens (Wilson, 1969). A *Curvularia* species may also have potential (Wilson, 1969).

A selected strain of *Colletotrichum destructivum* was shown to affect *C. campestris* and *C. epithymum* selectively in lucerne (Leach, 1958). A *forma specialis* of *C. gloeosporioides* has been tested and used for many years now in a formulation known as 'Lu-bao No. 1' for control of *C. australis, C. chinensis* and *C. maritima* on soyabean in China. Degradation of the fungus strain occurred in the 1970s but a new strain ('No. 2') is now in use (Zhang, 1985; Wang, 1986; Julien, 1987).

Integrated Control of *Cuscuta*

Prevention

The most severe infestations by *Cuscuta* species are often the result of planting contaminated seed. This is very well known for lucerne and it is largely as a seed contaminant of this crop that *C. campestris* has become so very widely distributed around the world. But it is also true for flax and linseed in many countries and for niger seed both in India and in Ethiopia. In Ethiopia the contamination of niger seed has also been indirectly responsible for infestation of coffee, where contaminated niger seed was used for interplanting in new coffee estates.

A first defence against *Cuscuta* is therefore to ensure that only clean seed is sown. Where necessary, there are procedures for separating seed of *Cuscuta* from that of lucerne or other crop seeds of similar size. Several different techniques have been used, all based on the fact that seeds of the parasite are less smooth than those of the crop. Dusting with a fine magnetic powder ('Purex') leaves a deposit on the rough seeds of *Cuscuta* so that the seeds can be separated and so extracted by magnets. It is uncertain, however, that this technique is still is use. The standard equipment now available has felt or velvet rollers to which the rough *Cuscuta* seeds stick much more readily than those of the crop. Even with this equipment, however, separation is not 100% and where possible seed should be obtained from an uncontaminated area.

Strasser (1988) has shown that dry heat (100°C for 15 minutes) devitalizes the seed of *Cuscuta* in niger seed to be used as bird-seed. He comments that a proportion of the niger seed is still viable after a very much longer exposure, but it is not altogether clear how safe this would be with seed to be used for sowing.

Another source of introduction to new areas is via farmyard manure as seeds of *Cuscuta* are known to survive passage through domestic animals (e.g. Gaertner, 1950; Bloomfield and Ruxton, 1977). As seed coats are partially digested in the gut of the animal, the seeds often have lower dormancy and are ready to germinate immediately. If contamination is suspected, it may therefore be worth delaying crop planting for a few days while any germinated seedlings are destroyed.

Other general methods of control

Where infestations are known to be serious **rotation** into crops known to be less susceptible will be advisable. Cereals and grass crops are ideal but broadleaved weeds may need to be controlled to prevent further seed production and build-up of the population.

The **removal of alternate wild hosts** in and around crop fields can be a valuable way of reducing the risks of build-up and spread of *Cuscuta*. Although most species have a wide host range, they generally have relatively few primary hosts on which they are fully vigorous. Kasasian (1977) concluded that the natural decline of *C. campestris* as a problem on his project area in Saudi Arabia was directly attributable to the control of *Convolvulus arvensis* around field edges.

Mechanical methods of **hand-pulling** or **hoeing**, or **spot treatment** with flame gun or non-selective herbicide or oil are all suitable for light infestations where whole-field treatments are not necessary or justified.

In temperate areas there is inevitably a flush of germination of *Cuscuta* following chilling over winter, and **timing of planting** to avoid this peak period can be valuable, ideally advancing planting so that the crop is established before temperatures are optimal for germination of the *Cuscuta*, to the extent of planting the previous autumn if this is feasible, or delaying planting until after the main flush.

Biological control is generally non-crop-specific and it is to be hoped that further progress will be made in the development and availabilty of both insects and mycoherbicide methods.

Crop-specific methods

Seedling lucerne and other forage crops Among the general methods already listed, rotation and clean seed are especially important, but where there is a risk on land with a known infestation, planting in summer or autumn is the main recommendation in the USA. For light infestations spot treatment can be made with a selective contact herbicide such as diquat for minimum crop damage. For a more general infestation, glyphosate is recommended.

Established lucerne and other forage crops In lucerne being grown **for seed**, it is especially important to control *Cuscuta* completely and minimize the risk of contamination of the harvested crop seed. This may require control methods lasting four months or more. As most of the available herbicides have limited persistence of effect it is necessary to use them more than once and to integrate them with other methods. In the USA it is recommended that the early flushes of germination are controlled mechanically, by repeated harrowing with skew treader, finger weeder or fine-toothed harrow. Several herbicides are available for use in seed crops, especially pendimethalin, but also trifluralin, chlorpropham, chlorthal-dimethyl, diquat and glyphosate. For spot treatment, hand-pulling, paraquat, oils or fire are all feasible.

In lucerne **for forage**, the range of herbicides registered for use tends to be smaller than for seed crops and in the USA chlorthal-dimethyl is not permitted in this situation. Otherwise the options are similar and a combination of chemical, cultural and physical methods are available with which to build an integrated programme. Cudney *et al.* (1992) specifically recommend a combination of pre-*Cuscuta*-emergence granular trifluralin, followed when necessary by flail-mowing and finishing the season with spot burning to ensure a complete prevention of seed production of *C. indecora*.

In **lespedeza** and **clovers**, the range of recommended herbicides is smaller, but chlorpropham can be used for both in the USA.

Niger seed Clean seed is especially important with this crop but if clean seed is not available, or there is a known infestation of the field, chlorpropham or propyzamide can be considered for pre-emergence or early post-emergence treatment. If reliance has to be placed on manual or mechanical control, then row-planting rather than broadcasting may help to minimize the number of crop plants that have to be destroyed.

Flax/linseed If there is no history of *C. epilinum* on the farm then the only risk is from contaminated seed and this should be avoidable by obtaining clean seed or, if necessary, by thorough sieving of seed before planting. If infestation occurs from soil infestation there is no known selective herbicide, and hand-pulling or spot-treatment by non-selective herbicide will be necessary. Row-planting may be preferable to broadcasting as a means of reducing the spread and hence the number of crop plants that have to be destroyed.

Tomato Tomatoes are usually regarded as a non-susceptible crop, at least once established, but in California, and presumably some other areas, they can be seriously attacked by *C. campestris*. Where infestation is expected, pre-emergence herbicides may prove useful. CDEC has been recommended in the past for direct control of *Cuscuta* in both transplanted and direct-sown tomato, but may no longer be available. Chlorpropham and chlorthal-dimethyl are also recommended but only in the established crop. Chlorpropham is applied before *Cuscuta* emergence either as granules or directed to the base of the crop plants. Chlorthal-dimethyl is applied 4–6 weeks after transplanting either to the soil or over the top of the crop. Other herbicides including pebulate can also be recommended if only to control broadleaved weeds on which the *Cuscuta* can establish and spread. Hand-weeding and hand-pulling may of course be all that is needed.

As for *Orobanche* control there will almost certainly be an option in due course to use genetically engineered varieties with resistance to herbicides such as glyphosate, sulphonyl ureas or imidazolinones.

Carrot and onion Chlorpropham can be used in both these crops, only pre-emergence in carrots, but also early post-emergence over the top of onions. Chlorthal-dimethyl can also be used pre- or post-emergence in onions. Pendimethalin is widely used in these crops in the USA.

Potato Pendimethalin is suitable for use in potato, where *Cuscuta* is expected.

Vegetable pulses/legumes Chlorpropham and chlorthal-dimethyl can be used in some legume crops but local availability and recommendations will need to be checked. The idea of mixed cropping with guar bean has been shown to work in black gram and should be considered in this and many other vegetable crops.

Sugarbeet Several herbicides have been shown to control *Cuscuta* in sugarbeet but ethofumesate is generally considered to be the herbicide of choice. Chlorpropham and propyzamide may not be sufficiently selective as pre-emergence treatments but can be used once the crop is established. Maleic hydrazide is a further possibility but has not been used outside eastern Europe.

Blueberries and cranberries Recommendations in the USA include chlorpropham, propyzamide and dichlobenil as pre-emergence treatments in both crops, but their use needs to be carefully timed. Bewick *et al.* (1988b) have developed a model based on degree days to predict the onset of *C. gronovii* emergence for this purpose. Even so, herbicide needs to be supplemented by hand-pulling and by cultivation as possible to destroy alternate weed hosts and germinating *Cuscuta* in and around the crop.

Tree crops A number of tree crops, including citrus, coffee, fruit and forest species, can be seriously affected by the large-vined *C. reflexa* and *C. monogyna* but also sometimes by *C. campestris*. In many cases the problems can be prevented by adequate hygiene in the plantations, including control of alternate weed hosts and spot treatment by hand-pulling where necessary. Once trees have become infested there are few remedies. Glyphosate has proved selective as an overall spray in citrus but there is little experience of this herbicide on other tree species.

Conclusion

Cuscuta problems are generally less widespread and intractable than the root parasites discussed in earlier chapters. There is also more scope for prevention by suitable care in selection or treatment of crop seed. Where the problem is acute in the more valuable horticultural crops, there are fortunately a number of herbicide options. Where these are not relevant, various forms of physical removal can be applied and may not have to be sustained for many years to achieve a substantial reduction in the problem. While further major advances may be achieved by research, they are not so badly needed for this parasite group as for those already discussed.

LAURACEAE

Cassytha filiformis L.

Although unrelated, the genus *Cassytha* has a strong superficial resemblance to *Cuscuta*, demonstrating convergent evolution in its simple twining stems and lack of leaves. Apart from the important morphological differences from *Cuscuta* which include the six-part perianth (three small sepals and three larger white petals) and fleshy fruit (a drupe), *Cassytha* species have much higher levels of chlorophyll and are generally yellow-green to green in colour (see Plates 13B, 13C). Non-flowering material can sometimes be difficult to distinguish from *Cuscuta* but *Cassytha* usually has at least some hairs, especially on and around the scales, while *Cuscuta* has none.

There are about 20 species of *Cassytha* among which ***C. melantha* R.Br.** causes severe damage to *Eucalyptus* spp. in Australia (Pederick and Zimmer, 1961) and ***C. ciliolata* L.** can completely overgrow host trees in South Africa (Visser, 1981). Otherwise *Cassytha filiformis* is by far the most widespread and common species of the genus.

Distribution *C. filiformis* occurs throughout the tropics, mainly near to coasts but occasionally also inland, e.g. at least 500 km from the sea in Burkina Faso.

Host range The host range of *C. filiformis* is very wide and Werth *et al.* (1979) list 81 species affected in the Bahamas, including grasses, ferns and gymnosperms as well as broadleaved angiosperms. They observed some evidence for host preference based perhaps on factors such as exfoliation of host bark and growth habit. Low-growing shrubs were favoured, though elsewhere it can be seen on quite tall trees such as the roadside neem trees (*Azadirachta indica*), conspicuously infested in and around Mombasa, Kenya.

Importance It has a wide range of medicinal uses (Watt and Breyer-Brandwijk, 1962) but is also responsible for occasional problems in agriculture or forestry, e.g. on *Pinus massonia* in China (Zhang, 1988), and on citrus in India and Tanzania. It has been referred to as 'public enemy No. 1' in Puerto Rico (King, 1966) and has also been noted to dominate the natural vegetation to an undesirable degree on the island of Diego Garcia in the Indian Ocean.

Biology The fruit is fleshy with a single seed about 5 mm long, hard-coated and needing to be degraded by microorganisms or other agency to break its dormancy (Visser, 1981). The seedlings also circumnutate in the same way and in the same anti-clockwise direction as *Cuscuta* but differ in being able to live for up to two months, reaching a length of 30 cm or more, without a host. This is attributable to a better developed root system. Although the primary root makes little or no growth, a number of laterals develop which provide anchorage and presumably water and mineral absorption. It also has a somewhat higher content of chlorophyll. Chloroplasts of *C. ciliolata*, studied by de la Harpe *et al.* (1980), were found to have well developed thylakoid system but rates of photosynthesis apeared to be low. Visser (1981) notes that the parasite varies in greenness according to the vigour of the host, and may rely more on its own photosynthesis when the host is weak. Kuijt (1969) emphasizes the perennial nature of *Cassytha*.

Ecology *C. filiformis* appears not to tolerate shade (Werth *et al.*, 1979). The need for and mechanism of pollination are uncertain. Werth *et al.* (1979) observed no insect visitors and the flowers open only slightly. Thrips were observed in most flowers and could have a role in pollination. The fruits are fleshy and are apparently distributed by birds.

Control No specific control methods are known for *Cassytha* species. It can only be suggested that infestations are removed manually as early as possible to prevent spread. This may have to involve the pruning of infested branches or the use of non-selective herbicide or fire for spot application to herbaceous crop species. In a few situations, shading might be used to reduce the vigour of the parasite. Herbicides which are selective against *Cuscuta* should be worth testing, especially glyphosate which could prove safe to certain tree crops, preferably when they are dormant.

Uses of *Cassytha* and *Cuscuta* species

Both of these genera are of value in the study of plant viruses being used to transmit infection from one host to another.

6 LORANTHACEAE AND VISCACEAE: THE MISTLETOES

Mistletoe is a name originally referring only to the European species *Viscum album*, but is now used as a general term for woody shoot parasites in several plant families, especially Loranthaceae and Viscaceae. *V. album* has always been of great mystical significance in northern Europe and played a role in pagan religious ceremony. Some of this significance persists in its use as a decoration at Christmas time in Europe and use of the corresponding species *Phoradendron serotinum* in North America. The history and uses of *V. album* and other mistletoes have been the subject of whole books (e.g. Becker and Schmoll, 1986) and are also usefully reviewed by Calder (1983) in the invaluable volume *The Biology of Mistletoes* (Calder and Bernhardt, 1983). These and many other mistletoes can, however, have undesirable effects on forest, fruit or ornamental trees, most notably the dwarf mistletoes (*Arceuthobium* species). Reviews that have been of value in preparing this chapter include those by Gill and Hawksworth (1961) and Hawksworth (1974).

TAXONOMY AND IDENTIFICATION

The two families, Loranthaceae and Viscaceae, are sometimes merged into Loranthaceae but are more usually kept separate. They are distinguished most conclusively on details of embryology (embryo sacs of the *Allium* type in Viscaceae and of the *Polygonum* type in Loranthaceae) but the simpler of some other distinguishing features are:

- Loranthaceae: flowers hermaphrodite (rarely dioecious), perianth petaloid, filaments distinct, anthers opening by slits.

- Viscaceae: flowers unisexual (monoecious or dioecious), perianth calyx-like and anthers sessile, opening by pores.

Loranthaceae: the Main Genera and Species

The larger family, Loranthaceae, is predominantly tropical in distribution with about 1000 species in 74 genera, according to Barlow *et al.* (1989) but the breakdown into genera has been a matter of controversy, still not completely resolved. Almost all species previously in *Loranthus* have been re-allocated to other genera.

Characteristics by which the genera and species are separated are often quite difficult and no attempt is made here to provide a detailed key. The simpler characters to look for include the number of corolla lobes (four v. five), the relative lengths of corolla tube and corolla lobes, hairiness of the corolla and foliage, the form of the anthers after anthesis (straight or gently curved v. reflexed or coiled).

At least 30 genera of Loranthaceae have been recorded as occurring on tree crops of economic importance. Information on their relative importance has been drawn especially from King (1966), Kuijt (1969), Knutson (1983) and Hawksworth (1974, 1983), supplemented by our own field observations. Some notes are given below on selected genera and species, especially those for which there is relatively recent evidence of economic importance, but it must be realized that other species and even genera could also occur locally and confirmation of identifications should always be sought from local or international authorities. The range of possibilities is somewhat simplified by the fact that each genus is restricted to either the Old World or the New, none in this family occurring in both.

Amyema

An Old World genus of about 150 species, restricted to Southeast Asia and Australasia. The generally colourful flowers, about 3 cm long, are arranged in small clusters, and have five petals free to the base. The leaves are simple, flat or needle-like and often closely mimic those of the host. Two species of significance are listed by Hawksworth (1974): ***Amyema congener* Tiegh.** on a wide range of hosts in Australia, including citrus, apple, pear, olive, plum, peach, oak and willow; and ***Amyema pendula* (Spreng.) Tiegh.** on eucalyptus and *Pinus radiata* in Australia.

Dendrophthoe

A genus of about 45 species in Africa and Asia with dense inflorescences of long, curved flowers with five reflexed lobes. Hawksworth (1974) lists eight species occurring on economic hosts, especially citrus. The most important species are probably the following:

***Dendrophthoe falcata* (L.f.) Ettingsh.** (=*Loranthus longiflorus* Desrousseaux) This is a common species in India recorded from several hundred hosts including teak, mango, citrus, custard apple, eucalyptus, apple, peach and guava. It is generally regarded as a damaging parasite where it occurs on fruit trees in India, while in teak in Kerala it has been reported to cause 37% and 42% reduction in annual growth of 34 and 12 year old teak respectively (Ghosh *et al.*, 1984).

***D. pentandra* (L.) Miq.** (= *Loranthus pentandrus* L.) This occurs on rubber and kapok (*Ceiba pentandra*) in Indonesia.

Englerina

An African genus of about 20 species, characterized by consistently 4-merous flowers and coiled filaments. Two species have been noted on crops: ***E. gabonensis* (Engl.) Balle** (= *Loranthus gabonensis* Engl.), on cocoa in West Africa; and ***E. woodfordioides* (Schweinf.) M. Gilbert** (= *Loranthus woodfordioides* Schweinf.) on peach in Ethiopia.

Helixanthera

An Old World genus of about 50 species, characterized by racemes of many small 4-merous flowers up to 2 cm long. Three species have been noted on crops including: ***H. mannii* (Oliver) Danser** (= *Loranthus mannii* Oliver) on citrus and coffee in West Africa; and ***H. parasitica* Loureiro** (= *Loranthus pentapetalus* Roxb.) on citrus in the Himalayas, including Bhutan.

Loranthus

This genus, which once included several hundred species, is curiously now reduced to just one. ***L. europaeus* Jacq.** has short spikes of very small yellow flowers with six petals 3–4 mm long, obovate leaves 4–5 cm long and fruits up to 1 cm diameter. It occurs on many species of oak in southern Europe, causing quite serious damage. Although most Loranthaceae are hermaphrodite, *L. europaeus* is dioecious, and this species also differs from most in being deciduous. The foliage has a yellow-green colour and is probably the origin of the legendary 'Golden Bough'.

Macrosolen

A genus of about 40 species restricted to South and East Asia with straight, five or six-lobed, corolla tubes. Three species have been recorded on crops, including: ***M. cochichinensis* (Lour.) Tiegh.** (= *Loranthus ampullaceus* Roxb.) on citrus, rubber, guava and cocoa in Malaya and India; and ***M. psilanthus* (Hook. f.) Danser** (= *Loranthus psilanthus* Hook. f.) on citrus in Bhutan.

Oryctanthus

A New World genus of about 20 species in tropical America. Two species have been recorded on crops, namely: ***O. occidentalis* (L.) Eichl.** on coffee and citrus in Costa Rica; and ***O. cordifolius* (Presl) Urban** on citrus in Costa Rica and Belize.

Phragmanthera

A genus of about 30 species in Africa and Arabia, characterized by flowers with five lobes, which are much shorter than the tube, and anthers transversely septate. The leaves are often densely covered in reddish or greyish hairs. At least five species have been recorded on crops including ***Phragmanthera regularis* (Sprague) M. Gilbert** (= *Loranthus regularis* Sprague), a robust shrubby species with leathery, ovate leaves up to 20 cm long and an inflorescence of dull orange flowers up to 5 cm long, densely covered in reddish-brown hairs. The berries are ovoid, bright red. *P. regularis* has been recorded in Kenya and Ethiopia on citrus, coffee, peach and a wide range of other tree species, including those used for shade over coffee. Its importance has not been clearly established, but it is certainly of local concern in Ethiopia.

Scurrula

A genus of about 50 species in South and East Asia with 4-merous flowers, leaves often with stellate hairs and ovary and fruit tapering at the base. At least six species occur on economic hosts, including: ***S. elata* (Edgeworth) Danser** (= *Loranthus elatus* Edgeworth) (see Plate 14A) on a range of deciduous fruit at higher elevations, above 2000 m, in Bhutan but apparently not on citrus or conifers; ***S. ferruginea*** **(Jacks) Danser** (= *Loranthus ferrugineus* Jacks) on many hosts in Malaysia (Lim, 1984) and elsewhere on citrus, peach, guava, etc.; ***S. parasitica*** **L.** (= *Loranthus scurrula* L.) on conifers as well as on deciduous fruit and citrus in India, Bhutan and Sri Lanka.

Struthanthus

A New World genus of about 75 species in tropical America, of which at least 11 species are recorded on economic hosts including ***Struthanthus orbicularis*** **(H.B.K.) Blume.** This is an especially common and serious parasite in Central America and the Caribbean, often highly conspicuous due to its growth habit which may result in a complete canopy being formed over the host tree, as on citrus in Costa Rica and Belize (see Fig. 6.1). Other host crops include rubber, coffee, peach and many forest species.

Tapinanthus

A genus of about 250 species in Africa, having straight 5-merous flowers, filaments elongate, with an apical tooth, coiling after anthesis. More than 30 species have been recorded on crops, including:

Tapinanthus bangwensis **(Engl. and Krause) Dans.** (= *Loranthus bangwensis* Engl. and Krause) This species has bright green broadly ovate or ovate-lanceolate leaves and red tubular flowers. It is the commonest of at least seven Loranthaceae on cocoa in Ghana (Phillips, 1977). There is direct damage to the trees by death of the distal parts of branches; indirect effects include damage to the associated shade trees, hence reduction of shade and encouragement of more competitive weed growth in the understorey; also increased capsid infestation and increased spread of swollen shoot disease resulting from encouragement of ants and associated mealybug vectors. It also occurs on cola (*Cola nitida*) in Ghana, where more than 40 infections per tree have been recorded. Distal parts of infected branches do not necessarily die off in this tree, but more than 25 per tree usually causes tree death (Clerk, 1978; Egbe and Oladokun, 1987).

Among other *Tapinanthus* species recorded by Hawksworth (1974), ***Tapinanthus globiferus*** **(A. Rich.) van Tiegh.** (= *Loranthus globiferus* A. Rich.) also has colourful red flowers (see Plates 14B, 14C, 14D) and occurs on coffee, citrus, other fruit and ornamental trees in Ethiopia and on shea butter nut (*Butryospermum paradoxum*) (Sallé *et al.*, 1987b) in West Africa. ***T. dodoneifolius*** **(DC.) Danser** (= *Loranthus dodoneifolius* DC.) is the most

Fig. 6.1. Mass of *Struthanthus orbicularis* covering citrus tree in Belize (courtesy Mr T.G. Denham).

important of the four *Tapinanthus* species on shea butter nut (Sallé *et al.*, 1987b). Crops affected by other species include especially citrus but also coffee, rubber, peach, cocoa, papaya, almond, mulberry, mango, avocado, teak and pecan. There are no records on conifers.

Taxillus

A genus of about 60 species in Africa and Asia, similar to *Scurrula* but the leaves are always glabrous when mature and the fruit is rounded at the base. At least six species are recorded on crop hosts including ***T. kaempferi* (DC.) Danser** (= *Viscum kaempferi* DC.) which is especially common and conspicuous on hemlock (*Tsuga* spp.) in the Himalayas, including Bhutan (see Plate 15A).

Other genera of Loranthaceae

Other genera in which at least one species has been recorded on economic hosts are ***Agelanthus, Amyltheca, Bakerella, Dendropemon, Elytranthe, Erianthemum, Globimetula, Helicanthes, Hyphear, Ileostylus, Lysiana, Muellerina, Nuytsia, Peraxilla, Phthirusa, Psittacanthus, Tolypanthus, Tripodanthus*** and ***Tristerix***.

Viscaceae: the Main Genera and Species

Viscaceae are also widespread in the tropics but extend further into the north temperate zone where the family is notably represented by the familiar *V. album* in Europe and by the notoriously serious *Arceuthobium* species in North America. According to Barlow (1983) there are about 400 species in seven genera, of which at least five can be of economic importance.

Arceuthobium

This is the only genus, in either mistletoe family, which is native in both the Old World and the New. The shoots arise from an extensive endophyte inside the host tissue and are often very numerous, but reduced to only a few centimetres long with inconspicuous scale leaves only 1–2 mm long. All species are dioecious and the flowers are equally inconspicuous, only 1–2 mm long. Hawksworth and Wiens (1984) list 34 species. All occur on conifers and almost all can therefore be of economic significance in forestry. Some species are subdivided into varieties or subspecies on the basis of morphology and/or host-specificity.

The major species include:

- ***Arceuthobium americanum* Nutt. ex Engelm.** on lodgepole and jack pine in western USA and Canada.
- ***Arceuthobium campylopodum* Engelm.** on ponderosa pine in western USA and Mexico.
- ***Arceuthobium douglasii* Engelm.** on Douglas fir in western and southern USA and Mexico (see Plate 16A).
- ***Arceuthobium pusillum* Peck** on black spruce (*Picea mariana*) in northeastern USA and in Canada. It has recently been reported as increasingly serious in Newfoundland (Singh and Carew, 1989).
- ***Arceuthobium tsugense* (Rosendahl) G.N. Jones** on western hemlock in northwestern USA.

- ***Arceuthobium vaginatum* (Willd.) Presl** on ponderosa pine in west and southwestern USA.
- ***Arceuthobium minutissimum* Hook.f.** on blue pine in the Himalayas of India, Pakistan and Bhutan (see Plate 16B).
- ***Arceuthobium pini* Hawksworth and Wiens** on *Pinus* and *Picea* species in southwestern China and the Himalayas, including Bhutan (see Plate 16C).

There are some further species of importance in Mexico and Central America, while juniper (*Juniperus*) is attacked by three more species in Europe, Africa and Asia.

Host range of *Arceuthobium* species For most species this is very narrow but rarely restricted to a single host species. Usually there are one to several major hosts, normally within one genus (most often *Pinus*), while infections can occasionally be found on a few further 'secondary' hosts. Such secondary hosts are rarely attacked except when growing in association with heavily infested 'primary host' (see Box 5.1, p. 198). The mechanism of host specificity, or resistance of non-host species, has not been clearly established.

Importance of *Arceuthobium* species The importance of the dwarf mistletoes in North American forestry has been repeatedly emphasized and it is often quoted that they represent the most serious of all 'disease' problems in the region. A very substantial proportion (often over 50%) of forests are infested over large regions (e.g. in the Rocky Mountains) and there is no doubt of the serious losses in timber volume and quality.

Damage includes reduced tree volume through decline in growth, tree and branch mortality, reduced cone and seed production and malformation of stems (see Plate 16D). Quantifying the losses has rarely been achieved with any precision, however, and some of the attempts that have been made can be criticized for not fully taking account of the tendency for *Arceuthobium* to be favoured by poor soil conditions and slow growth of the host trees (cf. *Striga hermonthica* on cereals). Thus in situations where reduced productivity of individual trees is associated with high parasite infestation, the reduced vigour and productivity of the tree may be at least partly the cause, and not altogether the effect, of the high parasite load. With this reservation we can refer to e.g. Gill (1937) who reported 33% loss of merchantable lodgepole pine in Colorado; Mathiasen *et al.* (1990), who reported 10, 25, 45 and 65% losses in volume growth from (respectively) class 3, 4, 5 and 6 ratings of *A. douglasii* on Douglas fir in southwestern USA; and Hawksworth and Zakaullah (1985), who reported that cone production ratings in blue pine in northern Pakistan were reduced from 2.9 in uninfected pines to 2.2 in moderately infected pines (6-class intensity rating 3), and to 0.8 and 0.06 at high intensity ratings of 5 and 6.

The 6-class intensity rating devised by Hawksworth (1977) has now become widely adopted in USA as a standard way of defining *Arceuthobium* infestation of individual trees. Each one third of the tree – top, middle and bottom – is scored 0, 1 or 2 according to the infestation being zero, light (less than half of branches infected) or heavy (more than half infected). The rating for each third is added to give a maximum of 6. At a rating of 6, 100% of ponderosa pine less than 23 cm in diameter are expected to die within seven years (Hawksworth and Geils, 1990).

Phoradendron

A genus of about 190 species in North and South America and the Caribbean. Foliage and general appearance are very similar to those of *Viscum*. Leaves are simple, opposite, often fleshy, on jointed brittle stems. The flowers are dioecious and have three very small perianth segments. The fruit is a viscid berry about 4 mm long. Five species occur on hardwoods in the USA (Scharpf and Hawksworth, 1974) and a different five species on conifers (Hawksworth and Scharpf, 1981). The most important are: ***Phoradendron serotinum*** **(Raf.) M.C. Johnst.** (= *P. flavescens* (Pursh.) Nutt.) with over 100 host species including pear, pecan (in Florida), walnut (in California) and citrus (in Texas), but no conifers; and ***P. piperoides*** **(H.B.K.) Trel.** on cocoa in Costa Rica. Others occur on fir (*Abies* spp.), cedar, etc. Plate 15B illustrates ***Phoradendron pauciflorum*** **Torr.** on *Abies alba* in California.

Host range of *Phoradendron* species This varies from very wide, as in *P. serotinum*, to very narrow with just a few hosts. Most species attack either hardwoods or conifers, not both. Lichter and Berry (1991) have shown that failure of *P. macrophyllum* to attack the non-host *Eucalyptus grandis* is due to the inducement of wound periderm by contact with the germinated seed (or the viscin surrounding it). The existence of variation within species has been shown for *P. tomentosum* in the form of significant variation in host preference, according to the source of seed samples from three different host species (Clay *et al.*, 1985). Samples from hackberry (*Celtis occidentalis*) and from elm (*Ulmus crassifolia*) developed best on those two hosts respectively.

Importance of *Phoradendron* species. Host trees are generally weakened by *Phoradendron* attack and are more subject to damage from drought. Infections in the trunk seriously reduce timber quality. Indirect damage includes that from increased fungal and insect attack, e.g. that to firs (*Abies* spp.) from fir engraver beetle (Scharpf and Hawksworth, 1981; Hawksworth and Scharpf, 1974).

Viscum

An Old World genus of about 60 species. Flowers may be monoecious or dioecious but are always very small and inconspicuous with three or four white or yellow lobes, 1–3 mm long, usually in the axils of the glabrous, opposite leaves. At least ten species occur on economic hosts including:

***Viscum album* L.** can be severe on apple, pear, pines, fir and poplars in Europe and on persimmon in China (see Plates 15C, 15D). Many thousands of papers have been published on *V. album*, but we doubt if there is any more entertaining description of the biology of any weed than that of *V. album* in two issues of the French forestry journal *La Hulotte* (the wood-owl) (Anon, 1981a,b) in which 'Monsieur Viscoglut' and his life history and control are described with the help of vivid and highly amusing cartoon drawings. The two articles are strongly recommended to anyone with a knowledge of French (see Fig. 6.2).

***V. cruciatum* Willd.** occurs mainly on olive in the Middle East, including Israel and Jordan.

Host range of *Viscum* species *V. album* has an extremely wide host range – 185 hosts are listed by Hawksworth (1974) – and these are only the introduced fruit, forest and ornamental species of some economic significance. They include conifers as well as hardwood species, but individual races may not be able to attack both of these groups. It is well known that three physiological races or subspecies of *V. album* can be distinguished with host preference for (i) deciduous tree species, (ii) pines, and (iii) fir (*Abies* spp.) and larch. Among the deciduous trees, apple and poplar are commonly attacked but oak and elm (*Ulmus* spp.) much more rarely. The occurrence of *V. album* on oak was always a matter of mystical importance in the past but is now of extra interest, as parasites from oak and elm are found to differ biochemically from those on other hosts and are of greater value for medicinal purposes (Grazi, 1982; Urech, 1987). *V. cruciatum* Willd. is said by Kuijt (1969) to occur only on olive, but Hawksworth (1974) also lists *Robinia pseudacacia* as a host in Israel.

Importance of *Viscum* species Damage from *V. album* is not generally regarded as serious, but up to 54% yield loss has been confirmed in the apple variety Discovery over the period 7–11 years after artifical inoculation (Preston, 1977). Yields were reduced by 47–54% on three different rootstocks but by only 7% when grafted onto the vigorous Malling 25 rootstock. The importance of *V. album* as a source of anti-cancer drugs is covered below.

Fig. 6.2. Cartoon drawing showing regrowth of *Viscum album* from the endophyte after pruning (from *La Hulotte*, Anon, 1981b).

Other genera of Viscaceae

Other genera in which at least one species is reported from economic hosts are ***Dendrophthora*** and ***Korthalsella.***

BIOLOGY, PHYSIOLOGY AND ECOLOGY

Loranthaceae and Viscaceae (Excluding *Arceuthobium*): the Leafy Mistletoes

Germination

Virtually all mistletoes have fruits surrounded by sticky viscin by which they adhere to the twigs or branches of the host tree where they have usually been deposited by birds. Contrary to some belief, passage though the gut of birds is not essential for germination. In most species, assuming favourable temperatures, germination is then almost immediate. In *V. album*, however, it may take up to four months. Some tropical species require high humidity or wet conditions but in many others this is not needed and in *Tapinathus bangwensis* germination and establishment on cocoa were both reduced by wet versus dry conditions (Phillips, 1991). In this study, germination of *T. bangwensis* occurred within 48 hours, shoot emergence within seven days and establishment within 14 days. Light is normally required for germination, perhaps because negative phototropism is the mechanism ensuring that the seedling immediately curves down and makes close contact with the host shoot surface. Elongation of the seedling occurs mainly by expansion of the cotyledons whose tips remain in the fruit. The tip of the seedling which makes contact and forms an appressorium on the host is referred to variously as the radicle or the hypocotyl. The seedling will keep growing to some extent if it does not immediately detect a suitably rough surface for attachment.

Attachment, penetration and development

The appressorium, or holdfast, develops into a disc shape, closely hugging the shape of the host stem, and becomes firmly attached around the perimeter. Intrusive papillae then develop from below the centre of the holdfast and penetrate the bark by a combination of mechanical pressure and enzymic action. The haustorium or sinker so formed reaches directly into the xylem to establish a bridge with the host vascular tissue. After successful establishment of the haustorium, the shoot of the parasite develops sometimes from the original apical tissue above the holdfast, but sometimes from the haustorial mass itself. Occasionally there are both types of shoot. An intercalary meristem develops at the level of the host cambium which serves to enlarge the haustorium subsequently as the host stem thickens.

Secondary runners develop from the primary haustorium to ramify either under or outside the bark (known as cortical strands or epicortical roots respectively). These form secondary wedge-shaped haustoria at intervals which penetrate the xylem in the same way as the primary haustorium. In *V. album* the cortical strands apparently extend no more than a few centimetres from the primary point of attachment. In many Loranthaceae, however, robust epicortical roots ramify extensively with secondary haustoria at intervals (see Fig. 6.3). Grazi (1984) showed that development of epicortical roots of *Scurrula pulverulenta* may depend on the host on which it is growing. They developed more strongly on oak (*Quercus ilex*), on which shoot development was poor, than on ivy (*Hedera helix*) on which shoot development was vigorous. Hence they may be a response to a less than optimal host. Shoots usually develop only from the original point of penetration, but many species, including *V. album*, have the ability to form additional shoots from the cortical strands especially after damage to, or removal of, the primary shoot, as amusingly illustrated in Fig. 6.2. Adventitious shoots are produced more extensively on a few species, including some *Oryctanthus* and *Amyema*. An unusual characteristic of all *Struthanthus* species is that their aerial branches and even the petioles of the parasite have the capacity to coil round host stems and sink further haustoria.

Fig. 6.3. Epicortical roots of *Scurrula parasitica* (in the foreground) on citrus branch (behind).

The nature of the contact between parasite and host at the cellular level has still been inadequately studied. It has often been assumed and stated that there is direct contact between the xylem of each, allowing a free flow of water and solutes from host to parasite. But Lamont (1983) points out that there is little direct evidence for such contact, while the quite different mineral constitution of the parasite from the host (such as the much higher ratio of potassium to calcium) is at total variance with the idea of a free flow of xylem fluids from one to the other. Most often it seems that there is at least some parenchyma tissue interposed between the xylem of host and parasite, leading to quite selective transfer of different components of the host xylem contents. Fineran (1987) presents interesting evidence for the lack of direct tracheary element connections between host and parasite (*Korthalsella linsayi* (Oliver ex Hook. f.) Engl.) but the existence of an effective apoplastic continuum via the walls of the parasite parenchyma cells in contact with host xylem.

A characteristic of most if not all established mistletoes is a brittleness of the stems, perhaps associated with the lack of need for structural strength.

Dependence on the host

Most species of Loranthaceae and Viscaceae have well developed photosynthetic capacity from an early stage and should not need to rely on the host as a source of carbon. Chlorophyll content is usually comparable to that of the host trees but can be lower, as in *Tapinanthus dodoneifolius* and *V. album* (Tuquet and Sallé, 1991). Photosynthesis was also relatively low in those two species, and in *Tieghemia bolusii* (Tuohy and Choinski, 1991), while in *Amyema nestor*, the net photosynthesis was somewhat higher in the parasite than in the host *Acacia grasbyi* (Hellmuth, 1971). Leonard (1973) concluded that hosts provided only traces of assimilated carbon to *Phoradendron* spp. However, a recent study using carbon isotopes (Marshall and Ehleringer, 1990) suggests that *P. juniperum* obtains 62% of its C from the host juniper; and de la Harpe *et al.* (1980) also found that two South African species of *Viscum* had apparently low rates of photosynthesis and are likely to be substantially dependent on their hosts for assimilated carbon. Richter and Popp (1992) estimate that *V. album* gets 23–45% of its carbon from the host, depending on host species. It seems that any generalization on this point is unwise. Where carbon is acquired from the host it may be quite largely in the form of organic nitrogen compounds.

The parasites are, however, completely dependent on the host for both water and minerals. A flow of water from host to parasite is maintained by a combination of high transpiration rate and high osmotic pressure (OP) in the parasite, the transpiration probably being the more important in most cases, as OP is not always higher than that in the host. High transpiration rates have been measured in *V. album*, *Phoradendron* species and in *Amyema nestor* (Gill and Hawksworth, 1961). Transpiration is maintained even under dry conditions through stomata which, as in other parasites, tend not to close under dry conditions. Transpiration may also be higher than the host at night (Fisher, 1983), though in some species at least the stomata close at night (Hellmuth, 1971; Davidson *et al.*, 1989; Tuohy and Choinski, 1991). Failure of stomata to close in response to the normal mechanism of raised levels of abscisic acid appears to be associated with the high levels of either or both potassium or cytokinins in the parasite (Fisher, 1983). Yan (1990) emphasizes the importance of the relative water potentials of host and parasite in host range. *Amyema preissii* can infect the evergreen shrub *Geijera linearifolia* but fails to continue growing because the water potential of this species is consistently lower than that of the parasite.

It was suggested by Knutson (1979) that the most important function of the high parasite transpiration is for transfer of nitrogen. This was supported by Ehleringer *et al.* (1985) who showed that there was less pronounced differential in transpiration rates when the host was rich in nitrogen.

Minerals are presumably obtained mainly from the host xylem, but selectively, such that the proportions are quite different from those in the host. Potassium levels can accumulate up to 13 times those in the host (Lamont, 1983). Calcium also accumulates but to a smaller extent so that K:Ca ratios in the parasite are higher. Other minerals including magnesium, manganese and several micronutrients are also generally higher in parasite than in host. The high levels of K are believed to be of significance in maintaining the higher OP of the parasite.

Nitrogen levels are often somewhat higher in the parasite, but not generally more than 1.5 times those in the host. The N obtained from the host is thought to be mainly in organic or ammonium form rather than nitrate, although Stewart and Orebamjo (1980) did confirm nitrate reductase activity in *Tapinanthus bangwensis*.

Effects on the host

Photosynthesis in the host tree is often found to be reduced as a result of infection. This is usually assumed to be a result of water loss and the closure of host stomata limiting access to atmospheric carbon dioxide (Hellmuth, 1971). Tuohy and Choinski (1991) claim evidence for some further effect on carbon assimilation but in general it is not thought that the depletion of carbohydrate in the host is as important as the effects on water balance.

The most important effect of almost all mistletoes on their host is their water uptake and the induction of water stress. Under moist conditions, damage is minimal, but at times of drought stress, the presence of the parasite may be critical and lead to death of branches, or of the whole host tree (see Plate 14D).

Some species block the whole vascular tissue of the host resulting in death of the distal parts of the branch. This occurs with *V. album* on apple (see Plate 15D), with *T. bangwensis* on cocoa (Room, 1971) and with many other species, though apparently less so on cola.

The high cytokinin levels often detected in parasites and sometimes in the adjacent host tissue can lead to witches' broom abnormality in trees infected by some Viscaceae other than *Arceuthobium*, e.g. in *Phoradendron* (Knutson, 1983).

Ecology

Climate The distribution of different mistletoe species is no doubt often limited by temperature, but the exact limits and the stage of growth which is most sensitive have not often been determined. Optimum temperatures for germination are usually well within the local ambient limits and it is probably the mature plant which is most critically affected. Frost almost certainly dictates the limits of some *Phoradendron* species in the USA (Gill and Hawksworth, 1961). By their ability to draw water from the host even under drought stress conditions, the mistletoes are able to survive under dry and even semi-desert conditions. However, they are still most abundant in the humid tropics. Weeraratna (1959) provides useful information on the distribution of a range of Loranthaceae and Viscaceae in Sri Lanka, as influenced by climatic zone and altitude.

Light Apart from the need for light at time of germination and attachment, most mistletoes other than *Arceuthobium* depend largely on their own photosynthesis and hence require adequate light. This is one of the more important factors in explaining the distribution and severity of mistletoes in an individual area, where they will tend to be most apparent on isolated trees, on the edge of forest and in the higher branches of the trees. Weeraratna (1959) confirms the association of heavy mistletoe attack with disturbed forest areas, and its almost total lack in dense jungle. In cocoa, only 6% of trees were infected by *T. bangwensis* under shade while 37% were infected in the absence of shade (Room, 1973). In a shading experiment on cocoa, Ahenkorah *et al.* (1974) concluded that cocoa yields were much higher under unshaded conditions but the life of the plantation was likely to be much shortened by various factors including increased attack by *T. bangwensis*. Preston (1977) also assumes that it was the shading effect of the very vigorous growth of apple on Malling 25 rootstock which was responsible for much reduced development of *V. album*. It is also important in relation to control measures (see below).

Soil type and fertility Gill and Hawksworth (1961) report little evidence for an association of mistletoes with particular soil types. They note that a number of authors have indicated a general association with poor soils but do not imply that this is a major factor. Our own observations, on the other hand, suggest that poor host growth, whether from soil or other conditions, is an important factor in predisposing a host to abundant and vigorous infection.

Birds Birds are intricately involved in the life history of mistletoes both as pollinators and in seed dispersal. Pollination of Loranthaceae is carried out mainly by sunbirds (Nectariniidae) and honeyeaters (Meliphagidae) or by the flowerpeckers or mistletoe birds (Dicaeidae). The last are also important in seed dispersal, along with many other less specialized seed-eating species. Flower structure is adapted to pollination by birds which take nectar from the half-open flowers, triggering an often explosive release of pollen on to their heads. Cross-pollination is assumed to be the norm in such cases. The Viscaceae are believed to be pollinated by insects and/or wind, but birds are still important in seed dispersal and the main species involved in dispersal of *V. album* is the European mistle thrush (*Turdus viscivorus*) whose English and Latin (specific) names both relate to this association. The way in which birds spread the seeds has been shown to be by defecation in the case of the mistletoe bird (*Dicaeum hirundinaceum*) in Australia (Liddy, 1983) but by regurgitation in the case of the yellow-fronted tinker bird (*Pogoniulus chrysoconus*) in South Africa (Godschalk, 1983). In neither case does the ingestion by the birds reduce viability, nor does it completely remove the viscin by which the seeds adhere to the tree stems after ejection.

The distance and rate of spread by birds has not often been measured, but the isolated infestation by *V. album* in California has been shown to have spread approximately 0.35 km per year over the past five years (Hawksworth *et al.*, 1991)

Mammals There have been observations of selective grazing of mistletoes by mammals, including gorillas and elephants. In the case of elephants it is attributed by Midgley and Joubert (1991) to the high nutritional status of mistletoes. Incidentally they note that the preferred spiny hosts of the parasites are also highly nutritious species. The brush-tail possum (*Trichosurus vulpecula*) apparently much prefers the foliage of the mistletoe *Amyema miquelii* (Lehm. ex Miq.) Tieghem to that of its host *Eucalyptus fasciculosa* but has difficulty distinguishing one from another. Choate *et al.* (1987) regard this as evidence for the importance of cryptic mimicry as a means of reducing predation by mammals. This was proposed originally by Barlow and Wiens (1977) to explain how mistletoe species do so often have quite strikingly similar leaf-shape to that of their hosts.

Insects Insects are believed to be involved in the pollination of many Viscaceae and smaller-flowered Loranthaceae but the relative importance of insects and wind is still not clarified.

Many insects are also known to damage mistletoes (Gill and Hawksworth, 1961), and a number have been studied as possible bio-control agents. The most damaging and interesting of these are noted under 'Biological Control' below.

Fungi Gill and Hawksworth (1961) note that over 100 fungi have been recorded on mistletoes including 24 species of rust, though none appears to be especially damaging.

Hyper-parasitism and epiparasitism Both Loranthaceae and Viscaceae often demonstrate hyper-parasitism, growing on other established mistletoe species, but not being restricted to parasitic hosts. Wiens and Calvin (1987) review the topic and suggest that the term epiparasitism should be used for those such as *Phoradendron calyculatum* Trel. and *Viscum capitellatum* Smith which are apparently restricted to other Loranthaceae as their sole hosts.

Uses of mistletoe

Among the mistletoes, *V. album* has been of exceptional interest for medicine, and in the bibliography of world literature on mistletoes (see Hawksworth and Scharpf, 1987) no less than 1300 of the 4400 citations to this species concern its medicinal uses. Many of the older qualities of *V. album* have now been discounted but it is still thought to be useful in curing warts and there is now very serious interest in the lectins and arginine from *V. album* in anti-cancer research and therapy (e.g. Urech, 1987).

Many mistletoes are also regarded as very nutritious for stock and are cut down for animal feed in times of drought. The sticky seeds are also a useful source of bird-lime.

Arceuthobium Species: the Dwarf Mistletoes

The dwarf mistletoes are dealt with separately here as their biology differs significantly from that of other mistletoes. They are distinct also in the seriousness of the damage they cause to conifer hosts. Apart from review sources already quoted, the proceedings of periodic symposia on the biology and control of dwarf mistletoes in North America have been of particular value in compiling this section (e.g. Scharpf and Parmeter, 1978; Hawksworth and Scharpf, 1984).

Germination

Seeds of *Arceuthobium* species are covered in viscin as in other mistletoes but their dispersal mechanism is quite different. Before they are released, a high pressure builds up inside the fruits and as the fleshy part of the fruit, plus seed, separates from the stalk, the seed is expelled upwards at a velocity approaching 100 km h^{-1}, commonly to a distance of 5–6 m and occasionally to 30 m (Hinds and Hawksworth, 1965). In spite of these dramatic statistics, the average rate of spread of an *Arceuthobium* infestation in North American forests is less than 1 m per year. The seeds adhere to host leaves or stem. In most species, germination is not immediate but occurs mainly in the spring, some months after release, apparently as a result of chilling (Knutson, 1974). *A. vaginatum*, however, germinates soon after release in the autumn. Light is required for germination of *Arceuthobium*, as for other genera of mistletoes.

Attachment, penetration and development

As for other mistletoes, negative phototropism ensures that the young seedling makes intimate contact with the host bark. The penetration and structure of the initial haustorium are not greatly different from other Viscaceae, but *Arceuthobium* species differ strikingly in the long delay between the initial infection and the eventual outgrowth of shoots, which varies from one to four years. During this period an endophyte is developed with spreading cortical strands and sinkers into host xylem. Some species develop only localized infections with endophyte in the secondary tissue only (e.g. *A. tsugense* and *A. occidentale*). Others ramify in the primary tissues and become extensively systemic (e.g. *A. douglasii* and *A. minutissimum*). The anatomy and morphology of these different endophyte systems are described in detail by Alosi and Calvin (1984). Direct xylem–xylem contact is not apparently common and the main contact is via the parasite parenchyma. Phloem is not usually detectable but has been seen in *A. globosum*. After 1–2 years, infection is often apparent from swelling of the host shoot around the point of penetration.

Dependence on the host

Unlike most of their relatives, *Arceuthobium* species depend very heavily on their hosts for elaborated carbon. Until the shoots emerge they are presumably photosynthesizing little if at all and are virtual holo-parasites. Even when the shoots emerge they have a low chlorophyll content and their photosynthesis is estimated to provide only part of their carbohydrate requirement. Estimates of their self-sufficiency vary from about 50% for *A. oxycedri* (Rey *et al.*, 1991) to no more than 30% for *A. tsugense* (Miller and Tocher, 1975) or even as little as 1% for *A. campylopodium* (Hull and Leonard, 1964; Leonard, 1973). In spite of this greater requirement for transfer of carbohydrate from the host there is still no evidence for close contact between phloem of host, and parasite. Indeed phloem tissue has not yet been detected in any *Arceuthobium* species other than *A. globosum* (Calvin *et al.*, 1984). This Mexican species is one of the largest in the genus, growing up to 70 cm and with stems up to 5 cm thick. It also has an unusually wide host range and is assumed to be a primitive representative of the genus, most others having become reduced in size and losing their phloem in the process.

Dependence on the host for water and minerals is similar to that of other mistletoes, and their mineral content is likewise high in potassium. Cytokinin content is especially high in the dwarf mistletoes (Schaffer *et al.*, 1981), but whether its role is in enhancing the ability of the parasite to attract water or carbohydrate from the host is not certain.

Effects on the host

The shoots of dwarf mistletoes transpire freely regardless of water stress conditions and can cause extra moisture stress in the host in spite of their relatively small surface area (Fisher, 1983). This additional stress is believed to contribute to making the host trees more susceptible to insect attack, particularly by reducing their resin exudation.

Arceuthobium species damage their hosts in two further important ways: (i) by causing a much greater drain on host carbohydrate; (ii) by causing abnormal host growth. The scale of carbohydrate drain has not been accurately measured and it is possible that this effect is exaggerated. As with *Striga* species, it is possible that other as yet unidentified physiological effects are more important than the carbohydrate drain into the parasite, which has a relatively small biomass. That the parasite can have marked physiological effect on the host is shown by the rise in the cytokinin and auxin levels and reductions in abscisic acid in tissue adjacent to the parasite (Livingston *et al.*, 1984). The increased cytokinin levels (possibly of parasite origin) are no doubt responsible for the typical witches' broom outgrowths which occur in response to most *Arceuthobium* species (see Plate 16D). The abnormal growth caused by *A. americanum* in lodgepole pine has been shown to be associated with locally increased starch content, from about 1% to about 2% (Brochot and Tinnin, 1986). The tree is damaged not only by the diversion of resources to these proliferating branches but, where they occur on the main trunk, by a reduction in timber quality. Reduced root-growth has also been recorded (Knutson and Toevs, 1972).

Among other indirect damaging effects is the predisposition of parasite infested forests to fire, as a result of dead branches and trees (Alexander and Hawksworth, 1975).

Ecology

Climate The northern spread of some *Arceuthobium* species in North America is known to be limited by low temperatures. Conversely, *A. minutissimum* occurs mainly at higher altitudes in the Himalayas but is thought to be limited by excess moisture rather than temperature.

Light Light is less important for *Arceuthobium* than for other mistletoes, owing to its weak photosynthesis and greater dependence on host carbohydrate. Hence infections often predominate in the lower, more shaded parts of the tree. Effects of light intensity on establishment and growth of *Arceuthobium* are not altogether consistent (Knutson, 1984) but there is some indication that growth of the parasite is best at moderate or low light intensities. Even so, there is evidence for greater vigour under less shade, and the association of severe dwarf mistletoe infestation with slow-growing trees is likely to be at least partly due to the better exposure to light.

Soil type and fertility There is evidence for association of at least some species of *Arceuthobium* with low fertility sites in North America (Mathiason and Blake, 1984). There is a similar tendency for greater importance of *A. minutissimum* on poor soils in Bhutan.

Fire Fire can have a varying effect on dwarf mistletoe populations, depending on whether it is localized or extensive. Localized fire can leave isolated trees or clumps which serve to infect regrowth, while extensive destructive fires can result in the establishment of clean, uninfested regrowth. Zimmerman and Laven (1984) usefully review the literature and describe the results of a detailed survey of *A. americanum* in lodgepole pine stands subject to varying history of fire over the past 200 years. Infestation was highest where fire had been least frequent.

Birds Although the primary mechanism of seed dispersal of *Arceuthobium* species is explosive, longer-distance spread of *A. vaginatum* in Colorado, USA has been attributed to birds (Hudler *et al.*, 1979).

Insects There is evidence for some involvement of insects (e.g. Hymenoptera) in the pollination of *Arceuthobium* species but wind pollination can certainly also occur. Their relative importance is still not clear (Stevens and Hawksworth, 1984).

The range of insects that have been recorded causing damage to dwarf mistletoes is reviewed by Stevens and Hawksworth (1984), but none appears to be of outstanding importance in limiting the range or intensity of *Arceuthobium* infestations.

Pathogens At least eight fungi are known to be parasitic on *Arceuthobium* species in North America (Hawksworth *et al.*, 1977) of which three are common and occur on most species in western USA. These are *Colletotrichum gloeosporioides*, *Cylindrocarpon gillii* (= *Septogloeum gillii*) and *Wallrothiella arceuthobii*. Their overall effect on the mistletoes in the field, however, is not thought to be great.

CONTROL OF LORANTHACEAE AND VISCACEAE (EXCLUDING *ARCEUTHOBIUM* SPECIES)

Cultural and Mechanical Control Methods

Pruning

This is virtually the only direct control method available for control of the majority of mistletoe problems, and is inevitably suitable only for relatively high value crops, especially fruit and ornamental species. Pruning generally has to involve the host tree branch itself rather than just the parasite, as cutting only the mistletoe allows regrowth from the embedded haustorium, as nicely illustrated in Fig. 6.2. To avoid such regrowth, it is necessary to cut the host branch some distance (15–20 cm) below the point of attachment. Wrapping the infected branch in opaque plastic or other material, after removal of the parasite only, has been suggested as a means of preventing regeneration but is not always successful.

As some mistletoes tend to attack only the smaller, younger branches, pruning may not be too seriously damaging to the host tree, but detecting and then reaching the infected branches with suitable cutting gear may still be difficult. In evergreen hosts, the time of parasite flowering is recommended, when the larger-flowered Loranthaceae at least are most readily seen. In deciduous hosts the best time is, of course, when the tree has dropped its leaves. Pruning can also be used on adjacent non-crop trees for indirect control, to reduce seed production and the threat of infestation via birds.

When continuously moist environments minimize the damage to the trees, the economic benefit from pruning may be quite marginal, even in relatively high value crops. Where moisture is regularly or at times seriously limiting, benefits are more likely to cover the costs.

Shade

Because most mistletoes rely mainly on their own photosynthesis, vigorous growth must depend on adequate exposure to light. Hence shading by host or other foliage is detrimental and forms the basis for one of the more important cultural control methods for crops which can be grown as an understorey. Shade from shade trees has been shown to significantly reduce the numbers and vigour of mistletoes such as *T. bangwensis* on cocoa (Room, 1973; Ahenkorah *et al.*, 1974). Unfortunately, it is extremely difficult to separate the beneficial effects of mistletoe-reduction from the other benefits of shading. Hence it cannot be certain that it is worth retaining or introducing shade trees, purely as a means of reducing mistletoe attack. Shading from the foliage of the host tree is dealt with in the next section.

Fertility/management

The previous section has dealt with the beneficial effects of shading by overstorey trees, but shading can be important within the host trees themselves. When parasites become established on young branches on the top or outside of the tree canopy, their success is likely to depend on how quickly and completely the host tends to outgrow and shade them. Hence a fast-growing tree will generally tend to suppress mistletoes, and conversely a slow-growing tree will allow more vigorous parasite growth. This was confirmed by Preston (1977) for *V. album* on different apple rootstocks. Attempts to assess yield loss due to mistletoes in a naturally infested site must inevitably be influenced by this phenomenon and it must be recognized that weak tree growth may be not so much a **result** of vigorous mistletoe infestation as its **cause**. Hence any management practices which improve tree growth should reduce the importance of mistletoes. Such practices could include fertilizer application, irrigation, pruning and adequate control of understorey weeds.

Prevention

Prevention is largely a matter of controlling mistletoe growth in trees adjacent to orchards or plantations, whether by pruning or selective felling. Theoretically, the main bird species responsible for seed dispersal might be targeted, but so far we know of no ecologically acceptable means of controlling or repelling them.

Resistant varieties

There are few if any reports of intra-specific variation in host susceptibility to Loranthaceae. However, distinct varietal resistance to *Viscum album* has been noted in poplar (*Populus* spp.). Sallé *et al.* (1991a) list three varieties of *P. nigra* (Blanc de Garonne, Vereecken and Italica) in which resistance has been attributed to a combination of higher pre-infection levels of flavonoids, with wound response involving further secretion of flavonoids and thickening of the inner periderm. Interestingly, there were high levels of phenolic acids in both the resistant and susceptible varieties. Sallé *et al.* (1991b) also refer to variations in varietal response of oak (*Quercus* spp.) to *V. album*.

Chemical Control Methods

Destruction of mistletoes by directed application of herbicide has been attempted as an alternative to pruning. The herbicides to have shown some suitability for this include **2,4-D** for control of *Phoradendron* in walnuts during the dormant phase in California (Graser, 1952), on *Amyema* in eucalypts (Hartigan, 1971), and against *Dendrophthoe falcata* on certain hosts in India (Seth, 1958). **Diesel oil** emulsion has also been used for control of *D. falcata* (Singh, 1958). In other studies in India, however, 2,4-D and a range of other compounds, including **copper sulphate**, have proved ineffective or damaging or both.

When one wishes to avoid pruning an ornamental tree, herbicide may be used to improve the effectiveness of pruning only the parasite. Michailides *et al.* (1987) showed that a **2,4-D/dicamba** mixture, painted onto the cut stumps of *P. tomentosum*, did not kill them but prevented their regrowth for 18 months.

A promising new treatment has been with **ethephon** applied as an overall spray to *Phoradendron tomentosum* in ornamental hardwoods in the dormant season. A 2% spray caused all parasite leaves and branchlets to fall off and there was little regrowth up to 150 days. Host trees were undamaged (Joyce *et al.*, 1987).

An alternative method is to apply herbicide to cuts in the branches just below the parasite. A range of herbicides including 2,4-D and ethephon were effective in this way against *V. album* in silver fir (*Abies alba*), poplar and willow (*Salix* spp.) (Delabraze and Lanier, 1972).

Trunk injection is an extension of this idea, used successfully in Australia by Greenham and Brown (1957) who achieved 70–100% control of *Amyema* spp. in eucalypts after 1–2 years using 2,4-D. Trunk injections of 2,4-D and copper sulphate have also been used successfully against *Scurrula pulverulenta* in India (Kadambi, 1954). In more recent work in Australia, Minko and Fagg (1989) obtained best results against *Amyema* species on eucalyptus using **glyphosate** (commercial product diluted 1:3 in water) and **triclopyr ester** (diluted 1:4) at 1 ml per cut, the cuts spaced at 10 cm intervals around the lower bole of the tree. **Clopyralid** and **hexazinone** were less satisfactory. Results varied according to eucalyptus species and time of year. Smaller trees (less than 30 cm diameter) were more liable to be damaged by the treatment.

Metribuzin has been shown to give promising control of *D. falcata* on teak (*Tectona grandis*) in southern India (Ghosh *et al.*, 1984). A volume of 600 ml 0.05% suspension, diffused into the trunk through an 8 mm nozzle, was sufficient to kill, within two months, an average of six clumps of parasite per tree of about 24 cm diameter at breast height. Most rapid diffusion occurred between 7 and 10 am and between June and October, but the treatment was also effective when the trees were leafless in February/March. This technique deserves wider testing in a range of tree and fruit crops.

Theoretically it should be possible to find soil-acting herbicides with selective action via the host roots and transpiration stream. It might be thought that the highly active substituted uracil herbicides such as **terbacil** and **bromacil** could be effective this way in citrus, but we are not aware of any experiments or observations to support this. This possibility would appear to be one of the obvious approaches to try for control in citrus, along with trunk-injection of the same compounds.

Biological Control Methods

By insects

The potential for control of *Dendrophthoe falcata* and other mistletoe species by a range of insect species has been studied over many years in Pakistan. Of these species, Mushtaque and Baloch (1979) believed at least four had potential as biocontrol agents: *Enzopherodes ephestialis* (Lepidoptera; Pyralidae); *Demarchus pubipennis* (Coleoptera; Chrysomelidae); *Anarsia sagmatica* (Lepidoptera; Gelachiidae); and *Ceratitella asiatica* (Diptera; Tephritidae). Samples of *A. sagmatica* and *C. asiatica* were subsequently released in Trinidad but failed to establish, and there is still no confirmation of their potential.

By pathogens

There have been no reported studies of pathogens for biocontrol of the leafy mistletoes.

By epiparasitic mistletoes

As *V. capitellatum* will only grow on other mistletoes, it has been suggested by Ghosh *et al.* (1984) as a possible bio-control measure for *D. falcata* on teak, but they did not propose practical ways of exploiting it. The use of *V. loranthi* Elmer was similarly proposed by Pundir (1981) for control of *Scurrula cordifolia* (Wall.) G.Don in northern India.

CONTROL OF *ARCEUTHOBIUM* SPECIES

The importance of the dwarf mistletoes in forestry in North America has resulted in very extensive research programmes and corresponding publications and symposia. In spite of all this study, there are still few if any simple direct control measures available and the dwarf mistletoes continue to pose one of the most difficult of all parasitic weed problems. The available methods, such as they are, include the following.

Cultural and Mechanical Control Methods

Pruning

Pruning is quite impracticable in the forestry context, and even for high value ornamental trees its value is limited by the systemic nature of many *Arceuthobium* species. Only for *A. americanum* and other species forming strictly localized infection will it be feasible. Even then, due allowance must be made for the hidden endophyte, and the branches be severed at least 20–30 cm below the lowest parasite shoots.

Forest management

As for the leafy mistletoes, management to encourage vigorous tree growth is likely to discourage development of dwarf mistletoes, but the lesser dependence of the parasite on light for its own photosynthesis reduces the scope for this approach. Hence there has to be greater dependence on destructive methods of crop felling, thinning and fire to prevent spread or reinfestation of new stands, rather than protection of the already infested trees.

The long time scales of both dwarf mistletoe spread (about 1 m per year) and forest growth (up to 100 years) mean that careful methods of prediction are required to ensure the most economical management systems. To help decision making, forestry workers have developed elaborate computer programs so that different strategies can be compared. Such programs have been in use for at least 20 years and continue to be developed with increasing sophistication (e.g. Geils and Mathiasen, 1990). Based on observed rates of spread of the *Arceuthobium* species both within and between trees and their estimated effect on yield, the models are used to predict the development and spread of the infestation and to calculate the optimum time and methods of thinning, harvesting and replanting.

Fire

The effects of fire have been mentioned under 'Ecology' and it is recognized that dwarf mistletoes are reduced by frequent fire. The use of fire as a deliberate control technique is supported by Alexander and Hawksworth (1975, 1976), but only in situations where residual infected trees need to be destroyed in logged-over areas or to destroy stands that are so heavily infested that the site is unproductive.

Prevention

Prevention of the short-distance dispersal of dwarf mistletoes is achieved by destroying infected trees by felling or by fire. It is also assisted by leaving or creating open corridors between infected and uninfected stands. A width of about 20 m is recommended (Hadfield and Russell, 1978). The same effect may be achieved more economically by planting strips of non-susceptible tree species.

Prevention of long-distance introduction requires quarantine legislation and *Arceuthobium* is unique in being the only genus of weeds specified in European Community quarantine rules as a prohibited import into community countries. This is a reflection of the present lack of any serious problem from the one species (*A. oxycedri*) already occurring in Europe and the severe threat that introduction of virulent species from North America would present.

Resistant varieties

Varietal resistance has been identified in Jeffrey pine (*Pinus jeffreyi*) (Scharpf, 1987) and in ponderosa pine (Scharpf and Roth, 1991). In the case of Jeffrey pine, a population from Placer County, northern California, has maintained almost total resistance over a 45 year period of exposure. No morphological differences are apparent and it is assumed that the mechanism of resistance is physiological. In the case of ponderosa pine, resistance has been less complete but there are significant differences among a number of selections (Scharpf, 1984). The mechanism was at one stage thought to be associated with a drooping needle character which resulted in shedding of parasite seed but this is now discounted.

Chemical Control Methods

Knutson (1978) and Hawksworth (1983) reviewed efforts at chemical control of dwarf mistletoes and concluded that, while theoretically it should be possible to find selective herbicides, and some **2,4,5-T** derivatives had shown activity, no practically useful technique had yet been developed. More recently, **ethephon** has been shown effective in causing kill of emerged shoots of *A. americanum*, thus preventing further seed dispersal. Attempts at aerial application, however, have not been as successful as mist-blower application from the ground (Baker *et al.*, 1989; Robbins *et al.*, 1989).

Injection of herbicides is hardly feasible in a forestry context, but is in any case problematic in conifers which exude resin and thus interfere with translocation from the point of injection.

Biological Control Methods

By insects

Knutson (1978), reviewing biological control approaches, mentions no insects at all in relation to bio-control. Stevens and Hawksworth (1984) review the large number of insects that have been recorded on *Arceuthobium* species, including those studied in Pakistan with a particular view to bio-control potential (Mushtaque and Baloch, 1979). They conclude that 'additional work on this management approach surely appears justified' but do not indicate which organisms present the greatest possibilities.

By pathogens

Several fungi are common on *Arceuthobium* species in North America, as noted above. Among the documented attempts to exploit them for bio-control, Knutson (1978) was not successful in introducing spores of *Cylindrocarpon gillii* into areas not naturally infested. Infection was established in most of the inoculated trees but failed to persist beyond three years, perhaps because of environmental and/or host specificity problems. Smith and Funk (1980) claimed more promising results with sprays of *Nectria macrospora* spores on *A. tsugense* in Vancouver, without damage to the western hemlock host trees, but admitted that improved inoculation techniques were needed.

Conclusion

Problems from mistletoes, especially in an extensive forest context, present enormous logistical problems of detection and accessibility. Fortunately, most of the mistletoes other than *Arceuthobium* species cause relatively mild damage and may not require intervention. In the less extensive context of fruit crops or ornamentals, there are a few options including pruning and spot-spraying by herbicide which could prove economic, while trunk-injection of herbicide may deserve more exploration. The *Arceuthobium*, dwarf mistletoe, problem continues to challenge the North American forester, but some management systems, backed up by computer modelling, are providing at least partial solutions.

Appendix: List of Crop Scientific Names

alfalfa = lucerne	*Medicago sativa* L.
almond	*Prunus dulcis* (Mill.) D.A. Webb
apple	*Malus sylvestris* (Borkh.) M. Ansf.
asparagus	*Asparagus officinalis* L.
avocado	*Persea americana* Mill.
bambara	*Vigna subterranea* (L.) Verdc.
barley	*Hordeum vulgare* L.
bean, common	*Phaseolus vulgaris* L.
bean, faba	*Vicia faba* L.
bean, moth	*Phaseolus aconitifolius* Jacq.
bean, mung	*Phaseolus radiata* (L.) R. Wilczek
bean, runner	*Phaseolus coccineus* L.
bean, tepary	*Phaseolus acutifolius* A. Gray
bean, velvet	*Mucuna pruriens* (L.) DC.
beet	*Beta vulgaris* L.
berseem	*Trifolium alexandrinum* L.
bird's foot trefoil	*Lotus corniculatus* L.
blueberry	*Vaccinium myrtillus* L.
cabbage	*Brassica oleracea* L.
carrot	*Daucus carota* L.
cedar	*Cedrus* spp.
celery	*Apium graveolens* L.
chickpea	*Cicer arietinum* L.
chilli	*Capsicum frutescens* L.
chrysanthemum	*Chrysanthemum* spp.
clover	*Trifolium* spp.
cocoa	*Theobroma cacao* L.
coffee	*Coffea arabica* L.
cola	*Cola acuminata* (P.B.) Schott & Endl.

cotton	*Gossypium* spp.
cowpea	*Vigna unguiculata* (L.) Walp.
cranberry	*Vaccinium macrocarpon* Ait.
cucumber	*Cucumis sativus* L.
cumin	*Cuminum cyminum* L.
custard apple	*Annona cherimola* Mill.
eggplant	*Solanum melongena* L.
fennel	*Foeniculum vulgare* Mill.
fir, douglas	*Pseudotsuga menziesii* (Mirbel) Franco
flax	*Linum usitatissimum* L.
garlic	*Allium sativum* L.
grape	*Vitis vinifera* L.
groundnut	*Arachis hypogaea* L.
guar	*Cyamopsis tetragonoloba* (L.) Taub.
guava	*Psidium guajava* L.
hemlock, western	*Tsuga heterophylla* (Raf.) Sarg.
hemp	*Cannabis sativa* L.
juniper	*Juniperus* spp.
jute	*Corchorus* spp.
kapok	*Ceiba pentandra* (Stick.) Gaertn.
kenaf	*Hibiscus cannabinus* L.
lablab	*Dolichos lablab* L.
larch	*Larix* spp.
lentil	*Lens culinaris* Medick.
lespedeza	*Lespedeza striata* (Thunb.) Hook. & Arn.
lettuce	*Lactuca sativa* L.
linseed	*Linum usitatissimum* L.
litchi	*Litchi chinensis* Sonn.
lucerne	*Medicago sativa* L.
lupin	*Lupinus* spp.
maize	*Zea mays* L.
mango	*Mangifera indica* L.
melon	*Cucumis melo* L.
millet, finger	*Eleusine coracana* (L.) Gaertn.
millet, Italian	*Setaria italica* (L.) P. Beauv.
millet, pearl	*Pennisetum americanum* (L.) K. Schum.
mint	*Mentha* spp.
mulberry	*Morus alba* L.
mustard	*Brassica* spp.
niger seed	*Guizotia abyssinica* (L.f.) Cass.
oak	*Quercus* spp.
olive	*Olea europaea* L.
onion	*Allium cepa* L.
orange	*Citrus aurantium* L.
papaya	*Papaya carica* Gaertn.
parsnip	*Pastinaca sativa* L.
pea	*Pisum sativum* L.
peach	*Prunus persica* Batsch

pear	*Pyrus communis* L.
pecan	*Carya illinoensis* (Wangenh.) K. Koch
pepper, sweet	*Capsicum annuum* L.
persimmon	*Diospyros kaki* L.
pigeon pea	*Cajanus cajan* (L.) Huth
pine, blue	*Pinus wallichiana* A.B. Jackson
pine, lodgepole	*Pinus contorta* Douglas ex Loudon
pine, jack	*Pinus banksiana* Lamb.
pine, Jeffrey	*Pinus jeffreyi* Grev. & Balf.
pine, ponderosa	*Pinus ponderosa* Douglas ex London & P. Lawson
plum	*Prunus domestica* L.
poplar	*Populus* spp.
potato	*Solanum tuberosum* L.
rapeseed	*Brassica napus* L.
rice	*Oryza sativa* L.
rubber	*Ficus elastica* Roxb.
safflower	*Carthamus tinctorius* L.
sesamum	*Sesamum indicum* L.
shea butter nut	*Butryospermum paradoxum* (Gaertn. f.) Hepper
siratro	*Macroptilum atropurpureum* (DC.) Urb.
sorghum	*Sorghum bicolor* (L.) Moench
soyabean	*Glycine max* (L.) Merr.
spruce	*Picea* spp.
squash	*Cucurbita* spp.
sugarbeet	*Beta vulgaris* L.
sugarcane	*Saccharum officinarum* L.
sunflower	*Helianthus annuus* L.
sweet potato	*Ipomoea batatas* (L.) Poir.
tea	*Camellia sinensis* L.
teak	*Tectona grandis* L.f.
tobacco	*Nicotiana tabacum* L.
tomato	*Lycopersicon esculentum* Mill.
vetch	*Vicia sativa* L.
walnut	*Juglans regia* L.
wheat	*Triticum aestivum* L.
willow	*Salix* spp.

REFERENCES

Abdel Halim, M.A., Amar, M.A., Hazem, A. and Abdel Hafeez, S. (1975) Contributions to the germination of *Orobanche* seeds with reference to root exudates of host. *Annals of Agricultural Science* 20, 127–139.

Abdel-Rahim, M.F., Satour, M.M., Mickail, K.Y., El-Eraki, S.A., Grinstein, A., Chen, Y. and Katan, J. (1988) Effectiveness of soil solarization in furrow-irrigated Egyptian soils. *Plant Disease* 72, 143–146.

Aber, M. (1984a) Les Orobanches: distribution, dégats et modalité de lutte. In: *Proceedings, European Weed Research Society Third Symposium on Weed Problems in the Mediterranean Area, Lisbon, 1984*, pp. 285–291.

Aber, M. (1984b) Cyto-physiological aspects of *Orobanche crenata* Forssk. parasitizing *Vicia faba* L. In: Parker, C., Musselman, L.J., Polhill, R.M. and Wilson, A.K. (eds) *Proceedings, 3rd International Symposium on Parasitic Weeds, Aleppo, Syria, 1984*. ICARDA, Aleppo, pp. 200–209.

Abou-Raya, M.A. and El-Sharkway, M.A. (1979) Observations on the parasitic behaviour of *Thesium humile* (Santalaceae): Fruit and its uptake of water. In: Musselman, L.J., Worsham, A.D. and Eplee, R.E. (eds) *Supplement to Proceedings, 2nd International Symposium on Parasitic Weeds, Raleigh, USA, 1979*. North Carolina State University, Raleigh, pp. 6–16.

Abou-Raya, M.A., Radi, A.F. and Heikel, M.M. Darwish (1973) Host–parasite relationship of *Orobanche* species. In: *Proceedings, Symposium on Parasitic Weeds, Malta, 1973*. European Weed Research Council, Wageningen, pp. 167–176.

Abou-Raya, M.A., El-Sharkway, M.A. and Garaboli, F.R. (1976) Observations on the parasitic behaviour of *Thesium* Vahl. (Santalaceae), (a) hosts and mode of attachment of haustorial structure. In: *Comptes Rendus du 5e Colloque International sur l'Ecologie et la Biologie des Mauvaises Herbes, Dijon, 1976*, pp. 367–376.

Abu-Gharbieh, W.I., Makkouk, K.M. and Saghir, A.R. (1978) Response of different tomato cultivars to the root-knot nematode, tomato yellow leaf curl virus and *Orobanche* in Jordan. *Plant Disease Reporter* 62, 263–266.

Abu-Irmaileh, B.E. (1979a) Occurrence of parasitic flowering plants in Jordan. *Plant Disease Reporter* 63, 1025–1028.

Abu-Irmaileh, B.E. (1979b) Effect of various fertilizers on broomrape (*Orobanche ramosa*) infestation of tomatoes. In: Musselman, L.J., Worsham, A.D. and Eplee, R.E. (eds) *Proceedings, 2nd International Symposium on Parasitic Weeds, Raleigh, 1979*. North Carolina State University, Raleigh, pp. 278–283.

Abu-Irmaileh, B.E. (1981) Response of hemp broomrape, *Orobanche ramosa* infestation to some nitrogenous compounds. *Weed Science* 29, 8–10.

Abu-Irmaileh, B.E. (1984) Effect of planting flax on the subsequent infestation of tomato by *Orobanche ramosa*. In: Parker, C., Musselman, L.J., Polhill, R.M. and Wilson, A.K. (eds) *Proceedings, 3rd International Symposium on Parasitic Weeds, Aleppo, Syria, 1984*. ICARDA, Aleppo, pp. 250–255.

Abu-Irmaileh, B.E. (1991) Soil solarization controls broomrapes (*Orobanche* spp.) in host vegetable crops in the Jordan Valley. *Weed Technology* 5, 575–581.

Abu-Irmaileh, B.E. and Fucik, J.E. (1989) Using glyphosate to control eastern dodder on citrus in Jordan. *HortScience* 24, 311–312.

Acimovic, M. (1980) Physiological races of *Orobanche cumana* Wallr. on sunflower in Yugoslavia. In: *Proceedings, 9th International Sunflower Conference*, pp. 162–167.

Adu-Tutu, K.O. and Drennan, D.S.H. (1991) Effect of sulphonylurea herbicides on *Striga*. In: Ransom, J.K., Musselman, L.J., Worsham, A.D. and Parker, C. (eds) *Proceedings, 5th International Symposium on Parasitic Weeds, Nairobi, 1991*. CIMMYT, Nairobi, pp. 361–371.

Agbobli, C.A. (1991a) Effect of nitrogen rates on *Striga asiatica* emergence in maize culture in Togo. In: Ransom, J.K., Musselman, L.J., Worsham, A.D. and Parker, C. (eds) *Proceedings, 5th International Symposium on Parasitic Weeds, Nairobi, 1991*. CIMMYT, Nairobi, pp. 28–30.

Agbobli, C.A. (1991b) *Striga gesnerioides* – a new threat to cowpea culture in southern Togo. In: Ransom, J.K., Musselman, L.J., Worsham, A.D. and Parker, C. (eds) *Proceedings, 5th International Symposium on Parasitic Weeds, Nairobi, 1991*. CIMMYT, Nairobi, pp. 26–27.

Agbobli, C.A. and Huguenin, B. (1987) Evaluation agronomique du problème du parasitisme du *Striga asiatica* sur maiz dans la sud Togo. In: Weber, H.C. and Forstreuter, W. (eds) *Proceedings, 4th International Symposium on Parasitic Flowering Plants, Marburg, Germany, 1987*. Philipps-Universität, Marburg, pp. 11–26.

Aggarwal, V.D. (1985) Cowpea *Striga* research. In: Singh, S.R. and Rachie, K.O. (eds) *Cowpea, Research, Production and Utilization*. John Wiley and Sons, London, pp. 335–340.

Aggarwal, V.D. (1991) Research on cowpea-*Striga* resistance at IITA. In: Kim, S.K. (ed.) *Combating Striga in Africa. Proceedings, International Workshop, Ibadan, 1988*. IITA, Nigeria, pp. 90–95.

Aggarwal, V.D. and Ouedraogo, J.T. (1989) Estimation of cowpea yield loss from *Striga* infestation. *Tropical Agriculture* 66, 91–92.

Aggarwal, V.D., Muleba, N., Drabo, I., Souma, J. and Mbewe, M. (1984) Inheritance of *Striga gesnerioides* resistance in cowpea. In: Parker, C., Musselman, L.J., Polhill, R.M. and Wilson, A.K. (eds) *Proceedings, 3rd*

International Symposium on Parasitic Weeds, Aleppo, Syria. ICARDA, Aleppo, pp. 143–147.

Aggarwal, V.D., Haley, S.D. and Brockman, F.E. (1986) Present status of breeding cowpea for resistance to *Striga* at IITA. In: ter Borg, S.J. (ed.) *Proceedings, Workshop on Biology and Control of* Orobanche, *Wageningen, 1986*. LH/VPO, Wageningen, pp. 176–180.

Ahenkorah, Y., Akrofi, G.S. and Adri, A.K. (1974) The end of the first cocoa shade and manurial experiment at the Cocoa Research Institute of Ghana. *Journal of Horticultural Science* 49, 43–51.

Aigbokhan, E.I. and Hussain, H.S.N. (1991) Reduction of leaf area in cowpeas due to attack by *Striga gesnerioides*. In: Ransom, J.K., Musselman, L.J., Worsham, A.D. and Parker, C. (eds) *Proceedings, 5th International Symposium on Parasitic Weeds, Nairobi, 1991*. CIMMYT, Nairobi, pp. 287–292.

Alexander, M.E. and Hawksworth, F.G. (1975) *Wildland Fires and Dwarf Mistletoes: A Literature Review of Ecology and Prescribed Burning*. USDA Forest Service General Technical Report RM-14, 12 pp.

Alexander, M.E. and Hawksworth, F.G. (1976) Fire and dwarf mistletoes in North American Forests. *Journal of Forestry* 74, 446–449.

Allred, K.R. and Tingey, D.C. (1964) Germination and spring emergence of dodder as influenced by temperature. *Weeds* 12, 45–48.

Al-Menoufi, O.A. (1991) Crop rotation as a control measure of *Orobanche crenata* in *Vicia faba* fields. In: Wegmann, K. and Musselman, L.J. (eds) *Progress in* Orobanche *Research. Proceedings, International Workshop on* Orobanche *Research, Obermarchtal, 1989*. Eberhard-Karls-Universität, Tübingen, pp. 241–247.

Al-Menoufi, O.A. and Ashton, F.M. (1991) Studies on the parasitism of *Cuscuta* spp. series 8: susceptibility and resistance of some *Lycopersicon* species to *Cuscuta campestris* infection. In: Ransom, J.K., Musselman, L.J., Worsham, A.D. and Parker, C. (eds) *Proceedings, 5th International Symposium on Parasitic Weeds, Nairobi, 1991*. CIMMYT, Nairobi, pp. 293–297.

Al-Menoufi, O.A. and Hassan, M.T. (1976) Studies on the parasitism of *Cuscuta* spp. 1. Survey study on *Cuscuta* spp. and their hosts in Noubareya Region (El Tahrir Province). *Egyptian Journal of Phytopathology* 8, 25–29.

Alosi, M.C. and Calvin, C.L. (1984) The anatomy and morphology of the endophytic system of *Arceuthobium* spp. In: Hawksworth, F.G. and Scharpf, R.F. (eds) *Biology of Dwarf Mistletoes: Proceedings of the Symposium. Fort Collins, Colorado, 1984*. USDA Forest Service General Technical Report RM-111, pp. 40–52.

Americanos, P.G. (1991) *Control of* Orobanche *in celery*. Technical Bulletin 47. Agricultural Research Institute, Ministry of Agriculture and Natural Resources, Nicosia, Cyprus, 5 pp.

Americanos, P. and Vouzounis, N. (1992) Weed science: *Orobanche*. *Agricultural Research Institute Review for 1991*. Ministry of Agriculture and Natural Resources, Nicosia, Cyprus, pp. 36–37.

Andrews, F.W. (1945) The parasitism of *Striga hermonthica* Benth. on sorghum spp. under irrigation. I. Preliminary results and the effect of heavy and light irrigation on *Striga* attack. *Annals of Applied Biology* 32, 193–200.

Andrews, F.W. (1946) The parasitism of *Striga hermonthica* Benth. on leguminous plants. *Annals of Applied Biology* 34, 267–275.

Anon (1959) *Review of crop experiments, Bechuanaland Protectorate 1947–1959.* Department of Agriculture, Mahalapye. Reprinted by Government Printer, Gaborone, Botswana, 1978, 111 pp.

Anon (1981a) Monsieur Viscoglut. *La Hulotte* 48, 48 pp.

Anon (1981b) Les aventures de Monsieur Viscoglut, 2ème épisode: Viscoglut contre Fritzi Pauley. *La Hulotte* 49, 52 pp.

Antonova, T.S. (1978) The development of the haustoria of *Orobanche cumana* Wallr. in the roots of immune and susceptible forms of sunflower (in Russian). *Botanicheski Zhurnal* 63, 1025–1030.

Arjona-Berral, A., Mesa-García, J. and García-Torres, L. (1988) Herbicide control of broomrape in peas and lentils. *FAO Plant Protection Bulletin* 36, 175–178.

Arjona-Berral, A., Mesa-García, J. and García-Torres, L. (1990) Distribution of 14C-glyphosate in legumes parasitised by *Orobanche crenata*. *Weed Research* 30, 53–59.

Ashton, F.M. and Santana, D. (1976) Cuscuta *spp. (dodder): A Literature Review of its Biology and Control.* University of California, Division of Agricultural Sciences Bulletin 1880, 22 pp.

Avdeev, Y. (1986) Identification of a dominant gene for resistance to broomrape in tomato and its use in breeding (in Russian). *Referativnyi Zhurnal* 5.65.111, 89–100.

Avdeev, Y.I. and Shcherbinin, B.M. (1976) Tomato specimen resistant to Egyptian broomrape (tolerance breeding) (in Russian). *Kartofel'i Ovoschi* 11, 35–36.

Avdeev, Y.I. and Shcherbinin, B.M. (1977) Tomato resistance to the broomrape *Orobanche aegyptiaca* Pers. (in Russian). *Tsiologiya i Genetika* 11, 491–496.

Avdeev, Y.I. and Shcherbinin, B.M. (1978) Investigating tomato resistance to broomrape and the inheritance of this trait (in Russian). *Doklady Vsesoyuznoi Akademii Sel'skokhozyaistevennykh Nauk imeni V.I. Lenina* 1, 14–16.

Awad, A.E. (1989) Effects of dicamba, nitrogen and pre-sowing hardening of host seed with phenolic acids on witchweed control in sorghum. PhD Thesis, North Carolina State University, Raleigh, USA.

Ayensu, E.S., Doggett, H., Keynes, R.D., Marton-Lefevre, J., Musselman, L.J., Parker, C. and Pickering, A. (eds) (1984) *Striga biology and control.* ICSU Press, Paris, 216 pp.

Ba, A.T. (1979) Notes on the histochemistry of the haustorium of *Striga hermonthica*. In: Musselman, L.J., Worsham, A.D. and Eplee, R.E. (eds) *Proceedings, 2nd Parasitic Weeds Symposium, Raleigh, 1979*. North Carolina State University, Raleigh, pp. 128–131.

Ba, A.T. (1984) Morphology, anatomy and ultrastructure of some parasitic species of the genus *Striga* (Scrophulariaceae). In: Ayensu, E.S., Doggett, H., Keynes, R.D., Marton-Lefevre, J., Musselman, L.J., Parker, C. and Pickering, A. (eds) *Striga Biology and Control*. ICSU Press, Paris, pp. 47–61.

Ba, A.T. and Kahlem, G. (1979) Mise en évidence d'activités enzymatiques au niveau de l'haustorium d'une phanérogamie parasite: *Striga hermonthica* (Scrophulariaceae). *Canadian Journal of Botany* 57, 2564–2567.

Babiker, A.G.T. and Fasil Reda (1991) Recent advances in *Striga* research in Sudan and Ethiopia. In: Ransom, J.K., Musselman, L.J., Worsham, A.D. and

Parker, C. (eds) *Proceedings, 5th International Symposium on Parasitic Weeds, Nairobi, 1991*. CIMMYT, Nairobi, pp. 180–189.

Babiker, A.G.T. and Hamdoun, M. (1982) Factors affecting the activity of GR7 in stimulating germination of *Striga hermonthica* (Del.) Benth. *Weed Research* 22, 111–115.

Babiker, A.G.T. and Hamdoun, M. (1983) Factors affecting the activity of ethephon in stimulating germination of *Striga hermonthica* (Del.) Benth. *Weed Research* 23, 125–131.

Babiker, A.G.T., Hamdoun. A.M. and Mansi, M.G. (1987) Influence of some soil and environmental factors on response of *Striga hermonthica* (Del.) Benth. seeds to selected germination stimulants. In: Weber, H.C. and Forstreuter, W. (eds) *Proceedings, 4th International Symposium on Parasitic Flowering Plants, Marburg, Germany, 1987*. Philipps-Universität, Marburg, pp. 53–66.

Babiker, A.G.T., Parker, C. and Suttle, J.C. (1992) Induction of *Striga* seed germination by thidiazuron. *Weed Research* 32, 243–248.

Bagnall-Oakeley, H., Gibberd, V. and Nyongesa, T.E. (1991) The incidence and control of *A. vogelii* in Embu district, Kenya. In: Ransom, J.K., Musselman, L.J., Worsham, A.D. and Parker, C. (eds) *Proceedings, 5th International Symposium on Parasitic Weeds, Nairobi, 1991*. CIMMYT, Nairobi, pp. 340–345.

Baker, F.A., Knowles, K., Meyer, T.R. and French, D.W. (1989) Aerial applications of ethylene releasing chemicals fail to promote abscision of dwarf mistletoe aerial shoots on jack pine. *Forestry Chronicle* 65, 194–195.

Baldev, B. (1959) *In vitro* responses of growth and development in *Cuscuta reflexa* Roxb. *Phytomorphology* 9, 316–319.

Baloch, G.M. (1968) Possibilities for biological control of some species of *Cuscuta* (Convolvulaceae). *PANS* 14, 27–33.

Baloch, G.M., Din, I.M. and Ghani, M.A. (1967a) Biological control of *Cuscuta* spp. I. *Cuscuta* spp. and insects associated with these in West Pakistan. *Technical Bulletin, Commonwealth Institute of Biological Control* 8, 149–158.

Baloch, G.M., Mohyuddin, A.I. and Ghani, M.A. (1967b) Biological control of *Cuscuta* spp. II. Biology and host plant range of *Melanagromyza cuscutae* Hering (Dipt. Agromyzidae) *Entomophaga* 12, 481–489.

Baloch, G.M., Mohyuddin, A.I. and Ghani, M.A. (1969) Biological control of *Cuscuta* spp. III. Phenology, biology and host-specificity of *Herpystis cuscutae* Bradley (Lep., Tortricidae). *Entomophaga* 14, 119–128.

Barlow, B.A. (1983) Biogeography of Loranthaceae and Viscaceae. In: Calder, M. and Bernhardt, P. (eds) *The Biology of Mistletoes*. Academic Press, Sydney, pp. 19–46.

Barlow, B.A. and Wiens, D. (1977) Host–parasite resemblance in Australian mistletoes: The case for cryptic mimicry. *Evolution* 31, 69–84.

Barlow, B.A., Hawksworth, F.G., Kuijt, J., Polhill, R.M. and Wiens, D. (1989) Genera of mistletoes. *The Golden Bough* 11, 1–4.

Bashir, M.O. (1987) The potential for biocontrol of witchweeds. In: Musselman, L.J. (ed.) *Parasitic Weeds in Agriculture. Volume I. Striga*. CRC Press, Boca Raton, pp. 183–206.

Bayer, D.E., Hoffman, E.C. and Foy, C.L. (1965) DCPA in host–parasite relations of alfalfa and dodder. *Weeds* 13, 92–95.

Bcharova, R. (1978) Investigating the diversity and pathogenicity of sunflower broomrapes from five sources (in Bulgarian). *Rastenievudni Nauki* 15, 164–169.

Bcharova, R. (1979) Physiological specialization in broomrape (*Orobanche cernua* Wall.) and the resistance of inbred sunflower lines (in Bulgarian). *Rastenievudni Nauki* 17, 127–132.

Bebawi, F.F. (1987) Cultural practices in witchweed management. In: Musselman, L.J. (ed.) *Parasitic Weeds in Agriculture. Volume I. Striga.* CRC Press, Boca Raton, pp. 159–171.

Bebawi, F.F. and El-Hag, G.A. (1983) Nutritive value of the parasitic weed *Striga hermontheca. Tropical Agriculture* 60, 44–47.

Bebawi, F.F. and Eplee, R.E. (1986) Efficacy of ethylene as a germination stimulant of *Striga hermonthica* seed. *Weed Science* 34, 694–698.

Bebawi, F.F. and Farah, A.F. (1981) Effects of parasitic and non-parasitic weeds on sorghum. *Experimental Agriculture* 17, 415–418.

Bebawi, F.F., Eplee, R.E., Harris, C.E. and Norris, R.S. (1984) Longevity of witchweed (*Striga asiatica*) seed. *Weed Science* 32, 494–497.

Beck, B.D.A. (1987) The effect of attack by *Alectra vogelii* (Benth.) on the yield of njugo beans (*Vigna subterranea* (L.) Verdc.). In: Weber, H.C. and Forstreuter, W. (eds) *Proceedings, 4th International Symposium on Parasitic Flowering Plants, Marburg, Germany,* 1987. Philipps-Universität, Marburg, pp. 79–82.

Becker, H. and Schmoll, H. (1986) *Mistel, Arzneipflanze, Brauchtum, Kuntsmotiv in Jugenstil.* Wissenschaftliche Verlagsgesellschaft MBH, Stuttgart, 132 pp.

Bedi, J.S. and Donchev, N. (1991) Results on mycoherbicide control of sunflower broomrape (*Orobanche cumana* Wall.) under glasshouse and field conditions. In: Ransom, J.K., Musselman, L.J., Worsham, A.D. and Parker, C. (eds) *Proceedings, 5th International Symposium on Parasitic Weeds, Nairobi, 1991.* CIMMYT, Nairobi, pp. 76–82.

Beilen, I.G. (1948) *Orobanche mutelii* F.Schultz and its specialization (in Russian). *Comptes Rendues Academie de Science, USSR* 61, 943–944.

Belyaeva, A.V., Cherkasova, A.P., Schpanova, L.G. and Alfimova, R.A. (1978) Maleic hydrazide against dodder (in Bulgarian). *Sakharnaya Svekla* 23, 37–39.

Benkov, B. (1970) The effect of the herbicide 2,4-D on yellow rattle in winter wheat crops (in Bulgarian). *Rasteniev"dni Nauki* 7, 111–117.

Benkov, B. (1978) Chemical control of field cow-wheat (*Melampyrum arvense* L.) in winter wheat (in Bulgarian). *Rasteniev"dni Nauki* 15, 126–131.

Bewick, T.A., Binning, L.K. and Dana, M.N. (1988a) Post-attachment control of swamp dodder (*Cuscuta gronovii*) in cranberry (*Vaccinium macrocarpon*) and carrot (*Daucus carota*). *Weed Technology* 2, 166–169.

Bewick, T.A., Binning, L.K. and Yandell, B. (1988b) A degree day model for predicting the emergence of swamp dodder in cranberry. *Journal of the American Society for Horticultural Science* 113, 839–841.

Bewick, T.A., Binning, L.K. and Dana, M.N. (1989) Control of swamp dodder in cranberry. *HortScience* 24, 850.

Bharathalakshmi and Jayachandra (1980) Presowing hardening of the host with phenolic acids reduces induction of seed germination in the root parasite *Striga asiatica. Tropical Pest Management* 26, 309–312.

Bharathalakshmi and Jayachandra (1984) Studies on the germination of some Indian collections of *Striga asiatica* (L.) Kuntze. In: Parker, C., Musselman, L.J., Polhill, R.M. and Wilson, A.K. (eds) *Proceedings, 3rd International*

Symposium on Parasitic Weeds, Aleppo, Syria, 1984. ICARDA, Aleppo, pp. 108–117.

Bharathalakshmi, Werth, C.R. and Musselman, L.J. (1990) A study of genetic diversity among host specific populations of the witchweed *Striga asiatica* (Scrophulariaceae) in Africa. *Plant Systematics and Evolution* 172, 1–12.

Bharati, M.P. (1991) Occurrence of *Orobanche* in Nepal. In: Wegmann, K. and Musselman, L.J. (eds) *Proceedings, International Workshop on* Orobanche *Research, Obermarchtal, 1989*. Eberhard-Karls-Universität, Tübingen, pp. 25–29.

Bhattarai, T., Bhandary, H. and Shrestha, P. (1989) Host range of *Cuscuta reflexa* Roxb. in the Kathmandu Valley, Nepal. *Plant Protection Quarterly* 4, 78–80.

Bischof, F. and Foroughi, M. (1971) Influence of pH of soil on the attachment of *Orobanche aegyptiaca* L. to tomato and tobacco (Research note). *Iranian Journal of Plant Pathology* 7, 56–58.

Bischof, F. and Koch, W. (1974) Chemical and biological control of *Orobanche aegyptiaca*. Paper read at: Conference on Plant Protection in Tropical and Subtropical Areas, Manila, 1974, 13 pp.

Bloomfield, J.R. and Ruxton, I.B. (1977) *The Effect of Farmyard Manure on Weed Introduction and Crop Establishment during Land Reclamation in Arid Land Agriculture (Saudi Arabia) II. Effects on Weed Increase and Crop Yields*. Research and Development Project, University College of North Wales, Bangor, UK and Ministry of Agriculture and Water, Saudi Arabia, Publication No. 87, 19 pp.

Boorsma, P.A. (1980) Variability in *Vicia faba* for resistance to *Orobanche crenata*. *FAO Plant Protection Bulletin* 28, 39–42.

Botha, P.J. (1946) Die lewensgeskiedenis morfologie en ontkiemingsfisiologie van Alectra vogelii Benth. PhD Thesis Potcheefstroom University for C.H.E., Republic of South Africa.

Botha, P.J. (1948) The parasitism of *Alectra vogelii* Benth. with special reference to the germination of its seeds. *Journal of South African Botany* 14, 63–80.

Botha, P.J. (1950) The effect of time of pre-exposure, temperature of pre-exposure and concentration of the host factor on the germination of the seed of *Alectra vogelii* Benth. *Journal of South African Botany* 16, 23–29.

Braun, M., Burgstaller, H. and Walter, H. (1984) Critical evaluation of control methods for *Orobanche ramosa* L. occurring in smallholder vegetable farms of the Khartoum Province, Sudan. In: Parker, C., Musselman, L.J., Polhill, R.M. and Wilson, A.K. (eds) *Proceedings, 3rd International Symposium on Parasitic Weeds, Aleppo, Syria, 1984*. ICARDA, Aleppo, pp. 245–249.

Braun, M., Koch, W. and Stiefvater, M. (1987) Solarization for soil sanitation – possibilities and limitations demonstrated in trials in southern Germany and the Sudan (in German). *Gesunde Pflanzen* 39, 301–309.

Braun, M., Koch, W., Mussa, H.H. and Stiefvater, M. (1988) Solarization for weed and pest control – possibilities and limitations. In: *Weed Control in Vegetable Production. Proceedings of a meeting of the EC Experts' Group, Stuttgart 28-31 Oct. 1986*. pp. 169–178.

Brenchley, W.E. (1920) *Weeds of Farm Land*. Longmans, Green and Co., London, 239 pp.

Brenchley, W.E. (1940) The weed problems in non-rotational wheat growing. *Empire Journal of Experimental Agriculture* 8, 126–137.

Brochot, N.E. and Tinnin, R.O. (1986) The effect of dwarf mistletoe on starch concentration in the twigs and needles of lodgepole pine. *Canadian Journal of Forest Research* 16, 658–660.

Brown, R., Greenwood, A.D., Johnson, A.W. and Long, A.G. (1951a) The stimulant involved in the germination of *Orobanche minor* Sm. I. Assay technique and bulk preparation of the stimulant. *Biochemical Journal (UK)* 48, 559–564.

Brown, R., Greenwood, A.D., Johnson, A.W., Long, A.G. and Tyler, G.J. (1951b) The stimulant involved in the germination of *Orobanche minor* Sm. II. Chromatographic purification of crude concentrates. *Biochemical Journal (UK)* 48, 564–568.

Brown, R., Johnson, A.W., Robinson, E. and Tyler, G.J. (1952) The *Striga* germination factor. II. Chromatographic purification of crude concentrates. *Biochemical Journal (UK)* 50, 596–600.

Calder, M. (1983) Mistletoes in focus: An introduction. In: Calder, M. and Bernhardt, P. (eds) *The Biology of Mistletoes*. Academic Press, Sydney, pp. 1–18.

Calder, M. and Bernhardt, P. (eds) (1983) *The Biology of Mistletoes*. Academic Press, Sydney, 348 pp.

Calvin, M., Hawksworth, F.G. and Knutson, D.M. (1984) Phloem in *Areuthobium globosum* (Viscaceae). *Botanical Gazette* 145, 461–464.

Capderon, M., Fer, A. and Ozenda, P. (1985) An unreported system of expulsion of a parasite: Dodder on cotton (*Cuscuta lupuliformis* on *Gossypium hirsutum*) (in French). *Comptes Rendus de l'Academie des Sciences, III (Sciences de la Vie)* 300, 227–232.

Cardwell, K.F., Smith, J., Weber, G. and Awad, A. (1991) Framework for orientation of *Striga* research at IITA. In: Ransom, J.K., Musselman, L.J., Worsham, A.D. and Parker, C. (eds) *Proceedings, 5th International Symposium on Parasitic Weeds, Nairobi, 1991*. CIMMYT, Nairobi, pp. 503–511.

Carson, A.G. (1988) Development and testing of a control package for *Striga hermonthica* on small-scale holdings in the Gambia. *Tropical Pest Management* 34, 97–101.

Carson, A.G. (1989a) Research and development strategies for the control of *Striga hermonthica* in the Gambia. In: Robson, T.O. and Broad, H.R. (eds) *Striga–Improved Management in Africa. Proceedings, FAO/OAU All-Africa Government Consultation on Striga control, Maroua, 1986*. FAO Plant Production and Protection Paper 96, pp. 100–118.

Carson, A.G. (1989b) Effect of intercropping sorghum and groundnuts on density of *Striga hermonthica* in The Gambia. *Tropical Pest Management* 35, 130-132.

Castejón-Muñoz, M., Romero-Muñoz, F. and García-Torres, L. (1987) Phenology and control of *Orobanche cernua* in sunflower with glyphosate. In: Weber, H.C. and Forstreuter, W. (eds) *Proceedings, 4th International Symposium on Parasitic Flowering Plants, Marburg, Germany, 1987*. Philipps-Universität, Marburg, pp. 121–126.

Castejón-Muñoz, M., Romero-Muñoz, F. and García-Torres, L. (1988) Dispersal of seeds of the sunflower broomrape (in Spanish). *Agriculture (Madrid)* 58, 981–982.

Castejón-Muñoz, M., Romero-Muñoz, F. and García-Torres, L. (1990)

Control of broomrape (*Orobanche cernua*) in sunflower (*Helianthus annuus* L.) with glyphosate. *Crop Protection* 9, 332–336.

Castejón-Muñoz, M., Romero-Muñoz, M. and García-Torres, L. (1993) Effect of planting date on broomrape (*Orobanche cernua* Loefl.) infections in sunflower (*Helianthus annuus* L.). *Weed Research* 33, 171–176.

Cechin, I. and Press, M.C. (1993) Nitrogen relations of the sorghum–*Striga hermonthica* host–parasite association: germination, attachment and early growth. *New Phytologist* 124, 681–687.

Cetinsoy, S. (1980) Studies on the determination of effective chemicals against *Melampyrum arvense* L. harmful in cereals in Central Anatolia (in Turkish). *Arastuma Datresa Baskanligi Sayt* 15, 118–119.

Cézard, R. (1973) Some aspects of the biology of *Orobanche* (in French). In: *Proceedings, Symposium on Parasitic Weeds, Malta, 1973*. European Weed Research Council, Wageningen, pp. 55–67.

Chabrolin, C. (1965) The role of certain saprophytic fungi in the germination of seeds with sclerenchymatous coats (in French). Paper read at 2[e] Colloque sur la Biologie des Mauvaises Herbes, Grignon, 1965, 10 pp.

Chalakov, C., Galev, I. and Tincheva, T. (1986) Broomrape (*Orobanche* spp.) control by means of pesticides provoking seed germination. In: ter Borg, S.J. (ed.) *Proceedings, Workshop on Biology and Control of* Orobanche, *Wageningen, 1986*. LH/VPO, Wageningen, pp. 190–195.

Chancellor, R.J. (1973) Germination and dormancy of *Odontites verna* (Bell.) Dum. In: *Proceedings, Symposium on Parasitic Weeds, Malta, 1973*. European Weed Research Council, Wageningen, pp. 260–268.

Chancellor, R.J., Parker, C. and Teferedegn, T. (1971) Stimulation of dormant weed seed germination by 2-chloroethyl-phosphonic acid. *Pesticide Science* 2, 35–37.

Chang, M and Lynn, D.L. (1986) The haustorium and the chemistry of host recognition in parasitic angiosperms. *Journal of Chemical Ecology* 12, 561–579.

Chang, M., Netzley, D.H., Butler, L.G. and Lynn, D.G. (1986) Chemical regulation of distance: characterization of the first natural host germination stimulant for *Striga asiatica*. *Journal of the American Chemical Society* 108, 7858–7860.

Chauhan, J.S., Singh, N.K. and Singh, S.V. (1989) Screening of higher plants for specific herbicidal principle active against dodder, *Cuscuta reflexa* Roxb. *Indian Journal of Experimental Biology* 27, 877–884.

Chennaveraiah, M.S., Nataraja, K. and Chikkannaiah, P.S. (1971) *In vitro* culture of the seeds of a root parasite: *Aeginetia indica* Linn. (without the association of the host). *Current Science* 24, 668–669.

Cherif-Ali, O., Housley, T.L. and Ejeta, G. (1990) Sorghum root length density and the potential for avoiding *Striga* parasitism. *Plant and Soil* 121, 67–72.

Chidley, V.L. and Drennan, D.S.H. (1987a) In vitro culture of *Striga asiatica* (L.) Kuntze (Scrophulariaceae). In: Weber, H.C. and Forstreuter, W. (eds) *Proceedings, 4th International Symposium on Parasitic Flowering Plants, Marburg, Germany, 1987*. Philipps-Universität, Marburg, pp. 829–831.

Chidley, V.L. and Drennan, D.S.H. (1987b) Effect of sorghum root residue on *Striga asiatica* (L.) Kuntze infection (Scrophulariaceae). In: Weber, H.C. and Forstreuter, W. (eds) *Proceedings, 4th International Symposium on Parasitic*

Flowering Plants, Marburg, Germany, 1987. Philipps-Universität, Marburg, pp. 829–831.

Choate, J.H., Andrews, R.H. and Barlow, B.A. (1987) Herbivory and cryptic mimicry in Australian Loranthaceae. In: Weber, H.C. and Forstreuter, W. (eds) *Proceedings, 4th International Symposium on Parasitic Flowering Plants, Marburg, Germany, 1987.* Philipps-Universität, Marburg, pp. 127–135.

Chun, D., Wilhelm, S. and Sagen, J.E. (1979) Components of record germination *in vitro* of branched broomrape, *Orobanche ramosa* L. In: Musselman, L.J., Worsham, A.D. and Eplee, R.E. (eds) *Supplement to Proceedings, 2nd International Symposium on Parasitic Weeds, Raleigh, 1979.* North Carolina State University, Raleigh, pp. 18,2.

Clay, K., Dement, D. and Rejmanek, M. (1985) Experimental evidence for host races in mistletoe (*Phoradendron tomentosum*). *American Journal of Botany* 72, 1225–1231.

Clerk, G.C. (1978) *Tapinanthus bangwensis* in a cola plantation in Ghana. *PANS* 24, 57–62.

Cook, C.E., Whichard, L.P., Wall, M.E., Egley, G.H., Coggan, P., Luhan, P.A. and McPhail, A.T. (1972) Germination stimulants 2. The structure of strigol – a potent seed germination stimulant for witchweed (*Striga lutea* Lour.). *Journal of the American Chemical Society* 94, 6198–6199.

Cooke, D.A. and Black, I.D. (1987) Biology and control of *Cuscuta campestris* and other *Cuscuta* spp.: a bibliographic review. *Technical Paper, Department of Agriculture, South Australia* 18, 33 pp.

Cordas, D.I. (1973) Effects of branched broomrape in tomatoes in California fields. *Plant Disease Reporter* 57, 926–927.

Cubero, J.I. (1983) Parasitic diseases in *Vicia faba* with special reference to broomrape (*Orobanche crenata* Forsk.) In: Hebblethwaite, P.D. (ed.) *The Faba Bean (*Vicia faba *L.).* Butterworths, London, pp. 493–521.

Cubero, J.I. (1986) Breeding for resistance to *Orobanche* and *Striga*: a review. In: ter Borg, S.J. (ed.) *Proceedings, Workshop on Biology and Control of* Orobanche, *Wageningen, 1986.* LH/VPO, Wageningen, pp. 127–139.

Cubero, J.I. (1991) Breeding for resistance to *Orobanche* species: a review. In: Wegmann, K. and Musselman, L.J. (eds) *Proceedings, International Workshop on* Orobanche *Research, Obermarchtal, 1989.* Eberhard-Karls-Universität, Tübingen, pp. 257–277.

Cudney, D.W., Orloff, S.B. and Reints, J.S. (1992) An integrated weed management procedure for the control of dodder (*Cuscuta indecora*) in alfalfa (*Medicago sativa*). *Weed Technology* 6, 603–606.

Czimber, G. (1969) The hard-coated character of seeds of *Cuscuta campestris* Yuncker and their resistance to herbicides (in Hungarian). *Mosonmagyaróvéri Agrartudomanyi Föisk. Közleményei* 12, 3–12.

Dalela, G.G. and Mathur, R.L. (1971a) Resistance of varieties of eggplant, tomato and tobacco to broomrape. *PANS* 17, 482–483.

Dalela, G.G. and Mathur, R.L. (1971b) Resistance of varieties of rape and mustard to broomrape (*Orobanche aegyptiaca*). *Indian Phytopathology* 24, 417–418.

Danielson, L.L. (1976) Effects of DCPA on growth of dodder. In: *Proceedings, Northeastern Weed Science Society* 30, pp. 268–276.

Davidson, N.J., True, K.C. and Pate, J.S. (1989) Water relations of the parasite:host relationship between the mistletoe *Amyema linophyllum* (Fenzl) Tieghem and *Casuarina obesa* Miq. *Oecologia* 80, 321–330.

Dawit Mulugeta (1992) Effects of planting and weeding methods on the control of dodder (*Cuscuta campestris*) in noug. In: Rezene Fessehaie and Parker, C. (eds) *Problems and Control of Parasitic Weeds in Ethiopia. Second Ethiopian Weed Science Workshop, Addis Ababa, 1988.* Ethiopian Weed Science Committee, Addis Ababa, pp. 79–84.

Dawson, J.H. (1965) Prolonged germination of field dodder. *Weeds* 13, 373–374.

Dawson, J.H. (1966) Response of field dodder to shade. *Weeds* 14, 4–5.

Dawson, J.H. (1971a) Establishing alfalfa on dodder-infested soil. *Weed Science* 19, 222–225.

Dawson, J.H. (1971b) Dodder control in alfalfa with dinoseb and D(-)(3-chlorophenylcarbamoyloxy)-2N-isopropylpropionamide. *Weed Science* 19, 551–554.

Dawson, J.H. (1972) Inhibitor of microbial enzyme prolongs dodder control with chlorpropham. *Weed Science* 20, 465–467.

Dawson, J.H. (1984a) A vegetative character that separates species of *Cuscuta*. In: Parker, C., Musselman, L.J., Polhill, R.M. and Wilson, A.K. (eds) *Proceedings, 3rd International Symposium on Parasitic Weeds, Aleppo, Syria, 1984.* ICARDA, Aleppo, pp. 184–187.

Dawson, J.H. (1984b) Effect of carbaryl and PCMC on dodder (*Cuscuta* spp.) control with chlorpropham. *Weed Science* 32, 290–292.

Dawson, J.H. (1984c) Control of *Cuscuta* in alfalfa – a review. In: Parker, C., Musselman, L.J., Polhill, R.M. and Wilson, A.K. (eds) *Proceedings, 3rd International Symposium on Parasitic Weeds, Aleppo, Syria, 1984.* ICARDA, Aleppo, pp. 188–199.

Dawson, J.H. (1987) *Cuscuta* (Convolvulaceae) and its control. In: Weber, H.C. and Forstreuter, W. (eds) *Proceedings, 4th International Symposium on Parasitic Flowering Plants, Marburg, Germany, 1987.* Philipps Universität, Marburg, pp. 137–149.

Dawson, J.H. (1989a) Dodder (*Cuscuta* spp.) control in established alfalfa (*Medicago sativa*) with glyphosate and SC-0224. *Weed Technology* 3, 552–559.

Dawson, J.H. (1989b) Established forage alfalfa (*Medicago sativa*) tolerates glyphosate and SC-0224 applied to control dodder (*Cuscuta* spp.). *Weed Technology* 3, 560–565.

Dawson, J.H. (1990a) Dodder (*Cuscuta* spp.) control with dinitroaniline herbicides in alfalfa (*Medicago sativa*). *Weed Technology* 4, 341–348.

Dawson, J.H. (1990b) Dodder (*Cuscuta* spp.) control in newly seeded alfalfa (*Medicago sativa*) with glyphosate. *Weed Technology* 4, 880–885.

Dawson, J.H. (1990c) Newly seeded alfalfa (*Medicago sativa*) tolerates glyphosate and SC-0224 at doses that control dodder (*Cuscuta* spp.). *Weed Technology* 4, 876–879.

Dawson, J.H. and Saghir, A.R. (1983) Herbicides applied to dodder (*Cuscuta* spp.) after attachment to alfalfa (*Medicago sativa*). *Weed Science* 31, 465–471.

Dawson, J.H., Ashton, F.M., Welker, W.V., Frank, J.R. and Buchanan, G.A. (1984) *Dodder and its control.* United States Department of Agriculture Farmers' Bulletin 2276, 24 pp.

de Bock, F. and Fer, A. (1992) Effects of abscisic acid on the transfer of sucrose from host, *Pelargonium zonale* (L.) Aiton, to a phanerogamic parasite *Cuscuta reflexa* Roxb. *Australian Journal of Plant Physiology* 19, 679–691.

Delabraze, P. and Lanier, L. (1972) Chemical control of *Viscum album* (in French). *European Journal of Forestry* 2, 95–103.

de la Harpe, A.C., Visser, J.H. and Grobbelaar, N. (1979) The chlorophyll concentration and photosynthetic activity of some parasitic flowering plants. *Zeitschrift für Pflanzenphysiologie* 95, 83–87.

de la Harpe, A.C., Visser, J.H. and Grobbelaar, N. (1980) Photosynthesis of certain South African parasitic flowering plants. *Zeitschrift für Pflanzenphysiologie* 97, 277–281.

de la Harpe, A.C., Visser, J.H. and Grobbelaar, N. (1981) Photosynthetic characteristics of some South African parasitic flowering plants. *Zeitschrift für Pflanzenphysiologie* 103, 265–275.

Delft, G.-J. and Stroud, A. (1992) The *Striga* problem in Welo Region; its relation to farmers and their farming techniques. In: Rezene Fessehaie and Parker, C. (eds) *Problems and Control of Parasitic Weeds in Ethiopia. Second Ethiopian Weed Science Workshop, Addis Ababa, 1988*. Ethiopian Weed Science Committee, Addis Ababa, pp. 10–14.

Deodikar, G.B. (1951) *Sorghum versicolor* Anderss – a species highly resistant to *Striga*. *Current Science (India)* 20, 135–136.

Dixon, N.H. and Parker, C. (1984) Aspects of the resistance of sorghum varieties to *Striga* species. In: Parker, C., Musselman, L.J., Polhill, R.M. and Wilson, A.K. (eds) *Proceedings, 3rd International Symposium on Parasitic Weeds, Aleppo, Syria, 1984*. ICARDA, Aleppo, pp. 123–132.

Doggett, H. (1965) *Striga hermonthica* on sorghum in East Africa. *Journal of Agricultural Science* 65, 183–194.

Doggett, H. (1984) *Striga* – advances in control. In: Hawksworth, D.L. (ed.) *Advancing Agricultural Production in Africa, Proceedings, CAB's First Scientific Conference, Arusha, Tanzania, 1984*. CAB, Slough, pp. 81–85.

Doggett, H. (1988) *Sorghum*. 2nd edition. Longman, London, 512 pp.

Doidge, E.M. (1924) A preliminary check list of plant diseases occurring in South Africa. *South Africa Botanical Survey Memoirs* 6, 56 pp.

Donald, W.W. and Fawcett, R.S. (1975) Growth inhibitors from red bartsia (*Bartsia odontites*), a minor weed in alfalfa. *Plant Growth Regulator Bulletin* 3, 37.

Dörr, I. (1972) Contact of *Cuscuta* hyphae with sieve tubes of its host plants (in German). *Protoplasma* 75, 167–184.

Dörr, I. (1987) The haustorium of *Cuscuta* – new structural results. In: Weber, H.C. and Forstreuter, W. (eds) *Proceedings, 4th International Symposium on Parasitic Flowering Plants, Marburg, 1987*. Philipps-Universität, Marburg, pp. 163–173.

Dörr, I. and Kollman, R. (1974) Structural features of the parasitism of *Orobanche*. I. Growth of the haustorial cells within the host tissue (in German). *Protoplasma* 80, 245–259.

Dörr, I. and Kollman, R. (1975) Structural features of parasitism of *Orobanche*. II. Differentiation of assimilate conducting elements within the haustorium (in German). *Protoplasma* 83, 185–201.

Dörr, I. and Kollman, R. (1976) Strukturelle Grundlagen des Parasitismus bei *Orobanche*. III Die Differentzierung des Xylemanschlusses bei *O. crenata*.

Protoplasma 89, 235–249.

Dörr, I., Visser, J.H. and Albers, F. (1977) On the parasitism of *A. vogelii* Benth. (Scrophulariaceae) II. Origin of lateral roots in the contact area of the haustorium. *Zeitschrift für Pflanzenphysiologie* 85, 349–359.

Dörr, I., Visser, J.H. and Kollmann, D. (1979) On the parasitism of *A. vogelii* Benth. (Scrophulariaceae) III. The occurrence of phloem between host and parasite. *Zeitschrift für Pflanzenphysiologie* 94, 427–439.

Doughty, L.R. (1941) Plant genetics. *East African Agricultural Research Station, Amani, Annual Report* 13, (1940), 13–15.

Dowler, C.C. and Robinson, E.L. (1990) Early herbicide evaluation for control of witchweed from 1957 through 1964. In: Sand, P.F., Eplee, R.E. and Westbrooks, R.G. (eds) *Witchweed Research and Control in the United States of America*. Weed Science Society of America, Champaign, pp. 91–98.

Drennan, D.S.H. and El-Hiweris, S.O. (1979) Changes in growth substances in *Sorghum vulgare* infected by *Striga hermonthica*. In: Musselman, L.J., Worsham, A.D. and Eplee, R.E. (eds) *Proceedings, 2nd International Symposium on Parasitic Weeds, Raleigh, 1979*. North Carolina State University, Raleigh, pp. 144–155.

Drennan, D.S.H. and Mohamed-Ahmed, A.G. (1992) Some effects of temperature on germination of *Orobanche ramosa* L. In: *IXème Colloque International sur la Biologie des Mauvaises Herbes, Dijon, 1992*. COLUMA/EWRS, pp. 117–120.

Edwards, W.G.H. (1972) *Orobanche* research at the Royal University of Malta. *PANS* 18, 475–476.

Edwards, W.G.H., Hiron, R.P. and Mallett, A.I. (1976) Aspects of the germination of *Orobanche crenata* seed. *Zeitschrift für Pflanzenphysiologie* 80, 105–111.

Efron, Y., Kim, S.K., Parkinson, V. and Bosque-Perez, N.A. (1989) IITA's strategies to develop *Striga* resistant maize germplasm. In: Robson, T.O. and Broad, H.R. (eds) *Striga – Improved Management in Africa. Proceedings, FAO/OAU All-Africa Government Consultation on* Striga *Control, Maroua, 1986*. FAO Plant Production and Protection Paper 96, pp. 141–153.

Egbe, N.E. and Oladokun, M.A.O. (1987) Factors limiting high yields in kola (*Cola nitida*) production in Nigeria. *Café Cacao Thé* 31, 303–310.

Egley, G.H. (1971) Mineral nutrition and the parasite–host relationship of witchweed. *Weed Science* 19, 528–533.

Egley, G.H. (1972) Influence of the seed envelope and growth regulators upon seed dormancy in witchweed (*Striga lutea* Lour.). *Annals of Botany* 36, 755–770.

Egley, G.H. and Dale, J.E. (1970) Ethylene, 2-chloroethylphosphonic acid, and witchweed germination. *Weed Science* 18, 586–589.

Egley, G.H., Eplee, R.E. and Norris, R.S. (1990) Discovery and testing of ethylene as a witchweed seed germination stimulant. In: Sand, P.F., Eplee, R.E. and Westbrooks, R.G. (eds) *Witchweed Research and Control in the United States of America*. Weed Science Society of America, Champaign, pp. 56–67.

Ehleringer, J.R., Schulze, E.D., Ziegler, H., Lange, O.A., Farquar, G.D. and Cowar, I.R. (1985) Xylem-tapping mistletoes: water or nutrient parasites? *Science* 227, 1479–1481.

El-Ghamrawy, N. and Neumann, K.-H. (1991) Some physiological aspects of *Orobanche crenata* parasitism on *Vicia faba*. In: Wegmann, K. and Musselman, L.J. (eds) *Proceedings, International Workshop on* Orobanche *Research, Obermarchtal, 1989*. Eberhard-Karls-Universität, Tübingen, pp. 134–146.

El-Ghamrawy, N., Salem, S.M. and Neumann, K.H. (1990) Nature of root exudates of *Vicia faba* plants in relation to induction of *Orobanche crenata* seed germination. *Angewandte Botanik* 64, 215–224.

Elia, Y.I., Al-Hahaf, N.A. and Al-Khesraji, T.O. (1987) Preliminary studies on *Orobanche aegyptiaca* Pers. in tomato (in Arabic). *Iraqi Journal of Agricultural Sciences 'Zanco'* 5, 233–244.

Emechebe, A.M. (1981) Cowpea pathology: notes on cowpeas and grain legume improvement program. Paper presented at Cropping Scheme Meeting, Institute for Agricultural Research, Samaru, Zaria, Nigeria.

Emechebe, A.M., Singh, B.B., Leleji, O.I., Atokple, I.D.K. and Adu, J.K. (1991) Cowpea–*Striga* problems and research in Nigeria. In: Kim, S.K. (ed.) *Combating Striga in Africa. Proceedings, International Workshop, Ibadan, 1988*. IITA, Nigeria, pp. 18–28.

Emiroglu, U., Nemli, Y. and Kucukozden, R. (1987) The resistance of Aegean tobacco lines to broomrape (*Orobanche ramosa* L.) and the effect of that parasite on yield and quality. In: Weber, H.C. and Forstreuter, W. (eds) *Proceedings, 4th International Symposium on Parasitic Flowering Plants, Marburg, Germany, 1987*. Philipps-Universität, Marburg, pp. 175–182.

Eplee, R.E. (1984) *Orobanche ramosa* in the United States. In: Parker, C., Musselman, L.J., Polhill, R.M. and Wilson, A.K. (eds) *Proceedings, 3rd International Symposium on Parasitic Weeds, Aleppo, Syria*, 1984. ICARDA, Aleppo, pp. 40–42.

Eplee, R.E. (1992) Witchweed (*Striga asiatica*): an overview of management strategies in the USA. *Crop Protection* 11, 3–7.

Eplee, R.E. and Norris, R.S. (1987a) Chemical control of *Striga*. In: Musselman, L.J. (ed.) *Parasitic Weeds in Agriculture. Volume I. Striga*. CRC Press, Boca Raton, pp. 173–182.

Eplee, R.E. and Norris, R.S. (1987b) Control of *Striga asiatica* (L.) Kuntze with systemic herbicides. In: Weber, H.C. and Forstreuter, W. (eds) *Proceedings, 4th International Symposium on Parasitic Flowering Plants, Marburg, Germany, 1987*. Philipps-Universität, Marburg, pp. 183–186.

Eplee, R.E. and Norris, R.S. (1990) Soil sampling collection equipment and equipment to separate seeds from the soil. In: Sand, P.F., Eplee, R.E. and Westbrooks, R.G. (eds) *Witchweed Research and Control in the United States*. Weed Science Society of America, Champaign, pp. 136–140.

Ernst, W.H.O. (1986) Mineral nutrition of *Nicotiana tabacum* cv, Bursana during infection by *Orobanche ramosa*. In: ter Borg, S.J. (ed.) *Proceedings, Workshop on Biology and Control of* Orobanche, *Wageningen, 1986*. LH/VPO, Wageningen, pp. 80–85.

Evans, D.C. (1960) Preliminary investigations into broomrape in the Cootamundra district. Paper No 60 (4 pp.). In: *Proceedings, 2nd Australian Weeds Conference, Australia, 1960*.

Evans, H.C. (1987) Fungal pathogens of some subtropical and tropical weeds and the possibilities for biological control. *Biocontrol News and Information* 8, 7–30.

Evtushenko, G.A., Iskhakova, N.A., Chepkasova, A.P. and Trushnikhina, E.L. (1973) The effect of MH on the productivity of sugarbeet and dodder (in Russian). In: Evtushenko, G.A. (ed.) *Vliyane Fiziologicheski Aktivnykh Soedinenii na Obmen Veschetstv i Produktivnost' Rastenii*, pp. 3–12.

EWRC (1973) *Proceedings, Symposium on Parasitic Weeds, Malta, 1973.* European Weed Research Council, Wageningen, 295 pp.

FAO (1985) *Proceedings, OAU/FAO Workshop on* Striga, *Yaounde, Cameroon, 23–27 September, 1985.* Food and Agriculture Organization of the United Nations, Rome, 136 pp.

Farina, M.P., Thomas, P.E.L. and Channon, P. (1985) Nitrogen, phosphorus and potassium effects on the incidence of *Striga asiatica* (L.) Kuntze in maize. *Weed Research* 25, 443–447.

Fasil Reda (1991) Laboratory and glasshouse studies on screening for *Striga* resistance in sorghum. In: *Proceedings, 7th East African Sorghum and Millet Regional Workshop, Nairobi, Kenya, 1990*, pp. 207–219.

Fasil Reda and Parker, C. (1990) A host range study of different *Striga* samples and *Alectra vogelii*. In: Parker, C. and Rezene Fessehaie (eds) *Proceedings, EWSC Sixth Annual Meeting, Addis Ababa, 1988.* Ethiopian Weed Science Committee, Addis Ababa, pp. 75–81.

Fate, G., Chang, M. and Lynn, D.G. (1990) Control of germination in *Striga asiatica*: chemistry of spatial definition. *Plant Physiology* 93, 201–207.

Feinbrun, N. (1970) A taxonomic review of European *Cuscutae. Israeli Journal of Botany* 19, 16–29.

Fer, A. (1976a) Photosynthesis and respiration in *Cuscuta lupuliformis* Krock seedlings during their preparasitic stage (in French). *Physiologie Végétale* 14, 357–365.

Fer, A. (1976b) Absorption, translocation and metabolism of phosphates by *Cuscuta gronovii* Willd. seedlings during their preparasitic stage (in French). *Biologia Plantarum* 18, 450–459.

Fer, A. (1979) Contribution to the physiology of the nutrition of phanerogamous parasites: study of the genus *Cuscuta* L. (Convolvulaceae) (in French). Doctor of Science Thesis, l'Université Scientifique et Médicale de Grenoble, France, 105 pp.

Fer, A. (1981a) Knowledge of the biology and physiology of *Cuscuta* spp. as a basis for research into a rational method of controlling these parasites (in French). In: *Compte Rendu de la 11e Conference du COLUMA*, pp. 330–337.

Fer, A. (1981b) Research on the pathway of transport involved in the nutrition of a phanerogamous parasite. A study on isolated leaves parasitised by *Cuscuta* (in French). *Physiologie Végétale* 19, 177–196.

Fer, A. (1983) Investigation of a new strategy for dodder using systemic herbicides. Results with labelled molecules in the laboratory (in French). In: *Compte Rendu de la 12e Conference du COLUMA* 1, pp. 179–186.

Fer, A. (1984) Physiological approach to the chemical control of *Cuscuta*: experiments with ^{14}C-labelled herbicides. In: Parker, C., Musselman, L.J., Polhill, R.M. and Wilson, A.K. (eds) *Proceedings, 3rd International Symposium on Parasitic Weeds, Aleppo, 1984.* ICARDA, Aleppo, pp. 164–174.

Fer, A., Benharrat, H., Rey, L. and Renaudin, S. (1990) A study of some effects of *Cuscuta lupuliformis* Krock parasitism on *Vicia sativa* L.: impact on biomass and the content of several important mineral elements (in French). *Comptes Rendus de L'Academie des Sciences, Series 3, Science de la Vie* 310, pp. 113–120.

Fetvadzhiev, V. (1970) Possibilities of controlling tobacco broomrape by the pesticide Basudin Granular (in Bulgarian). *Bulgarski Tyntyun* 15, 25–29.

Fineran, B.A. (1987) A structural approach towards investigating transport systems between host and parasite, as exemplified by some mistletoes and root parasites. In: Weber, H.C. and Forstreuter, W. (eds) *Proceedings, 4th International Symposium on Parasitic Flowering Plants, Marburg, Germany, 1987*. Philipps-Universität, Marburg, pp. 201–220.

Fisher, J.T. (1983) Water relations of mistletoes and their hosts. In: Calder, M. and Bernhardt, P. (eds) *The Biology of Mistletoes*. Academic Press, Sydney, pp. 161–184.

Fitzgerald, C.H. (1975) Early root development and anatomical parasite–host relationships of black senna with slash pine. *Forest Science* 21, 239–242.

Fitzgerald, C.H. and Terrell, S. (1975) Control of *Seymeria cassioides* with phenoxyalkanoic acid herbicides. In: *Proceedings, 28th Annual Meeting, Southern Weed Science Society*, pp. 226–229.

Forbes, J.O. (1964) Two problem weeds of Manitoba. In: *Proceedings, 20th North Central Weed Control Conference, 1964*, p. 61.

Foschi, S. and Rapparini, G. (1977) The control of *Cuscuta campestris* Yunck. and *Cuscuta europaea* L. (in Italian). In: *Proceedings, European Weed Research Society Symposium on Different Methods of Weed Control and their Integration, Uppsala, 1977*, pp. 129–137.

Foy, C.L., Jacobsohn, R. and Jain, R. (1988). Screening of *Lycopersicon* spp. for glyphosate and/or *Orobanche aegyptiaca* Pers. resistance. *Weed Research* 28, 383–391.

Foy, C.L., Jain, R. and Jacobsohn, R. (1989) Recent approaches for chemical control of broomrape (*Orobanche* spp.). *Reviews in Weed Science* 4, 123–152.

Foy, C.L., Jacobsohn, R., Bohlinger, B. and Jacobsohn, M. (1991) Seasonal behaviour of broomrape species as determined by host range and environmental factors. In: Ransom, J.K., Musselman, L.J., Worsham, A.D. and Parker, C. (eds) *Proceedings, 5th International Symposium on Parasitic Weeds, Nairobi, 1991*. CIMMYT, Nairobi, pp. 454–457.

Frater, K.C. (1975) Broomrape control. *New Zealand Tobacco Growers' Journal* (November), 10.

Fratianne, D.G. (1965) The interrelationship between the flowering of dodder and the flowering of some long and short day plants. *American Journal of Botany* 52, 556–562.

French, R.C. and Sherman, L.J. (1976) Factors affecting dormancy, germination and seedling development of *Aeginetia indica* L. (Orobanchaceae). *American Journal of Botany* 63, 558–570.

Friesen, G.H. and Korwar, G.H. (1991) Effect of phenolic acids and mixed cropping on *Striga asiatica* infestations in sorghum. In: Ransom, J.K., Musselman, L.J., Worsham, A.D. and Parker, C. (eds) *Proceedings, 5th International Symposium on Parasitic Weeds, Nairobi, 1991*. CIMMYT, Nairobi, pp. 10–13.

Fuggles-Couchman, N.R. (1935) A parasitic weed of sorghums (*Rhamphicarpa veronicaefolia* Vatke). *East African Agriculture Journal* 1, 145–147.

Gaertner, E.E. (1950) Studies of seed germination, seed identification and host relationships in dodders, *Cuscuta* spp. *Cornell University Agricultural Experiment Station Memoirs* 294, 56 pp.

García-Torres, L. and López-Granados, F. (1991a) Progress of herbicide control of broomrape (*Orobanche* spp.) in legumes and sunflower (*Helianthus annuus* L.) In: Ransom, J.K., Musselman, L.J., Worsham, A.D. and Parker, C. (eds) *Proceedings, 5th International Symposium on Parasitic Weeds, Nairobi, 1991.* CIMMYT, Nairobi, pp. 306–309.

García-Torres, L. and López-Granados, F. (1991b) Control of broomrape (*Orobanche crenata* Forsk.) in broad bean (*Vicia faba* L.) with imidazolinones and other herbicides. *Weed Research* 31, 227–235.

García-Torres, L., López-Granados, F., Saavedra, M. and Mesa- García, J. (1989) Selection of herbicides for the control of broomrape (*Orobanche* spp.) in faba bean (*Vicia faba* L.). *Fabis Newsletter* 24, 32–36.

García-Torres, L., López-Granados, F. and Saavedra, M. (1991) New herbicides for broomrape (*Orobanche crenata*) control in faba bean (*Vicia faba*) In: Wegmann, K. and Musselman, L.J. (eds) *Proceedings, International Workshop on* Orobanche *Research, Obermarchtal, 1989.* Eberhard-Karls-Universität, Tübingen, pp. 200–208.

Geils, B.W. and Mathiasen, G.L. (1990) Intensification of dwarf mistletoe on southwestern Douglas-fir. *Forest Science* 36, 955–969.

Ghosh, S.K., Balasundaram, M. and Mohamed, A.M. (1984) Studies on the host–parasite relationship of phanerogamic parasites on teak and their possible control. *Kerala Forest Research Institute Research Report* 21, 39 pp.

Gibson, C.C. and Watkinson, A.R. (1989) The host range and selectivity of a parasitic plant: *Rhinanthus minor* L. *Oecologia* 78, 401–406.

Gil, J., Martin, L.M. and Cubero, J.I. (1982) Resistencia al jopo (*Orobanche crenata* Forsk.) en *Vicia faba* L. *Anales del Instituto Nacional de Investigaciones Agrarias, Serie Agricola* 21, 175–181.

Gil, J., Martin, L.M. and Cubero, J.I. (1984) Resistance to *Orobanche crenata* Forsk. in *Vicia sativa* L. II. Characterization and genetics. In: Parker, C., Musselman, L.J., Polhill, R.M. and Wilson, A.K. (eds) *Proceedings, 3rd International Symposium on Parasitic Weeds, Aleppo, Syria, 1984.* ICARDA, Aleppo, pp. 221–229.

Gill, L.S. (1937) *Dwarf mistletoe on lodgepole pine.* USDA Forest Service Pest Leaflet 18.

Gill, L.S. and Hawksworth, F.G. (1961) *The Mistletoes: A Literature Review.* USDA Forest Service Technical Bulletin 1424, 87 pp.

Gimesi, A. (1966) Selective control of dodder (*Cuscuta* spp.) in clover and lucerne. *Weed Research* 6, 81–82.

Gimesi, A. (1979) Protection against *Cuscuta* with herbicides. In: Musselman, L.J., Worsham, A.D. and Eplee, R.E. (eds) *Proceedings, 2nd International Symposium on Parasitic Weeds, Raleigh, 1979.* Carolina State University, Raleigh, pp. 292–296.

Gimesi, A. (1987) Chemical control of *Cuscuta* species. In: Weber, H.C. and Forstreuter, W. (eds) *Proceedings, 4th International Symposium on Parasitic Flowering Plants, Marburg, Germany, 1987.* Philipps-Universität, Marburg, pp. 249–252.

Giray, H. and Nemli, Y. (1983) Investigations on the morphological characters, brief life history and effectiveness of the natural enemy of *Orobanche, Phytomyza orobanchia* Kalt. (Diptera, Agromyzidae) in Izmir Province (in Turkish). *Turkiye Bitki Kormua Gergisi* 7, 183–192.

Girling, D.J., Greathead, D.J., Mohyuddin, A.I. and Sankaran, T. (1979) The potential for biological control in the suppression of parasitic weeds. *Biocontrol News Information* (Sample issue), 7–16.

Godschalk, S.K.B. (1983) Mistletoe dispersal by birds in South Africa. In: Calder, M. and Bernhardt, P. (eds) *The Biology of Mistletoes*. Academic Press, Sydney, pp. 117–128.

Gold, A.H., Sagen, J.E. and Wilhelm, S. (1979) California soils suppressive to branched broomrape (*Orobanche ramosa* L.). In: Musselman, L.J., Worsham, A.D. and Eplee, R.E. (eds) *Supplement to the Proceedings, 2nd International Symposium on Parasitic Weeds, Raleigh, 1979*. North Carolina State University, Raleigh, p. 1.

Gouws, J., Visser, J.H. and Grobbelaar, W.P. (1980) Some aspects of the bidirectional translocation of ^{14}C-labelled metabolites between *Alectra vogelii* Benth. and *Voandzeia subterranea* (L.) Thou. *Zeitschrift für Pflanzenphysiologie* 99, 225–233.

Govier, R.N. and Harper, J.L. (1964) Hemi-parasitic weeds. In: *Proceedings, 7th British Weed Control Conference, Brighton, 1984*, pp. 577–582.

Govier, R.N., Nelson, M.D. and Pate, J.S. (1967) Hemiparasitic nutrition in angiosperms. 1. The transfer of organic compounds from host to *Odontites verna* (Bell.) Dom. (Scrophulariaceae). *New Phytologist* 66, 285–297.

Govier, R.N., Brown, J.G.S. and Pate, J.S. (1968) Hemiparasitic nutrition in angiosperms. 2. Root haustoria and leaf glands of *Odontites verna* (Bell.) Dum. and their relevance to the abstraction of solutes from the host. *New Phytologist* 67, 963–971.

Graph, S., Herzlinger, G., Kleifeld, Y., Bargutti, A. and Retig, B. (1982) Control of dodder in chickpeas. *Phytoparasitica* 10, 267.

Graser, H.L. (1952) A progress report on mistletoe control in walnuts. *Western Fruit Grower* 6, 23–24.

Graves, J.D., Press, M.C. and Stewart, G.R. (1989) A carbon balance model of the sorghum–*Striga hermonthica* host–parasite association. *Plant, Cell and Environment* 12, 101–107.

Graves, J.D., Wylde, A., Press, M.C. and Stewart, G.R. (1990) Growth and carbon allocation in *Pennisetum typhoides* infected with the parasitic angiosperm *Striga hermonthica*. *Plant, Cell and Environment* 13, 367–373.

Graves, J.D., Press, M.C., Smith, S. and Stewart, G.R. (1992) The carbon canopy economy of the association between cowpea and the parasitic angiosperm *Striga gesnerioides*. *Plant, Cell and Environment* 15, 283–288.

Grazi, G. (1982) Research in *Viscum album* at Laboratorium Hiscia. *The Golden Bough* 1, 5.

Grazi, G. (1984) Possible host dependency of epicortical root formation in cultivated *Scurrula pulverulenta* (Loranthaceae). In: Parker, C., Musselman, L.J., Polhill, R.M. and Wilson, A.K. (eds) *Proceedings, 3rd International Symposium on Parasitic Weeds, Aleppo, Syria, 1984*. ICARDA, Aleppo, pp. 6–9.

Greathead, D.J. (1984) The natural enemies of *Striga* species, and the prospects for their utilization as biological control agents. In: Ayensu, E.S., Doggett, H.,

Keynes, R.D., Marton-Lefevre, J., Musselman, L.J., Parker, C. and Pickering, A. (eds) *Striga Biology and Control. Proceedings, International Workshop, Dakar, Senegal, 1983*. ICSU Press, Paris, pp. 133–160.

Greathead, D.J. and Milner, J.E.D. (1971) A survey of *Striga* spp. (Scrophulariaceae) and their insect natural enemies in East Africa with a discussion on the possibilities of biological control. *Tropical Agriculture* 48, 111–124.

Greenham, C.G. and Brown, A.G. (1957) The control of mistletoe by trunk injection. *Australian Institute of Agricultural Science Journal* 23, 308–318.

Grelen, H.E. and Mann, W.F. (1973) Distribution of senna seymeria (*Seymeria cassioides*), a root parasite on southern pines. *Economic Botany* 27, 339–342.

Grobbelaar, W.P. (1952) Kaffir corn seed. *Farming in South Africa* 27, 424, 435, 446.

Gworgwor, N.A. and Weber, H.C. (1991) Effect of N-application on sorghum growth, *Striga* infestation and the osmotic pressure of the parasite in relation to the host. *Journal of Plant Physiology* 139, 194–198.

Gwynn, T.M., Musselman, L.J. and Mann, W.F. (1978) The floral and seed biology of *Seymeria cassioides* (Scrophulariaceae). *Beiträge zur Biologie der Pflanzen* 54, 105–124.

Habib, S.A. and Rahman, A.A.A. (1988) Evaluation of some weed extracts against field dodder on alfalfa (*Medicago sativa*). *Journal of Chemical Ecology* 14, 443–452.

Hacquet, J., Dumont, R., Fer, A., Cadot, B. and Rousselot, J. (1983) Account of trials in 1982 and 1983 for dodder control in lucerne (in French). In: *Compte Rendu de la 12e Conference de COLUMA, Paris*, pp. 167–178.

Hadfield, J.S. and Russell, K.W. (1978) Dwarf mistletoe management in the Pacific northwest. In: Scharpf, R.F. and Parmeter, J.R. (eds) *Proceedings, Symposium on Dwarf Mistletoe Control through Forest Management, Berkeley, California, 1978*. General Technical Report PSW-31, Pacific Southwest Forest and Range Experiment Station, Forest Service, USDA, pp. 73–81.

Hakansson, S. (1956) Biology, ecology and control of wild onion and yellow rattle. In: *Proceedings, 3rd British Weed Control Conference, Blackpool, 1956*, pp. 207–210.

Harahap, Z., Olela, J.C., Ampong-Nyarko, K. and Saxena, R.C. (1992) Upland rice cultivars resistant to parasitic weed *Striga hermonthica*. *International Rice Research Newsletter* 17, 10–11.

Hartigan, D.T. (1971) *Hormone Spray Technique for the Control of Mistletoe*. Technical Publication, Forestry Commission of New South Wales 16, 4 pp.

Hassan, H.M., Kheir, N.F., Harb, E.Z., Badr, L.A. and Ghobashy, A.M. (1991) The effect of CCC and GA_3 on growth, yield, sugars and anatomical structure of tomato plants parasitized by *Orobanche ramosa*. In: Wegmann, K. and Musselman, L.J. (eds) *Proceedings, International Workshop on* Orobanche *Research, Obermarchtal, 1989*. Eberhard-Karls-Universität, Tübingen, pp. 70–79.

Hattingh, I.D. (1954) Geelblom control. *Farming in South Africa*, May edition, 525–526.

Hauck, C., Muller, S. and Schilknecht, H. (1992) A germination stimulant for parasitic flowering plants from *Sorghum bicolor*, a genuine host plant. *Journal of Plant Physiology* 139, 474–478.

Hawksworth, F.G. (1974) *Mistletoes on introduced trees of the world*. US Department of Agriculture Agricultural Handbook 469, 49 pp.

Hawksworth, F.G. (1977) *The Six-Class Dwarf Mistletoe Rating System*. USDA Forest Service General Technical Report RM-48, 7 pp.

Hawksworth, F.G. (1983) Mistletoes as forest parasites. In: Calder, M. and Bernhardt, P. (eds) *Biology of Mistletoes*. Academic Press, Sydney, pp. 317–333.

Hawksworth, F.G. and Geils, B.W. (1990) How long do mistletoe-infected ponderosa pines live? *Western Journal of Applied Forestry* 5, 47–48.

Hawksworth, F.G. and Scharpf, R.F. (1981) *Phoradendron on conifers*. USDA Forest Service Forest Insect and Disease Leaflet 164, 7 pp.

Hawksworth, F.G. and Scharpf, R.F. (eds) (1984) *Biology of Dwarf Mistletoes: Proceedings of the Symposium. Fort Collins, Colorado, 1984*. USDA Forest Service General Technical Report RM-111, 131 pp.

Hawksworth, F.G. and Scharpf, R.F. (1987) Mistletoe literature of the world. In: Weber, H.C. and Forstreuter, W. (eds) *Proceedings, 4th International Symposium on Parasitic Flowering Plants, Marburg, Germany, 1987*. Philipps-Universität, Marburg, p. 317.

Hawksworth, F.G. and Wiens, D. (1972) *Biology and classification of Arceuthobium*. USDA Forest Service Agriculture Handbook 401, 324 pp.

Hawksworth, F.G. and Wiens, D. (1984) Biology and classification of *Arceuthobium*: an update. In: Hawksworth, F.G. and Scharpf, R.F. (eds) *Biology of Dwarf Mistletoes: Proceedings of the Symposium. Fort Collins, Colorado, 1984*. USDA Forest Service General Technical Report RM-111, pp. 2–17.

Hawksworth, F.G. and Zakaullah. (1985) Observations on dwarf mistletoe on Himalayan blue pine in Pakistan. *The Golden Bough* 7, 1–2.

Hawksworth, F.G., Wicker, E.F. and Scharpf, R.F. (1977) *Fungal Parasites of Dwarf Mistletoes*. USDA Forest Service General Technical Report RM-36, 14 pp.

Hawksworth, F.G., Scharpf, R.F. and Marosy, M. (1991) European mistletoe continues to spread in Sonoma County. *California Agriculture* November–December, 1991, 39.

Hellmuth, E.O. (1971) Eco-physiological studies on plants in arid and semi-arid regions of W. Australia. IV. Comparison of the field physiology of the host *Acacia grasbyi* and its hemiparasite *Amyema nestor* under optimal and stress conditions. *Journal of Ecology* 59, 351–363.

Hemsley, W.B. and Skan, S.A. (1906) *Alectra* Thunb. In: Dyer, W.T. (ed.) *Flora of Tropical Africa*, Volume 4. L. Reeve, London, pp. 362–373.

Hepper, F.N. (1960) New and noteworthy Scrophulariaceae in Africa. *Kew Bulletin* 14, 402–416.

Hepper, F.N. and Gasson, P. (1986) A tuberous haustorium of *Thonningia sanguinea* (Balanophoraceae) growing on *Hevea brasiliensis*. *Haustorium* 16, 1–2.

Herb, R., Visser, J.H. and Schildknecht, H. (1987) Recovery, isolation and preliminary investigation of germination stimulants produced by *Vigna unguiculata* Walp. cv Saunders Upright. In: Weber, H.C. and Forstreuter, W. (eds) *Proceedings, 4th International Symposium on Parasitic Flowering Plants, Marburg, Germany, 1987*. Philipps-Universität, Marburg, Germany, pp. 351–366.

Herbaugh, H., Upton, N. and Eplee, R.E. (1980) *Striga gesnerioides* in the United

States of America. *Proceedings, Southern Weed Science Society* 33, pp. 187–190.

Hess, D.E. and Ejeta, G. (1987) Effect of cultural treatments on infestation of *Striga hermonthica* (Del.) Benth. (Scrophulariaceae). In: Weber, H.C. and Forstreuter, W. (eds) *Proceedings, 4th International Symposium on Parasitic Flowering Plants, Marburg, Germany, 1987*. Philipps-Universität, Marburg, pp. 367–375.

Hess, D.E., Ejeta, G. and Butler, L.G. (1991). Research into germination of *Striga* seed by sorghum root exudates. In: Ransom, J.K., Musselman, L.J., Worsham, A.D. and Parker, C. (eds) *Proceedings, 5th International Symposium on Parasitic Weeds, Nairobi, 1991*. CIMMYT, Nairobi, pp. 217–222.

Hinds, T.E. and Hawksworth, F.G. (1965) Seed dispersal velocity in four dwarf mistletoes. *Science* 148, 517–519.

Hiron, R.W.P. (1973) An investigation into the processes involved in germination of *Orobanche crenata* using a new bio-assay technique. In: *Proceedings, Symposium on Parasitic Weeds, Malta, 1973*. European Weed Research Council, Wageningen, pp. 76–88.

Hodosy, S. (1981) Biological control of broomrape, *Orobanche ramosa*, a tomato parasite: I. Occurrence and adaptibility of *Fusarium* species to control broomrape in Hungary (in Hungarian). *Zöldségtermesztesi Kutato Intézet Bulletinje* 14, 21–29.

Hofstra, J.J. and Klaren, C.H. (1973) Host–parasite relations in *Rhinathus serotinus*. 2. Physiological interactions between host and parasite. In: *Proceedings, Symposium on Parasitic Weeds, Malta, 1973*. European Weed Research Council, Wageningen, pp. 247–256.

Holdsworth, M. and Nutman, P.S. (1947) Flowering response in a strain of *Orobanche minor*. *Nature* 160, 223–224.

Holm, L., Pancho, J., Herberger, J. and Plucknett, D. (1979) *A Geographical Atlas of World Weeds*. Wiley-Interscience, New York, 440 pp.

Horowitz, M., Regev, Y. and Herzlinger, G. (1983) Solarization for weed control. *Weed Science* 31, 170–179.

Horvath, Z. and Bujaki, G. (1990) New data on the biology of *Smicronyx jungermanniae* Reich. (Col.: Curculonidae) (in Hungarian). *Novenyvedelem* 26, 346–351.

Hsiao, A.I., Worsham, A.D. and Moreland, D.E. (1981) Regulation of witchweed (*Striga asiatica*) conditioning and germination by dl-strigol. *Weed Science* 29, 101–104.

Hudler, G.W., Oshima, N. and Hawksworth, F.G. (1979) Bird dissemination of dwarf mistletoe on ponderosa pine in Colorado. *American Midland Naturalist* 102, 273–280.

Hughes, C.G. (1961) Flowering plants parasitizing sugar cane. In: Martin, J.P., Abbott, E.V. and Hughes, C.C. (eds) *Sugar Cane Diseases of the World*. Elsevier, pp. 479–493.

Hull, R.J. and Leonard, O.A. (1964) Physiological aspects of parasitism in mistletoes (*Arceuthobium* and *Phoradendron*). 2. The photosynthetic capacity of mistletoes. *Plant Physiology* 39, 1008–1017.

Hussain, H.S.N. and Aigbokhan, E.I. (1991) Growth stages of *S. gesnerioides* on cowpea roots. In: Ransom, J.K., Musselman, L.J., Worsham, A.D. and Parker, C. (eds) *Proceedings, 5th International Symposium on Parasitic Weeds,*

Nairobi, 1991. CIMMYT, Nairobi, pp. 279–286.

Hutchison, J.M. (1977) Control of dodder in tomatoes. In: *Proceedings, 49th Annual California Weed Conference*, pp. 48–50.

Hutchison, J.M. and Ashton, F.M. (1980) Germination of field dodder (*Cuscuta campestris*). *Weed Science* 28, 330–333.

Ibrahim, N.E., Babiker, A.G.T., Edwards, W.G. and Parker, C. (1985) Activity of extracts from *Euphorbia* species on the germination of *Striga* species. *Weed Research* 25, 135–140.

ICARDA (1992) *ICARDA Annual Report, 1991.* International Centre for Agricultural Research in Dry Areas, Aleppo, Syria, 120 pp.

ICRISAT (1979) *Annual Report for 1977/78.* International Crops Research Institute for the Semi-arid Tropics, Patancheru, India.

Igbinnosa, I. and Okonkwo, S.N.C. (1991) Studies on the germination of cowpea witchweed (*S. gesnerioides*) and its effect on cowpea (*Vigna unguiculata*). In: Ransom, J.K., Musselman, L.J., Worsham, A.D. and Parker, C. (eds) *Proceedings, 5th International Symposium on Parasitic Weeds, Nairobi, 1991.* CIMMYT, Nairobi, pp. 58–67.

Igbinnosa, I. and Okonkwo, S.N.C. (1992) Stimulation of the germination of seeds of cowpea witchweed (*Striga gesnerioides*) by sodium hypochlorite and some growth regulators. *Weed Science* 40, 25–28.

Ihl, B., Jabob, F. and Sembdner, G. (1984) Studies on *Cuscuta reflexa* Roxb. V. The level of endogenous hormones in the parasite, *Cuscuta reflexa* and its host, *Vicia faba* L. and a suggested role in the transfer of nutrients from host to parasite. *Plant Growth Regulation* 2, 77–90.

Ihl, B., Jabob, F., Meyer, A. and Sembdner, G. (1987) Investigation on the endogenous levels of abscisic acid in a range of parasitic phanerogams. *Journal of Plant Growth Regulation* 5, 191–205.

Ismail, A.M.A. and Obeid, M. (1976) A study of assimilation and translocation in *Cuscuta hyalina* Heyne ex Roth., *Orobanche ramosa* L. and *Striga hermonthica* Benth. *Weed Research* 16, 87–92.

Ivannikov, A.I. (1972) The dodder midge (*Melanagromyza cuscutae* Hg.) (in Russian). *Zashchita Rastenii* 7, 23–24.

Izard, C. (1959) L'*Orobanche* du tabac. *Annales de l'Institut Expérimental du Tabac de Bergerac* 3, 299–316.

Jackson, M.B. and Parker, C. (1991) Induction of germination by a strigol analogue requires ethylene action in *Striga hermonthica* but not in *S. forbesii. Journal of Plant Physiology* 138, 383–386.

Jacobsohn, R. (1986) Broomrape avoidance and control: agronomic problems and available methods. In: ter Borg, S.J. (ed.) *Proceedings, Workshop on Biology and Control of* Orobanche, *Wageningen, 1986*. LH/VPO, Wageningen, pp. 18–24.

Jacobsohn, R. (1989) *Orobanche*. In: Helevy, H.A. (ed.) *Handbook of Flowering*, Volume 6. CRC Press, Florida, pp. 490–493.

Jacobsohn, R. and Eldar, E. (1992) Imazethapyr for broomrape control in peas. *Phytoparasitica* 20, 345.

Jacobsohn, R. and Levy, D. (1986) Glyphosate for *Orobanche* control in various crops; problems and promises. In: ter Borg, S.J. (ed.) *Proceedings, Workshop on Biology and Control of* Orobanche, *Wageningen, 1986*. LH/VPO, Wageningen, pp. 171–175.

Jacobsohn, R., Greenberger, A., Katan, J., Levi, J. and Alon, H. (1980) Control of Egyptian broomrape (*Orobanche aegyptiaca*) and other weeds by means of solar heating of the soil by polyethylene mulching. *Weed Science* 28, 312–316.

Jacobsohn, R., Ben-Ghedalia, D. and Marton, K. (1987) Effect of the animal's digestive system on the infectivity of *Orobanche* seeds. *Weed Research* 27, 87–90.

Jacobsohn, R., Bohlinger, B., Eldar, E. and Agrawal, V.P. (1991a) Crop host range of *Orobanche* species in an experimental field. In: Ransom, J.K., Musselman, L.J., Worsham, A.D. and Parker, C. (eds) *Proceedings, 5th International Symposium on Parasitic Weeds, Nairobi, 1991*. CIMMYT, Nairobi, pp. 176–179.

Jacobsohn, R., Kleifeld, Y., Agrawal, V.P., Jha, P. and Marton, K. (1991b) Soil fumigation with Telone II for broomrape (*Orobanche* spp.) control. In: Wegmann, K. and Musselman, L.J. (eds) *Proceedings, International Workshop on* Orobanche *Research, Obermarchtal, 1989*. Eberhard-Karls-Universität, Tübingen, pp. 185–190.

Jain, R. and Foy, C.L. (1987) Influence of various nutrients and growth regulators on germination and parasitism of *Orobanche aegyptiaca*. In: Weber, H.C. and Forstreuter, W. (eds) *Proceedings, 4th International Symposium on Parasitic Flowering Plants, Marburg, Germany, 1987*. Philipps-Universität, Marburg, pp. 427–436.

Jain, R. and Foy, C.L. (1992) Nutrient effects on parasitism and germination of Egyptian broomrape (*Orobanche aegyptiaca*). *Weed Technology* 6, 269–275.

James, R.W. (1976) A preliminary note on the control of broomrape (*Orobanche minor*) in flue-cured tobacco. *New Zealand Tobacco Growers' Journal* May 1976, 7–8.

James, R.W. and Frater, K.C. (1977) The control of broomrape (*Orobanche minor*) in flue-cured tobacco. II. The results of experimental work with metham-sodium in 1976–1977. *New Zealand Tobacco Growers' Journal* October 1977, 10–12.

Joel, D.M. (1987a) Detection and identification of *Orobanche* seeds using fluorescence microscopy. *Seed Science and Technology* 15, 119–124.

Joel, D.M. (1987b) Identification of *Orobanche* seeds. In: Weber, H.C. and Fostreuter, W. (eds) *Proceedings, 4th International Symposium on Parasitic Flowering Plants, Marburg, Germany, 1987*. Philipps-Universität, Marburg, pp. 437–444.

Joel, D.M. (1988a) A key for the identification of *Orobanche* seeds. Abstract of paper presented at the 10th Conference of the Weed Science Society of Israel. *Phytoparasitica* 16, 376.

Joel, D.M. (1988b) *Orobanche cumana* – a new adventive weed in Israel. Abstract of paper presented at the 10th Conference of the Weed Science Society of Israel. *Phytoparasitica* 16, 375.

Joel, D.M. (1992) Control of broomrape (*Orobanche aegyptiaca* Pers.) with chlorsulfuron on a transgenic sulfonylurea-resistant crop. In: *Abstracts, 1992 Meeting of the Weed Science Society of America*, p. 65.

Joel, D.M., Back, A., Kleifeld, Y. and Gepstein, S. (1991a) Seed conditioning and its role in *Orobanche* seed germination: inhibition by paclobutrazole. In: Wegmann, K. and Musselman, L.J. (eds) *Proceedings, International Workshop on* Orobanche *Research, Obermarchtal, 1989*. Eberhard-Karls-Universität, Tübingen, pp. 147–157.

Joel, D.M., Kleifeld, Y. and Bucsbaum, H. (1991b) *Osyris alba* causing damage in orchards. In: Ransom, J.K., Musselman, L.J., Worsham, A.D. and Parker, C. (eds) *Proceedings, 5th International Symposium on Parasitic Weeds, Nairobi, 1991.* CIMMYT, Nairobi, pp. 378–381.

Johnson, A.W., Roseberry, G. and Parker, C. (1976) A novel approach to *Striga* and *Orobanche* control using synthetic germination stimulants. *Weed Research* 16, 223–227.

Johnson, A.W., Gowda, G., Hassanali, A,. Knox, J., Monaco, S. and Razawi, Z. (1981) The preparation of synthetic analogues of strigol. *Journal of the Chemical Society Perkin Transactions* I 6, 1734–1743.

Jones, B.M.G. and Safa, S.B. (1982) Variation of seed-coat ornamentation in *Striga hermonthica* (Scrophulariaceae). *Annals of Botany* 50, 629–634.

Joyce, D.C., Rein, K., Berry, A.M. and Reid, M.S. (1987) Control of broadleaf mistletoe (*Phoradendron tomentosum*) with dormant season ethephon sprays. *Acta Horticulturae* 201, 141–145.

Julien, M.H. (ed.) (1987) *Biological Control of Weeds: A World Catalogue of Agents and their Target Weeds.* 2nd Edition. CABI, Slough, 150 pp.

Kabambe, V.H. (1991) The development of cultural methods for control of *Striga* in maize in Malawi. In: Ransom, J.K., Musselman, L.J., Worsham, A.D. and Parker, C. (eds) *Proceedings, 5th International Symposium on Parasitic Weeds, Nairobi, 1991.* CIMMYT, Nairobi, pp. 46–50.

Kabulov, D.T. and Kabulova, A.D. (1977) The anatomical structure of broomrape and the vegetative organs of parasitized plants (in Russian). *Trudy Samarkandskogo Gosudarstvennogo Universiteta imeni A. Navoi* 336, 92–108.

Kabulov, D.T. and Khalimov, M.K. (1974) Some biological characteristics of Egyptian broomrape and ways of controlling it by means of the fungus *Fusarium* (in Russian). *Nauchnye Trudy Biologicheskogo Fakulteta Samarkandski Gosudarstvennyi Universitet imeni A. Navoi* 207, 167–173.

Kabulov, D.T. and Tashpulatova, S.Z. (1974) The effects of Egyptian broomrape on tomato yields (in Russian). *Nauchnye Trudy Samarkandski Gosudarstvennyi Universitet imeni A. Navoi* 207, 141–145.

Kachelriess, S. (1988) A method for the detection of *Orobanche* seeds in the soil (in German). *Zeitschrift für Pflanzenkrankheiten and Pflanzenschutz* 198, 55–60.

Kaculas, K. and Tarziev, Z. (1965) Control of dodder (in Russian). *Zashchita Rastenii* 10, 47–48.

Kadambi, K. (1954) On *Loranthus* control. *Indian Forester* 80, 493–495.

Kadry, A. and Tewfic, H. (1956) Seed germination in *Orobanche crenata* Forssk. *Svensk Botanik Tidskrift* (Sweden) 50, 270–286.

Kasasian, L. (1971) *Orobanche* spp. *PANS* 17, 35–41.

Kasasian, L. (1973a) Miscellaneous observations on the biology of *Orobanche crenata* and *O. aegyptiaca*. In: *Proceedings, Symposium on Parasitic Weeds, Malta, 1973.* European Weed Research Council, Wageningen, pp. 68–75.

Kasasian, L. (1973b) The chemical control of *Orobanche crenata* in *Vicia faba* and the susceptibility of 53 cultivars of *V. faba* to *O. crenata*. In: *Proceedings, Symposium on Parasitic Weeds, Malta, 1973.* European Weed Research Council, Wageningen, pp. 224–230.

Kasasian, L. (1973c). Control of *Orobanche*. *PANS* 19, 368–371.

Kasasian, L. (1977) *The Control of Grass and Broadleaf Weeds and of* Cuscuta *in Young and Established Hasawi Alfalfa.* Joint Agricultural Research and Development

Project, University College of North Wales, Bangor, UK and Ministry of Agriculture and Water, Saudi Arabia. Publication No. 106, 9 pp.

Kasasian, L. and Parker, C. (1971) The effect of numerous herbicides on the germination of *Orobanche aegyptiaca* and *Striga hermonthica*. *PANS* 17, 471–481.

Kaswari, M.A. and Abu-Irmaileh, B.E. (1989) Resistance to branched broomrape (*Orobanche ramosa*) in tomato germplasm. *HortScience* 24, 822–824.

Katan, J, Greenberger, A., Alon, H. and Grinstein, A. (1976) Solar heating by polyethylene mulching for the control of diseases caused by soil-borne pathogens. *Phytopathology* 66, 683–688.

Kaufman, D.D. and Kearney, P.C. (1965) Microbial degradation of isopropyl-*N*-3-chlorophenylcarbamate and 2-chloroethyl-*N*-3-chlorophenylcarbamate. *Applied Microbiology* 13, 443–446.

Kaufman, D.D., Kearney, P.C., Von Endt, D.W. and Miller, D.E. (1970) Methylcarbamate inhibition of phenylcarbamate metabolism in soil. *Journal of Agricultural and Food Chemistry* 18, 513–519.

Kearney, P.C. (1965) Purification and properties of an enzyme responsible for hydrolyzing phenylcarbamates. *Journal of Agricultural and Food Chemistry* 13, 561–564.

Kelili, D., Zemragh, M. and Marliere, A.M. (1983) Broomrape control in Morocco: assessment and future prospects (in French). In: *Compte Rendu de la 12e Conference du COLUMA*, Volume 1, pp. 195–202.

Khalaf, K.A.H.A. (1982) Some studies on the control of *Orobanche crenata* parasitism of *Vicia faba*. PhD Thesis, University of London, 233 pp.

Khalaf, K.A. (1992) Evaluation of the biological activity of flax as a trap crop against *Orobanche* parasitism of *Vicia faba*. *Tropical Agriculture (Trinidad)* 69, 35–38.

Khatri, G.B., Jha, P.K., Agrawal, V.P. and Jacobsohn, R. (1991) Distribution, host range and phenology of *Orobanche* spp. in Nepal. In: Wegmann, K. and Musselman, L.J. (eds) *Proceedings, International Workshop on* Orobanche *Research, Obermarchtal, 1989*. Eberhard-Karls-Universität, Tübingen, pp. 18–24.

Khot, R.S., Bhat, B.N., Angadi, S.V., Rao, S. and Patil, S. (1989) Effect of trap crops on the incidence of *Orobanche* and yield of Bidi tobacco. *Farming Systems* 5, 101–103.

Kim, S.K. (ed.) (1991a) *Combating Striga in Africa. Proceedings, International Workshop, Ibadan, 1988.* International Institute of Tropical Agriculture, Ibadan, 145 pp.

Kim, S.K. (1991b) Breeding maize for *Striga* tolerance and the development of a field infestation technique. In: Kim, S.K. (ed.) *Combating Striga in Africa. Proceedings, International Workshop, Ibadan, August 1988.* IITA, Nigeria, pp. 96–108.

Kim, S.K. and Winslow, M.D. (1991) Progress in breeding maize for *Striga* tolerance/resistance at IITA. In: Ransom, J.K., Musselman, L.J., Worsham, A.D. and Parker, C. (eds) *Proceedings, 5th International Symposium on Parasitic Weeds, Nairobi, 1991.* CIMMYT, Nairobi, pp. 494–499.

Kim, S.K. and Winslow, M.D. (1992) Breeding maize for *Striga* resistance. *IITA Research* 4, 9–12.

Kim, S.K., Khadr, F., Parkinson, V., Fajemesin, J. and Efron, Y. (1985) Maize breeding for *Striga* resistance in Africa. In: *Proceedings, FAO/OAU Workshop on Striga, Yaounde, Cameroon, 1985*, pp. 58–74.

Kim, S.K., Efron, Y., Khadr, F., Mareck, J.H., Fajemisin, J. and Lee, M.H. (1987) Registration of 16 maize-streak virus resistant tropical maize parental inbred lines. *Crop Science* 27, 824–825.

King, L.J. (1966) *Weeds of the World*. Interscience Publishers, New York, 526 pp.

King, S.B. (1975) Screening sorghum for resistance to witchweed (*Striga hermonthica*) in Nigeria. In: *Proceedings, American Phytopathological Society, Southern Division*, 2, 136.

King, S.B. and Zummo, N. (1977) Physiologic specialisation in *Striga hermonthica* in West Africa. *Plant Disease Reporter* 61, 770–773.

Klaren, C.H. and van der Dijk, S.J. (1976) Water relations of the hemi-parasite *Rhinanthus serotinus* before and after attachment. *Physiologia Plantarum* 38, 121–125.

Klaren, H.C. and Janssen, G. (1978) Physiological changes in the hemiparasite *Rhinanthus serotinus* before and after attachment. *Physiologia Plantarum* 42, 151–155.

Kleifeld, Y., Goldwasser, Y., Herzlinger, G., Golan, S. and Baragutti, A. (1991) Broomrape control with metham-sodium. In: Ransom, J.K., Musselman, L.J., Worsham, A.D. and Parker, C. (eds) *Proceedings, 5th International Symposium on Parasitic Weeds, Nairobi, 1991*. CIMMYT, Nairobi, pp. 382–390.

Kleifeld, Y., Goldwasser, Y., Herzlinger, G., Joel, D.M., Golan, S. and Kahana, D. (1993) Flax (*Linum usitatissimum* L.) as a host plant for Egyptian broomrape (*Orobanche aegyptiaca* Pers.). *Weed Research* (in press).

Klyueva, M.P. and Pamuchki, G.V. (1982) Technology of the use of *Phytomyza* (in Russian). *Zashchita Rastenii* 1, 33–34.

Klyueva, M.P. and Pamuchki, G.V. (1983) How to determine the technical effectiveness of *Phytomyza* (in Russian). *Zashchita Rastenii* 6, 41–42.

Knepper, D.A., Creager, R.A. and Minn, L.J. (1990) Identifying dodder seed as contaminants in seed shipments. *Seed Science and Technology* 18, 731–741.

Knepper, D., Obilana, A.T. and Musselman, L.J. (1991) Morphology of *Striga forbesii* and preliminary screening for resistance in sorghum. In: Ransom, J.K., Musselman, L.J., Worsham, A.D. and Parker, C. (eds) *Proceedings, 5th International Symposium on Parasitic Weeds, Nairobi, 1991*. CIMMYT, Nairobi, pp. 241–246.

Knutson, D.M. (1974) Infection technique and seedling response to dwarf mistletoes. *Plant Disease Reporter* 58, 235–238.

Knutson, D.M. (1978) Biological and chemical control of dwarf mistletoe. In: Scharpf, R.F. and Parmeter, J.R. (eds) *Proceedings, Symposium on Dwarf Mistletoe Control through Forest Management, Berkeley, 1978*. General Technical Report PSW-31, Pacific Southwest Forest and Range Experiment Station, Forest Service, USDA, pp. 151–155.

Knutson, D.M. (1979) How parasitic seed plants induce disease in other plants. In: Horsefall, J.G. and Cowling, E.B. (eds) *Plant Disease: An Advanced Treatise. Vol. IV. How Pathogens Induce Disease*. Academic Press, New York, pp. 293–312.

Knutson, D.M. (1983) Physiology of mistletoe parasitism and disease responses in the host. In: Calder, M. and Bernhardt, P. (eds) *The Biology of Mistletoes*. Academic Press, Sydney, pp. 295–316.

Knutson, D.M. (1984) Seed development, germination behaviour and infection characteristics of several species of Arceuthobium. In: Hawksworth, F.G.

and Scharpf, R.F. (eds) *Biology of Dwarf Mistletoes: Proceedings of the Symposium. Fort Collins, Colorado, 1984.* USDA Forest Service General Technical Report RM-111, pp. 77–84.

Knutson, D.M. and Toevs, W.J. (1972) Dwarf mistletoe reduces root growth of ponderosa pine seedlings. *Forest Science* 18, 323–324.

Kotoula-Syka, E. and Eleftherohorinos, I.G. (1991) *Orobanche ramosa* L. (broomrape) control in tomato (*Lycopersicon esculentum* Mill.) with chlorsulfuron, glyphosate and imazaquin. *Weed Research* 31, 19–27.

Kott, S.A. (1969) Cultural control of broomrape (in Russian). In: *Weeds and their Control*. Kolos, Moscow, pp. 96–98, 177–178.

Kovalev, O.V. (1973) Modern outlooks of biological control of weed plants in the USSR and the international phytophagous exchange. In: *Proceedings, 2nd International Symposium on Biological Control of Weeds, Rome, 1971*, pp. 166–172.

Krishnaiah, M.S., Rao, C.R. and Rao, G.N.S. (1968) *Striga lutea* resistance of male sterile lines and hybrids of pearl millet (*Pennisetum typhoides*). *Andhra Agricultural Journal* 15, 180.

Krishnamurthy, G.V.G. and Chari, M.S. (1991) Further studies on post-emergence control of broomrape on tobacco with plant oils. In: Ransom, J.K., Musselman, L.J., Worsham, A.D. and Parker, C. (eds) *Proceedings, 5th International Symposium on Parasitic Weeds, Nairobi, 1991.* CIMMYT, Nairobi, pp. 108–110.

Krishnamurthy, G.V.G. and Nagarajan, K. (1991) Post-emergence control of *Orobanche cernua* on tobacco with oils. *Tropical Pest Management* 37, 149–151.

Krishnamurthy, G.V.G., Nagarajan, K. and Chandwani, G.H. (1976) Studies on the control of *Orobanche* on tobacco. *Tobacco Research* 2, 58–62.

Krishnamurthy, G.V.G., Nararajan, K. and Ramji Lal (1977a) Some studies on *Orobanche cernua* Loefl., a parasitic weed on tobacco in India. In: *Program and Abstracts of Papers, Weed Science Conference and Workshop in India, 1977,* pp. 113–114.

Krishnamurthy, G.V.G., Lal, R. and Nagarajan, K. (1977b). Further studies on the effect of various crops on the germination of *Orobanche* seed. *PANS* 23, 206–208.

Krishnamurthy, G.V.G., Lal, R. and Nagarajan, K. (1982) Preliminary studies on the control of *Orobanche* in tobacco. *Indian Journal of Weed Science* 14, 115–120.

Krishnamurthy, G.V.G., Nararajan, K. and Ramji Lal (1987) Studies on longevity and depth of emergence of broomrape in tobacco fields. *Indian Journal of Weed Science* 17 (1985), 38–41.

Krishnamurthy, S. and Rao, U.M. (1976) Control of *Orobanche* through crop rotation. *Indian Farming* 25, 23.

Kuijt, J. (1969) *The Biology of Parasitic Flowering Plants.* University of California Press, Berkeley, 346 pp.

Kuijt, J. (1977) Haustoria of phanerogamic parasites. *Annual Review of Phytopathology* 17, 91–118.

Kuijt, J. (1991) The haustorial interface: what does it tell us? In: Ransom, J.K., Musselman, L.J., Worsham, A.D. and Parker, C. (eds) *Proceedings, 5th International Symposium on Parasitic Weeds, Nairobi, 1991.* CIMMYT, Nairobi, pp. 1–5.

Kukula, S.T. and Masri, H. (1984) Integrated control practices and chemical control of *Orobanche crenata* in faba bean. In: Parker, C., Musselman, L.J., Polhill, R.M. and Wilson, A.K. (eds) *Proceedings, 3rd International Symposium on Parasitic Weeds, Aleppo, Syria, 1984*. ICARDA, Aleppo, pp. 256–261.

Kumar, L.S.S. (1940) *Striga* research in Bombay. *Indian Farming* 1, 609–610.

Kumar, L.S.S. and Abraham, A. (1941) Cytological studies in Indian parasitic plants. I. The cytology of *Striga*. *Indian Academy of Science Proceedings, Section B* 14, 509–516.

Kumar, L.S.S. and Solomon, S. (1940) The influence of light on the germination of species of *Striga*. *Current Science* 9, 541.

Kumar, L.S.S. and Solomon, S. (1941) A list of hosts of some phanerogammic root-parasites attacking economic crops in India. *Indian Academy of Sciences Proceedings, Series B* 13, 151–156.

Kumar, R.M., Kondap, S.M., Rao, P.C. and Reddy, C.N. (1989) Chemical control of *Cuscuta* in mungbean (*Vigna radiata* L.). In: *Abstracts of the Twelfth Asian-Pacific Weed Science Society Conference, Seoul, August 21–26, 1989*, p. 69.

Kumar, U. (1977) Morphogenetic regulation of seed germination in *Orobanche aegyptiaca* Pers. *Canadian Journal of Botany* 55, 2613–2621.

Kunisch, M., Linke, K-H., Richter, O. and Koch, W. (1991) Inclusion of conceptual modeling in studies on the population dynamics of the genus *Striga*. *Angewandte Botanik* 65, 45–57.

Kust, C.A. (1964) Effect of photoperiod on the growth of witchweed. *Agronomy Journal* 56, 93.

Kwak, M.M. (1979) Variation in *Rhinanthus* (Rhinanthoideae; Scrophulariaceae) in relation to bumblebee pollination. In: Musselman, L.J., Worsham, A.D. and Eplee, R.E. (eds) *Proceedings, 2nd International Symposium on Parasitic Weeds, Raleigh, USA, 1979*. North Carolina State University, Raleigh, pp. 48–58.

Labrada, R. and Perez, R. (1988) Non-chemical control methods for *Orobanche ramosa* (in Spanish). *Agrotecnica de Cuba* 20, 35–40.

Lagoke, S.T.O. (1989) *Striga* in Nigeria. In: Robson, T.O. and Broad, H.R. (eds) *Striga – Improved Management in Africa. Proceedings, FAO/OAU All-Africa Government Consultation on Striga Control, Maroua, 1986*. FAO Plant Production and Protection Paper 96, pp. 68–73.

Lagoke, S.J.O., Shebayan, J.Y., Webber, G., Olufajo, O., Elemo, K., Adu, J.K., Emechebe, A.M., Singh, B.B., Zaria, A., Ngawa, L., Olaniyan, G.O., Olafare, S.O. and Adeoti, A.A. (1990) Survey of *Striga* problem and evaluation of *Striga* control methods and packages in crops in the Nigerian savanna. Paper presented at: *Workshop of Pan-African Striga control Network, Ibadan, 1990*. 43 pp.

Lamont, B. (1983) Mineral nutrition of mistletoes. In: Calder M. and Bernhardt, P. (eds) *The Biology of Mistletoes*. Academic Press, Sydney, pp. 185–204.

Lane, H.C. and Kasperbauer, M.J. (1965) Photomorphogenic responses of dodder seedlings. *Plant Physiology* 40, 109–116.

Lane, J.A. and Bailey, J.A. (1991) Resistance of cowpea (*Vigna unguiculata*) to *Striga gesnerioides*. In: Wegmann, K. and Musselman, L.J. (eds) *Proceedings, International Workshop on* Orobanche *Research, Obermarchtal, 1989*. Eberhard-Karls-Universität, Tübingen, pp. 344–350.

Lane, J.A. and Bailey, J.A. (1992) Resistance of cowpea and cereals to the parasitic angiosperm *Striga*. *Euphytica* 63, 85–93.

Lane, J.A., Bailey, J.A. and Terry, P.J. (1991a) An *in-vitro* growth system for studying the parasitism of cowpea (*Vigna unguiculata*) by *Striga gesnerioides*. *Weed Research* 31, 211–217.

Lane, J.A., Kershaw, M.J., Moore, T.H.M., Child, D.V., Terry, P.J. and Bailey, J.A. (1991b) An *in vitro* system to investigate the parasitism of cereals by *Striga* species and resistance of cowpea to *A. vogelii*. In: Ransom, J.K., Musselman, L.J., Worsham, A.D. and Parker, C. (eds) *Proceedings, 5th International Symposium on Parasitic Weeds, Nairobi, 1991*. CIMMYT, Nairobi, pp. 237–240.

Lang, D.J., Sun, R.W., Gu, H.Y. and Ma, G.C. (1989) Trials on the application of butralin to control Chinese dodder (*Cuscuta chinensis*) in soybean fields (in Chinese). *Ningxia Journal of Agro-Forestry Science and Technology* 1, 18–19.

Langston, M.A. and English, T.J. (1981) Corn varietal resistance or tolerance to witchweed (*Striga asiatica*). Abstract in: *Proceedings, 34th Southern Weed Control Conference*, p. 41.

Langston, M.A. and English, T.J. (1990a) Vegetative control of witchweed and herbicide evaluation of techniques. In: Sand, P.F., Eplee, R.E. and Westbrooks, R.G. (eds) *Witchweed Research and Control in the United States of America*. Weed Science Society of America, Champaign, pp. 107–125.

Langston, M.A. and English, T.J. (1990b) Depletion of witchweed (*Striga asiatica*) in the soil. In: Sand, P.F., Eplee, R.E. and Westbrooks, R.G. (eds) *Witchweed Research and Control in the United States of America*. Weed Science Society of America, Champaign, pp. 114–125.

Langston, M.A., English, T.J. and Eplee, R.E. (1991) Herbicides for control of *Striga asiatica* in the USA. In: Ransom, J.K., Musselman, L.J., Worsham, A.D. and Parker, C. (eds) *Proceedings, 5th International Symposium on Parasitic Weeds, Nairobi, 1991*. CIMMYT, Nairobi, pp. 400–406.

LARS (1988) *Improved Weed Management for Developing Countries. Final Report of EEC-Funded Project TSD.A.198*. Long Ashton Research Station, UK, 165 pp.

Last, F.T. (1960a) Incidence of *Striga hermonthica* (Del.) Benth. on two varieties of irrigated sorghum differently manured, spaced and thinned. *Tropical Agriculture* 37, 309–319.

Last, F.T. (1960b) Effect of cultural treatments on the incidence of *Striga hermonthica* (Del.) Benth. and yields of sorghum in the Sudan: field experiments 1957/58. *Annals of Applied Biology* 48, 207–229.

Laycock, D. (1989) Recent results of *Striga* investigations in the CILSS Integrated Pest Management Project with special reference to Niger. In: Robson, T.O. and Broad, H.R. (eds) *Striga – Improved Management in Africa*, FAO Plant Production and Plant Protection Paper 96, FAO Rome, 119–136.

Leach, C.M. (1958) A disease of dodder caused by the fungus *Colletotrichum destructivum*. *Plant Disease Reporter* 42, 827–829.

Lee, K.B. and Lee, C.D. (1989) The structure and development of the haustorium in *Cuscuta australis*. *Canadian Journal of Botany* 67, 2975–2982.

Lee, W.O. and Timmons, F.L. (1956) Evaluation of pre-emergence and stubble treatments for control of dodder in alfalfa seed crops. *Agronomy Journal* 48, 6–10.

Lekic, M.B. (1970) Phytophagous insects observed on parasitic phanerogams of the genera *Orobanche* and *Cuscuta* in 1968. In: Simmonds, F.J. (ed.) *Proceedings, 1st International Symposium on Biological Control of Weeds, Delemont,*

1969. Commonwealth Institute of Biological Control, Miscellaneous Publication No. 1, pp. 21–24.

Leonard, O.A. (1973) Translocation in and between mistletoes and their hosts and the significance of this in relation to chemical control. In: *Proceedings, Symposium on Parasitic Weeds, 1973*. European Weed Research Council, Wageningen, pp. 188–193.

Lichter, J.M. and Berry, A.M. (1991) Establishment of the mistletoe *Phoradendron macrophyllum:* phenology of early stages and host compatibility studies. *Botanical Gazette* 152, 468–475.

Liddy, J. (1983) Dispersal of Australian mistletoes: the Cowiebank study. In: Calder M. and Bernhardt, P. (eds) *The Biology of Mistletoes*. Sydney, Academic Press, pp. 101–116.

Lim, W.C. (1984) Incidence of parasitic higher plants and strangling figs on ornamental trees in Penang, Malaysia. In: Parker, C., Musselman, L.J., Polhill, R.M. and Wilson, A.K. (eds) *Proceedings, 3rd International Symposium on Parasitic Weeds, Aleppo, Syria, 1984*. ICARDA, Aleppo, pp. 1–9.

Linke, K.-H. (1987) Untersuchungen über Keimung und Jugendwicklung von *Striga* und *Orobanche*. *PLITS* 5 (3), 95 pp.

Linke, K.-H. (1992) Biology and control of *Orobanche* in legume crops. *PLITS* 10 (2), 62 pp.

Linke, K.-H. and Saxena, M.C. (1991a) Study on viability and longevity of *Orobanche* seed under laboratory conditions. In: Wegmann, K. and Musselman, L.J. (eds) *Proceedings, International Workshop* on Orobanche *Research, Obermarchtal, 1989*. Eberhard-Karls-Universität, Tübingen, pp. 110–114.

Linke, K.-H. and Saxena, M.C. (1991b) Towards an integrated control of *Orobanche* spp. in some legume crops. In: Wegmann, K. and Musselman, L.J. (eds) *Proceedings, International Workshop on* Orobanche *Research, Obermarchtal, 1989*. Eberhard-Karls-Universität, Tübingen, pp. 248–256.

Linke, K.-H., Sauerborn, J., and Saxena, M.C. (1989) *Orobanche Field Guide*. University of Hohenheim/International Center for Agricultural Research in Dry Areas, 42 pp.

Linke, K.-H., Vorlaender, C. and Saxena, M.C. (1990) Occurrence and impact of *Phytomyza orobanchia* (Diptera: Agromyzidae) on *Orobanche crenata* (Orobanchaceae) in Syria. *Entomophaga* 35, 633–639.

Linke, K.-H., Schnell, H. and Saxena, M.C. (1991a) Factors affecting the seed bank of *Orobanche crenata* in fields under lentil based cropping systems in northern Syria. In: Ransom, J.K., Musselman, L.J., Worsham, A.D. and Parker, C. (eds) *Proceedings, 5th International Symposium on Parasitic Weeds, Nairobi, 1991*. CIMMYT, Nairobi, pp. 321–327.

Linke, K.-H., Saxena, M.C., Sauerborn, J. and Masri, H. (1991b) Effect of soil solarization on the yield of food legumes and on pest control. In: *Proceedings, 1st International Conference on Soil Solarization, Amman, Jordan, 1990*. FAO Plant Production and Protection Paper 109, pp. 139–154.

Linke, K.-H., Scheibel, C., Saxena, M.C. and Sauerborn, J. (1992) Fungi occurring on *Orobanche* spp. and their preliminary evaluation for *Orobanche* control. *Tropical Pest Management* 38, 127–130.

Liu, L. and Li, X. (1988) The geographical distribution of sunflower diseases in

China. *Plant Pathology* 37, 470–474.

Liu, Z.Q., Fer, A. and Lecocq, F.M. (1991) L'imazaquine: un herbicide prometteur pour la lutte curative contre la cuscute (*Cuscuta* spp.) dans les culture de soja (*Glycine max*). *Weed Research* 31, 33–40.

Livingston, W.H., Brenner, M.L. and Blanchette, R.A. (1984) Altered concentrations of abscisic acid, indole-3-acetic acid and zeatin riboside associated with eastern dwarf mistletoe infections on black spruce. In: Hawksworth, F.G. and Scharpf, R.F. (eds) *Biology of Dwarf Mistletoes: Proceedings of the Symposium. Fort Collins, Colorado, 1984.* USDA Forest Service General Technical Report RM-111, pp. 53–61.

Lo, T.T. (1955) NCo310 highly resistant to the root parasite Bunga (*Aeginetia indica*). *Taiwan Sugar* 2, 18–20.

Logan, D.C. and Stewart, G.R. (1991) Role of ethylene in the germination of the hemiparasite *Striga hermonthica*. *Plant Physiology* 97, 1435–1438.

Logan, D.C. and Stewart, G.R. (1992) Germination of the seeds of parasitic angiosperms. *Seed Science Research* 2, 179–190.

Lolas, P.C. (1986) Control of broomrape (*Orobanche ramosa*) in tobacco (*Nicotiana tabacum*). *Weed Science* 34, 427–430.

López, M.E. and Barile, R.L. (1964) The effect of different herbicides in the control of *Aeginetia indica* Roxb. *Philippine Sugar Institute Quarterly* 10, 19–31.

López, M.E. and Barile, R.L. (1965) The effect of different herbicides in the control of *Aeginetia indica* Roxb. *Sugar News* 41, 71–83, 91.

Losner-Goshen, D., Joel, D.M., Mayer, A.M. and Steffens, J.C. (1991) Surface characteristics and internal structure of *Orobanche* seedlings. In: Ransom, J.K., Musselman, L.J., Worsham, A.D. and Parker, C. (eds) *Proceedings, 5th International Symposium on Parasitic Weeds, Nairobi, 1991*. CIMMYT, Nairobi, pp. 356–360.

Lukovin, S. and Kitenko, V. (1974) An important method for dodder control (in Russian). *Sel'skoe Khozysistvo Kirgizii* 4, 20.

Lukovin, S. and Rudenko, V. (1975) Dodder destruction by flaming (in Russian). *Zashchita Rastenii* 6, 47.

Luther, P. and Beck, H. (1987) *Die Mistel. Botanik, Lektine, medizinische Anwendung.* Springer-Verlag, Berlin, 188 pp.

Lynn, D.G. and Chang, M. (1990) Phenolic signals in cohabitation: implications for plant development. *Annual Review of Plant Physiology and Plant Molecular Biology* 41, 497–526.

Mabasa, S. and Lagoke, S.T.O. (1991) *Alectra vogelii* attacks bambara groundnut in Zimbabwe. *Striga Newsletter* 3, 11.

McGrath, H., Shaw, W.C., Jansen, L.L., Lipscomb, B.R., Miller, P.R. and Ennis, W.B. (1957) *Witchweed* (Striga asiatica) *– a new parasitic plant in the United States.* USDA Special Publication 10, 142 pp.

Machado, M.A. and Zetsche, K. (1990) A structural, functional and molecular analysis of plastids of the holoparasites *Cuscuta reflexa* and *C. europaea*. *Planta* 181, 91–96.

McNeely, G.H., Hoffman, E.C., Bayer, D.E. and Foy, C.L. (1966) Control of dodder in alfalfa with DCPA. *California Agriculture* 20, 14–16.

Macqueen, M. (1984) Haustorial initiating activity of several simple phenolic compounds. In: Parker, C., Musselman, L.J., Polhill, R.M. and Wilson, A.K.

(eds) *Proceedings, 3rd International Symposium on Parasitic Weeds, Aleppo, Syria, 1984.* ICARDA, Aleppo, pp. 118–122.

Maganin, I., Lagoke, S. and Adu, A. (1992) Nitrogen reduces *Alectra* emergence but depresses cowpea yield. *Striga Newsletter* 3, 13. FAO, Accra, Ghana.

Maiti, R.K., Ramaiah, K.V., Bisen, S.S. and Chidley, V.L. (1984) A comparative study of the haustorial development of *Striga asiatica* (L.) Kuntze on sorghum cultivars. *Annals of Botany* 54, 447–457.

Mallet, A.I. (1973) Studies in the chemistry of the *Orobanche crenata* germination factor present in the roots of *Vicia faba* and other host plants. In: *Proceedings, Symposium on Parasitic Weeds, Malta, 1973.* European Weed Research Council, Wageningen, pp. 89–97.

Malykhin, I.I. (1974) The time to restore sunflower to its former place in the crop rotation (in Russian). *Zernovya i Maslichnye Kul'tury* 10, 36–37.

Mangnus, E. (1992) Strigol analogues: design, synthesis and biological activity. PhD Thesis. University of Nijmegen, Netherlands, 135 pp.

Mangnus, E. and Zwanenburg, B. (1991) Design and synthesis of germination stimulants for seeds of *Striga* and *Orobanche* spp. In: Wegmann, K. and Musselman, L.J. (eds) *Progress in* Orobanche *Research. Proceedings, International Workshop on Orobanche Research, Obermarchtal, Germany, 1989.* Eberhard- Karls-Universität, Tübingen, pp. 157–165.

Mangnus, E.M. and Zwanenburg, B. (1992) Tentative molecular mechanisms for germination stimulation of *Striga* and *Orobanche* seeds by strigol and its synthetic analogues. *Journal of Agricultural and Food Chemistry* 40, 1066–1070.

Mann, W.F., Grelen, H.E. and Williamson, B.C. (1969) *Seymeria cassioides,* a parasitic weed on slash pine. *Forest Science* 15, 318–319.

Mann, W.F. and Musselman, L.J. (1979) *Senna seymeria* parasitises western conifers. *Economic Botany* 33, 338–339.

Marikovskii, P. and Ivannikov, A.I. (1966) Natural enemies of dodder in Kazakhstan (in Russian). *Zashchita Rastenii* 4, 27–28.

Markham, R.H. (1985) *Possibilities for the Biological Control of* Striga *Species in the Sahel.* Commonwealth Institute of Biological Control, Nairobi, Kenya, 16 pp.

Marshall, J.D. and Ehleringer, J.R. (1990) Are xylem-tapping mistletoes partially heterotrophic? *Oecologia* 84, 244–248.

Marudarajan, D. (1950) Note on *Orobanche cernua* Loefl. *Current Science (India)* 19, 64–65.

Mathiasen, R.L. and Blake, E.A. (1984) Relationships between dwarf mistletoes and habitat types in western coniferous forests. In: Hawksworth, F.G. and Scharpf, R.F. (eds) *Biology of Dwarf Mistletoes: Proceedings of the Symposium. Fort Collins, Colorado, 1984.* USDA Forest Service General Technical Report RM-111, pp. 111–116.

Mathiasen, R.L., Hawksworth, F.G. and Edminster, C.B. (1990) Effects of dwarf mistletoe on growth and mortality of douglas-fir in the southwest. *Great Basin Newsletter* 50, 173–179.

Mathur, R.L. and Bhargava, L.P. (1971) Screening of pearl millet (*Pennisetum typhoides*) varieties against *Striga lutea. Indian Phytopathology* 24, 804.

Matthews, L.J. (1989) Improved package for *Striga* control for the small scale farmer. In: Robson, T.O. and Broad, H.R. (eds) Striga *– Improved Management in Africa. Proceedings, FAO/OAU All-Africa Government Consultation on* Striga *control, Maroua, 1986*. FAO Plant Production and Protection Paper 96, pp. 183–186.

Matthies, D. (1989) The effect of killing the host plant on the parasites *Striga hermonthica* (Scrophulariaceae) and *Orobanche crenata* (Orobanchaceae). *Beiträge zur Biologie Pflanzen* 64, 1–11.

Mazaheri, A., Moazami, N., Vaziri, M. and Moayed-Zadeh, N. (1991) Investigations on *Fusarium oxysporum* a possible biological control of broomrape (*Orobanche* spp.). In: Ransom, J.K., Musselman, L.J., Worsham, A.D. and Parker, C. (eds) *Proceedings, 5th International Symposium on Parasitic Weeds, Nairobi, 1991*. CIMMYT, Nairobi, pp. 93–95.

Mbogo, J.O. and Osoro, M.O. (1991) Reaction of sugarcane clones to *Striga hermonthica* infection in Kenya. In: Ransom, J.K., Musselman, L.J., Worsham, A.D. and Parker, C. (eds) *Proceedings, 5th International Symposium on Parasitic Weeds, Nairobi, 1991*. CIMMYT, Nairobi, pp. 310–312.

Melchior, H. (1941) Die Gattung *Alectra* Thunb. *Notizblätt der Botanischen Garten und Museum Berlin-Dahlem* 15, 423–447.

Melero-Vara, J.M., Dominguez, J. and Fernandez-Martinez, J.M. (1989) Evaluation of differential lines and a collection of sunflower parental lines for resistance to broomrape (*Orobanche cernua*) in Spain. *Plant Breeding* 102, 322–326.

Menke, H.F. (1954) Dodder infestation can halt certified seed production. *Western Feed and Seed* 9, 24, 36–37.

Mesa-García, J. and García-Torres, L. (1982) Effects of bean (*Vicia faba* L.) planting dates on broomrape (*Orobanche crenata* Forsk.) phenology and competition. In: *Proceedings, British Crop Protection Conference, Weeds, Brighton, 1982*, pp. 757–764.

Mesa-García, J. and García-Torres, L. (1984) A competition index for *Orobanche crenata* Forsk. effects on broad bean (*Vicia faba* L.). *Weed Research* 24, 379–382.

Mesa-García, J. and Vasquez-Cobo, A. (1985) Problems in the use of glyphosate in faba beans (*Vicia faba* L.) for the control of *Orobanche crenata* Forsk (in Spanish). *Información Tecnica Economica Agraria* 16, 52–56.

Mesa-García, J., de Haro, A. and García-Torres, L. (1984) Phytotoxicity and yield response of broad bean (*Vicia faba*) to glyphosate. *Weed Science* 32, 445–450.

Michailides, T.J., Ogawa, J.M., Parmeter, J.R. and Yoshimine, S. (1987) Survey for and chemical control of leafy mistletoe (*Phoradendron tomentosum* ssp. *macrophyllum*) on shade trees in Davis, California. *Plant Disease* 71, 533–536.

Midgley, J.J. and Joubert, D. (1991) Mistletoes, their host plants and the effects of browsing by large mammals in Addo Elephant National Park. *Koedoe* 34, 149–152.

Mihajlovic, L. (1986) Results of investigation on *Orobanche* spp. entomofauna in Yugoslavia and the possibility of using insects for biological control. In: ter Borg, S.J. (ed.) *Proceedings, Workshop on Biology and Control of* Orobanche, *Wageningen, 1986*. LH/PVO, Wageningen, pp. 118–126.

Mikhailova, A. and Genov, G. (1978) Paarlan for dodder control on capsicum seedlings (in Bulgarian). *Rastitelna Zashchita* 26, 30–32.

Miller, J.R. and Tocher, R.D. (1975) Photosynthesis and respiration of *Arceuthobium tsugense* (Loranthaceae). *American Journal of Botany* 62, 765–769.

Milvain, H. (1990) Golden dodder. *Farmers' Newsletter, Large Areas* 135, 25–26.

Minko, G. and Fagg, P.C. (1989) Control of some mistletoe species on eucalypts by trunk injection with herbicides. *Australian Forestry* 52, 94–102.

Misra, A., Tosh, G.C., Mohanty, D.C. and Patro, G.K. (1981) Herbicidal and selective effect of pronamide for control of dodder in niger seed. In: *Proceedings, 8th Asian Pacific Weed Science Society Conference, Bangalore, 1981*, pp. 255–257.

Misra, A., Tosh, G.C. and Tosh, B.N. (1982) Preliminary studies on broomrape (*Orobanche* spp.) – a parasitic weed on brinjal. In: *Abstracts of papers, Annual Conference of Indian Society of Weed Science, 1982*, p. 45.

Miusov, I.N. and Bashaeva, E.G. (1968) Mass production of the fungus *Alternaria* spp., a parasite of field dodder (in Russian). *Vestnik sel'Skokhozyaistvennoi Nauki Alma-Ata* 11, 87–90.

Moshalenko, G.P. and Dement'ev, P.E. (1991) Control of *Cuscuta* species (in Russian). *Zashchita Rastenii* 4, 48–49.

Movesian, T.B. and Azarian, K.A. (1971) Pathological changes in cattle poisoned by the dodder *Cuscuta campestris* Yuncker (in Russian). *Biologicheskii Zhurnal Armenii* 24, 67–70.

Mugabe, N.R. (1983) Effect of *Alectra vogelii* Benth. on cowpea (*Vigna unguiculata* (L.) Walp.) 1. Some aspects of reproduction of cowpea. *Zimbabwe Journal of Agricultural Research* 21, 135–147.

Mukumov, K.M. and Faizieva, S.Z. (1977) Results of container trials investigating the broomrape resistance of vegetable and melon crops (in Russian). *Trudy Samarkandskogo Gosudarstvennogo Universiteta imeni A. Navoi*. 336, 73–78.

Müller, F. and Distler, B. (1991) Translocation of glyphosate in the host/parasite system *Vicia faba* and *Orobanche crenata*. In: Wegmann, K. and Musselman, L.J. (eds) *Progress in* Orobanche *Research. Proceedings of the International Workshop on* Orobanche *Research, Obermarchtal, 1989*. Eberhard-Karls-Universität, Tübingen, pp. 226–233.

Müller, S., Hauck, C. and Schildknecht, H. (1992) Germination stimulants produced by *Vigna unguiculata* Walp cv Saunders Upright. *Journal of Plant Growth Regulation* 11, 77–84.

Mumera, L. (1983) *Striga* infestation in maize and sorghum relative to cultivar, herbicidal activity and nitrate. In: *Proceedings, 9th East African Weed Science Society Conference, Nairobi, 1983*, pp. 82–104.

Muromtsev, G.S. and Agnistikova, V.N. (1969) Inhibition of dodder seedling growth by synthetic and biosynthetic fungicides (in Russian). *Doklady Vsesoyuznoi Akademii Sel'skokhozyaistvennykh Nauki* 4, 16–17.

Mushtaque, M. and Baloch, G.M. (1979) Possibility of biological control of mistletoes, *Loranthus* spp., using oligophagous insects from Pakistan. *Entomophaga* 24, 73–81.

Musselman, L.J. (1973) Preliminary studies on the anatomy of seedling roots and haustoria of *Striga* lutea Lour. (Scrophulariaceae). In: *Proceedings, Symposium on Parasitic Weeds, Malta, 1973*. European Weed Research Council, Wageningen, pp. 149–153.

Musselman, L.J. (1979) Fertility and floral patterns in some species of *Striga* (Scrophulariaceae). *Research Reports, National Geographic Society*, 487–491.

Musselman, L.J. (1980) The biology of *Striga, Orobanche* and other root parasitic weeds. *Annual Review of Phytopathology* 18, 463–489.

Musselman, L.J. (1986) Taxonomy of *Orobanche*. In: ter Borg, S.J. (ed.) *Proceedings, Workshop on Biology and Control of* Orobanche, *Wageningen, 1986*. LH/VPO, Wageningen, pp. 2–10.

Musselman, L.J. (ed.) (1987) *Parasitic Weeds in Agriculture. Volume I.* Striga. CRC Press, Boca Raton, 317 pp.

Musselman, L.J. (1991) *Orobanche ramosa* and *Orobanche aegyptiaca* in Flora Palaestina. In: Wegmann, K. and Musselman, L.J. (eds) *Progress in* Orobanche *Research. Proceedings, International Workshop on* Orobanche *Research, Obermarchtal, 1989*. Eberhard-Karls-Universität, Tübingen, pp. 1–5.

Musselman, L.J. and Hepper, F.N. (1986) The witchweeds (*Striga*, Scrophulariaceae) of the Sudan Republic. *Kew Bulletin* 41, 205–221.

Musselman, L.J. and Mann, W.F. (1978) *Root Parasites of Southern Forests*. US Department of Agriculture, Forest Service, General Technical Report SO-20, 76 pp.

Musselman, L.J. and Parker, C. (1981a) Surface features of *Striga* seeds (Scrophulariaceae). *Adansonia* 20, 431–437.

Musselman, L.J. and Parker, C. (1981b) Studies on indigo witchweed, the American strain of *Striga gesnerioides* (Scrophulariaceae). *Weed Science* 29, 594–596.

Musselman, L.J. and Parker, C. (1982) Preliminary host ranges of some strains of economically important broomrapes (Orobanchaceae). *Economic Botany* 36, 270–273.

Musselman, L.J., Worsham, A.B. and Eplee, R.E. (eds) (1979) *Proceedings, 2nd International Symposium on Parasitic Weeds, Raleigh, 1979*. North Carolina State University, Raleigh, 296 pp.

Musselman, L.J., Parker, C. and Dixon, N. (1981) Notes on autogamy and flower structure in agronomically important species of *Striga* (Scrophulariaceae) and *Orobanche* (Orobanchaceae). *Beiträge zur Biologie der Pflanzen* 56, 329–343.

Musselman, L.J., Matteson, P.C. and Fortune, S. (1983) Potential pollen vectors of *Striga hermonthica* (Scrophulariaceae) in West Africa. *Annals of Botany* 51, 859–862.

Musselman, L.J., Bharathalakshmi, Safa, S.B., Knepper, D.A., Mohamed, K.I. and White, C.L. (1991) Recent research on the biology of *Striga asiatica, S. gesnerioides* and *S. hermonthica*. In: Kim, S.K. (ed.) *Combating* Striga *in Africa. Proceedings, International Workshop, Ibadan, 1988*. IITA, Nigeria, pp. 31–41.

Myatt, O.W.D. and Rehbein, C.A. (1954) A new cane-killing weed, *Thesium australe*. *Cane Growers' Quarterly Bulletin* 18, 51–52.

Nag Raj, T.R. (1966) Fungi occurring on witchweed in India. *Commonwealth Institute of Biological Control Technical Communication* 7, 75–80.

Nagar, N. and Sanwal, G.C. (1984) Biochemical aspects of parasitism in *Cuscuta reflexa*: inhibition of cell wall degrading enzymes of *Cuscuta* by non-susceptible plants. In: Parker, C., Musselman, L.J., Polhill, R.M. and Wilson, A.K. (eds) *Proceedings, 3rd International Symposium on Parasitic Weeds, Aleppo, Syria, 1984*. ICARDA, Aleppo, pp. 175–183.

Nassib, A.M. and Hussein, A.H.A. (1989) On-farm *Orobanche* control in faba bean in middle Egypt. *FABIS Newsletter* 24, 11–15.

Nassib, A.M., Ibrahim, A.A. and Saber, H.A. (1979) Broomrape (*Orobanche crenata*) resistance in broad beans – breeding work in Egypt. In: Hawtin, G.C.

and Chancellor, G.J. (eds) *Food Legume Improvement and Development*. IDRC, Ottawa, pp. 133–135.

Nassib, A.M., Ibrahim, A.A. and Khalil, S.A. (1982) Breeding for resistance to *Orobanche*. In: Hawtin, G. and Webb, C. (eds) *Faba Bean Improvement*. Martinus Nijhoff, The Hague, pp. 199–206.

Nassib, A.M., Hussein, A.H.A. and El Reyes, F.M. (1984) Effect of variety, chemical control, sowing date and tillage on *Orobanche* spp. infestation and faba bean yield. *FABIS Newsletter* 10, 11–15.

Nassib, A.M., Hussein, A.H.A. and El-Deeb, M.A. (1991a) Demonstration plots on *Orobanche* control in faba bean in Minia Governate. In: *Nile Valley Regional Program on Cool Season Food Legumes, Proceedings, 3rd Annual Regional Meeting, Giza, 1991*, pp. 9–16.

Nassib, A.M., Saber, H.A., Farrag, H.M. and Seada, A.Y. (1991b) Demonstration plots on *Orobanche* control in faba bean in Fayoum Governate. In: *Nile Valley Regional Program on Cool Season Food Legumes, Proceedings, 3rd Annual Regional Meeting, Giza, 1991*, pp. 45–51.

Nelson, R.R. (1958a) Preliminary studies on the host range of *Striga asiatica*. *Plant Disease Reporter* 42, 376–382.

Nelson, R.R. (1958b) The effect of soil type and soil temperature on the growth and development of witchweed (*Striga asiatica*) under controlled soil temperatures. *Plant Disease Reporter* 42, 152–155.

Nemli, Y. (1987) Preliminary studies on the resistance of some crops to *Cuscuta campestris* Yunck. In: Weber, H.C. and Forstreuter, W. (eds) *Proceedings, 4th International Symposium on Parasitic Flowering Plants, Marburg, Germany, 1987*. Philipps-Universität, Marburg, pp. 591–596.

Nemli, Y., Emiroglu, U. and Kücüdözden, R. (1991) Chemical control of broomrape (*Orobanche ramosa* L.) in tobacco. In: Wegmann, K. and Musselman, L.J. (eds) *Progress in* Orobanche *Research. Proceedings, International Workshop on* Orobanche *Research, Obermarchtal, Germany, 1989*. Eberhard-Karls-Universität, Tübingen, pp. 191–199.

Netzly, D.H., Riopel, J.L., Ejeta, G. and Butler, L.G. (1988) Germination stimulants of witchweed (*Striga asiatica*) from hydrophobic root exudate of sorghum (*Sorghum bicolor*). *Weed Science* 36, 441–446.

Norris, R.S. and Eplee, R.E. (1983) Effect of stimulant plus herbicide on *Striga* germination. In: Ramaiah, K.V. and Vasudeva Rao, M.J. (eds) *Proceedings, 2nd International Workshop on Striga, Oagadougou, 1981*. ICRISAT, Patancheru, pp. 43–46.

Nour, J., Tod, P., Yachmie, P., Panchal, G. and Stewart, G.R. (1984) The role of mannitol in *Striga hermonthica* and other parasitic angiosperms. In: Parker, C., Musselman, L.J., Polhill, R.M. and Wilson, A.K. (eds) *Proceedings, 3rd International Symposium on Parasitic Weeds, Aleppo, Syria, 1984*. ICARDA, Aleppo, pp. 81–89.

Nwoke, F.I.O. and Okonkwo, S.N.C. (1974) Facultative hemi-parasitism in *Buchnera hispida* Buch.-Ham ex D.Don. *Annals of Botany* 38, 993–1002.

Nwoke, F.I.O. and Okonkwo, S.N.C. (1978) Structure and development of the primary haustorium in *A. vogelii* Benth. (Scrophulariaceae). *Annals of Botany* 42, 447–454.

Nwoke, F.I.O. and Okonkwo, S.N.C. (1980) Photocontrol of seed germination in the hemiparasite *Buchnera hispida* (Scrophulariaceae). *Physiologia Plantarum* 49, 388–392.

Obilana, A.T. (1983) *Striga* studies and control in Nigeria. In: Ramaiah, K.V. and Vasudeva Rao, M.J. (eds) *Proceedings, 2nd International Workshop on* Striga, *Oagadougou, 1981*. ICRISAT, Patancheru, pp. 87–96.

Obilana, A.T., Knepper, D and Musselman, L.J. (1987) *Striga* (witchweeds) in Zimbabwe. Paper presented at Fourth Regional SADCC/ICRISAT Sorghum Millet Improvement Program Workshop, Matopos, Zimbabwe, 1987, 28 pp.

Obilana, A.T., de Milliano, W.A.J. and Mbwaga, A.M. (1991) *Striga* research in sorghum and millets in Southern Africa: status and host plant resistance. In: Ransom, J.K., Musselman, L.J., Worsham, A.D. and Parker, C. (eds) *Proceedings, 5th International Symposium on Parasitic Weeds, Nairobi, 1991*. CIMMYT, Nairobi, pp. 435–441.

Oganyan, E.A. (1976) The most susceptible growth stages of dodder to the action of tricothecin (in Russian). *Biologicheskii Zhurnal Armenii* 29, 92–94.

Ogborn, J.E.A. (1972a) The significance of the seasonal pattern of emergence of *Striga hermonthica* Benth. In: Rao, N.G.P. and House, L.R. (eds) *Proceedings, Sorghum in Seventies Workshop, New Delhi, 1972*. ICRISAT, Patancheru, pp. 562–571.

Ogborn, J.E.A. (1972b) The control of *Striga hermonthica* in peasant farming. In: *Proceedings, 11th British Weed Control Conference, Brighton, 1972*, pp. 1068–1077.

Ogborn, J.E.A. (1984) *Striga*: Research priorities with specific reference to agronomy. In: Ayensu, E.S., Doggett, H., Keynes, R.D., Marton-Lefevre, J., Musselman, L.J., Parker, C. and Pickering, A. (eds) Striga *biology and control. Proceedings, International Workshop, Dakar, Senegal, 1983*. ICSU Press, Paris, pp. 195–212.

Ogborn, J.E.A. (1987) *Striga* control under peasant farming conditions. In: Musselman, L.J. (ed.) *Parasitic Weeds in Agriculture. Volume I.* Striga. CRC Press, Boca Raton, pp. 145–158.

Okonkwo, S.N.C. (1966a) Studies on *Striga senegalensis* Benth. I. Mode of host-parasite union and haustorial structure. *Phytomorphology (India)* 16 (1966), 453–463.

Okonkwo, S.N.C. (1966b) Studies on *Striga senegalensis*. II. Translocation of ^{14}C-labeled photosynthate, urea-^{14}C and sulphur-35 between host and parasite. *American Journal of Botany* 53, 142–148.

Okonkwo, S.N.C. (1975) In vitro post-germination growth and development of embryos of *Alectra* (Scrophulariaceae). *Physiologia Plantarum* 34, 378–383.

Okonkwo, S.N.C. (1982) Nutrient factors for shoot development and seedling growth of *Striga gesnerioides* (Willd.) Vatke. *Zeitschrift für Pflanzenphysiologie* 106, 381–389.

Okonkwo, S.N.C. (1991) *In vitro* growth response of cultured germinated seeds of witchweed (*Striga asiatica*). In: Weber, H.C. and Forstreuter, W. (eds) *Proceedings, 4th International Symposium on Parasitic Flowering Plants, Marburg, Germany, 1987*. Philipps-Universität, Marburg, pp. 155–163.

Okonkwo, S.N.C. and Nwoke, F.I.O. (1974) Seed germination in *Buchnera hispida* Buch.-Ham. ex D.Don. *Annals of Botany* 38, 409–417.

Okonkwo, S.N.C. and Nwoke, F.I.O. (1975) Bleach-induced germination and breakage of dormancy of seeds of *Alectra vogelii*. *Physiologia Plantarum* 35, 175–180.

Okonkwo, S.N.C. and Nwoke, F.I.O. (1978) Initiation, development and structure of the primary haustorium in *Striga gesnerioides* (Scrophulariaceae). *Annals of Botany* 42, 455–463.

Okonkwo, S.N.C. and Okafor, D.N. (1988) Stimulation of germination of seeds of root parasitic weeds by strigol analogues GR7 and GR24. In: Okonkwo, S.N.C., Akobundu, I.O. and Ayeni, A.O. (eds) *Proceedings, 3rd Biennial Conference of the West African Weed Science Society, Ibadan, 1985*, pp. 31–41.

Okonkwo, S.N.C. and Raghavan, V. (1982) Studies on the germination of seeds of the root parasites, *A. vogelii* and *S. gesnerioides*. 1. Anatomical changes in the embryos. *American Journal of Botany* 69, 1636–1645.

Olivier, A., Benhamou, N. and Leroux, G.D. (1991a) Cell surface interactions between sorghum roots and the parasitic weed *Striga hermonthica*: cytochemical aspects of cellulose distribution in resistant and susceptible host tissues. *Canadian Journal of Botany* 69, 1679–1690.

Olivier, A., Ramaiah, K.V. and Leroux, G.D. (1991b) Selection of sorghum (*Sorghum bicolor* (L.) Moench) varieties resistant to the parasitic weed *Striga hermonthica* (Del.) Benth. *Weed Research* 31, 219–225.

Orloff, S.B. (1987) Biology of dodder and its control in alfalfa. In: *Proceedings, 39th Annual California Weed Conference*, pp. 161–168.

Orloff, S.B., Vargas, R.N., Cudney, D.W., Canevari, W.M. and Schmierer, J. (1989) Dodder control in alfalfa. *California Agriculture* 43 (4), 30–32.

Osman, M.A., Raju, P.S., Peacock, J.M. (1991) The effect of soil temperature, moisture and nitrogen on *Striga asiatica* (L.) Kuntze seed germination, viability and emergence on sorghum (*Sorghum bicolor* (L.) Moench.) roots under field conditions. *Plant and Soil* 131, 265–273.

Ozenda, P. and Capdepon, M. (1972) Recherches sur les phanérogames parasites. II. Une Scrophulariacée holoparasite méconnue, *S. gesnerioides. Phytomorphology* 22, 314–324.

Panchenko, V.P. (1974) Micro-organisms in the control of Egyptian broomrape parasitising water melons (in Russian). *Mikologia i Fitopathalogya* 8, 122–125.

Panchenko, V.P. (1981) The biological protection of water melons and tomatoes from broomrape in Astrakhan Province (in Russian). *Doklady Vsesoyuznoi Akademii Sel'skokhozyaistvennykh Naukimeni V.I. Lenina* 8, 25–27.

Pant, D.D. and Banerji, R. (1965) Epidermal structure and development of stomata in some Convolvulaceae. *Senkenbergiana Biologica* 46, 155–173.

Parker, C. (1974) Pot experiments with herbicides for the control of *Striga* spp. In: Ransom, J.K., Musselman, L.J., Worsham, A.D. and Parker, C. (eds) *Proceedings, 5th International Symposium on Parasitic Weeds, Nairobi, 1991*. CIMMYT, Nairobi, pp. 152–171.

Parker, C. (1984) The influence of *Striga* species on sorghum under varying nitrogen fertilization. In: Parker, C., Musselman, L.J., Polhill, R.M. and Wilson, A.K. (eds) *Proceedings, 3rd International Symposium on Parasitic Weeds, Aleppo, Syria, 1984*. ICARDA, Aleppo, pp. 90–98.

Parker, C. (1988) Parasitic plants in Ethiopia. *Walia* 11, 21–27.

Parker, C. (1990) Some observations on the prevention of *Striga* seed production by hand-pulling. In: Parker, C. and Rezene Fessehaie (eds) *Proceedings, Ethiopian Weed Science Committee 6th Annual Meeting, Addis Ababa, 1988*. EWSC, Addis Ababa, pp. 83–88.

Parker, C. (1991a) Flower morphology in *Striga hermonthica* and *S. aspera. Haustorium* 26, 4–5.

Parker, C. (1991b) Protection of crops against parasitic weeds. *Crop Protection* 10, 6–22.

Parker, C. and Polniaszek, T.I. (1990) Parasitism of cowpea by *Striga gesnerioides*: variation in virulence and discovery of a new source of host resistance. *Annals of Applied Biology* 116, 305–311.

Parker, C. and Reid, D.C. (1979) Host specificity in *Striga* species – some preliminary observations. In: Musselman, L.J., Worsham, A.D. and Eplee, R.E. (eds) *Supplement to Proceedings, 2nd International Symposium on Parasitic Weeds, Raleigh, USA*. North Carolina State University, Raleigh, pp. 79–90.

Parker, C. and Wilson, A.K. (1983) *Striga*-resistance identified in semi-wild 'shibra' millet (*Pennisetum* sp.). *Mededelingen van de Faculteit Landbouwweteschapen, Rijkuniversiteit, Gent* 48, 1111–1117.

Parker, C., Hitchcock, A-M and Ramaiah, K.V. (1977) The germination of *Striga* species by crop root exudates; techniques for selecting resistant crop cultivars. In: *Proceedings, Asian-Pacific Weed Science Society 6th Conference, Jakarta, 1977*, pp. 67–74.

Parker, C., Musselman L.J., Polhill, R.M. and Wilson, A.K. (eds) (1984) *Proceedings, 3rd International Symposium on Parasitic Weeds, Aleppo, Syria, 1984*. ICARDA, Aleppo, 265 pp.

Parkinson, V. (1985) *Striga*, serious threat to maize production in Africa and research being conducted by IITA. In: *Proceedings, OAU/FAO Workshop on Striga, Yaounde, Cameroon, 1985*, pp. 47–55.

Parkinson, V.O. (1989) A survey of the infestation of crops by *Striga* spp. in Benin, Nigeria and Togo. *Proceedings, Nova Scotian Institute of Science* 39, 1–9.

Patterson, D.T. (1990) Effects of environment on growth and reproduction of witchweed and the ecological range of witchweed. In: Sand, P.F., Eplee, R.E. and Westbrooks, R.G. (eds) *Witchweed Research and Control in the United States*. Weed Science Society of America, Champaign, pp. 68–80.

Pearson, H.H.W. (1913) *The Problem of the Witchweed*. South African Department of Agriculture Publication No. 40, 34 pp.

Pederick, L.A. and Zimmer, W.J. (1961) *The Parasitic Forest Dodder-laurel* Cassytha melantha *R.Br.* Bulletin of the Forestry Commission, Victoria 12, 16 pp.

Pennell, F.W. (1935) The Scrophulariaceae of eastern temperate North America. *Academy Natural Science Philadelphia Monographs* 1, 1–650.

Pesch, C. and Pieterse, A.H. (1982) Inhibition of germination in *Striga* by means of urea. *Experientia* 38, 559–560.

Petrov, D. (1970) A new physiological race of sunflower broomrape (*Orobanche cumana* Wallroth) in Bulgaria (in Bulgarian). In: Kovachevski, I. *et al.* (eds) *Plant Protection in the Service of Agriculture*. Bulgarian Academy of Science Press, Bulgaria, pp. 37–48.

Petzoldt, K. (1979) Bacterial nodules of *Rhizobium leguminosarum* and *Orobanche crenata* germination and penetration of broad beans with an integrated control program. In: Musselman, L.J., Worsham, A.D. and Eplee, R.E. (eds) *Proceedings, 2nd International Symposium on Parasitic Weeds, Raleigh, USA, 1979*. North Carolina State University, Raleigh, pp. 260–268.

Petzoldt, K. and Sneyd, J. (1986) *Orobanche cumana* control by breeding and glyphosate treatment in sunflowers. In: ter Borg, S.J. (ed.) *Proceedings, Workshop on Biology and Control of* Orobanche, *Wageningen, 1986*. LH/VPO, Wageningen, pp. 166–171.

Philcox, D. (1990) *Alectra* Thunb. In: Launert, E. (ed.) *Flora Zambeziaca* 8, Part 2, pp. 85–93.

Phillips, E.J. (1977) Mistletoe on cocoa in Ghana. *Ghana Journal of Agricultural Science* 10, 137–143.

Phillips, E.J. (1991) Germination of *Tapinanthus bangwensis* on different substrates and subsequent development on cocoa. In: Ransom, J.K., Musselman, L.J., Worsham, A.D. and Parker, C. (eds) *Proceedings, 5th International Symposium on Parasitic Weeds, Nairobi, 1991*. CIMMYT, pp. 415–419.

Pieters, R. and Aalders, A.J.G. (1986) The effect of *Vicia faba* sowing density on broomrape attack. *FAO Plant Protection Bulletin* 34, 209–211.

Pieterse, A.H. (1979) The broomrapes (Orobanchaceae) – a review. *Abstracts on Tropical Agriculture* 5, 9–35.

Pieterse, A.H. (1990) *Report of Mission on* Striga *Infestation and Control in Zaire, 4–23 November 1992 for Project UNDP/FAO ZAI 83/113*. Royal Tropical Institute, Amsterdam, 22 pp.

Pieterse, A.H. (1991) The effect of nitrogen fertilizers on the germination of seeds of *Striga hermonthica* and *Orobanche crenata*. In: Wegmann, K. and Musselman, L.J. (eds) *Progress in* Orobanche *Research. Proceedings, International Workshop on* Orobanche *Research, Obermarchtal, 1989*. Eberhard-Karls-Universität, Tübingen, pp. 115–124.

Pieterse, A.H. and Pesch, C.J. (1983) The witchweeds (*Striga* spp.) – a review. *Abstracts on Tropical Agriculture* 9, 9–37.

Pieterse, A.H. and Verkleij, J.A.C. (1991) Effect of soil conditions on *Striga* development – a review. In: Ransom, J.K., Musselman, L.J., Worsham, A.D. and Parker, C. (eds) *Proceedings, 5th International Symposium on Parasitic Weeds, Nairobi, 1991*. CIMMYT, Nairobi, pp. 329–339.

Pieterse, A.H., Ebbers, A.E.H. and Pesch, C.J. (1984) A study on wet dormancy in seeds of *Striga hermonthica* (Del.) Benth. *in vitro*. In: Parker, C., Musselman, L.J., Polhill, R.M. and Wilson, A.K. (eds) *Proceedings, 3rd International Symposium on Parasitic Weeds, Aleppo, Syria, 1984*. ICARDA, Aleppo, pp. 99–107.

Pieterse, A.H., García-Torres, L., Al-Menoufi, O.A., Linke, K-H. and ter Borg, S.J. (1992) Integrated control of the parasitic angiosperm *Orobanche* (broomrape). Paper presented at the International Food Legume Conference II, Cairo, 1992. 9 pp.

Polniaszek, T.I. and Parker, C. (1987) Variation in host specificity of *Alectra vogelii* Benth. (Scrophulariaceae). In: Weber, H.C. and Forstreuter, W. (eds) *Proceedings, 4th International Symposium on Parasitic Flowering Plants, Marburg, Germany, 1987*. Philipps-Universität, Marburg, pp. 613–619.

Polniaszek, T.I., Parker, C. and Riches, C.R. (1991) Variation in the virulence of *Alectra vogelii* populations on cowpea. *Tropical Pest Management* 37, 152–154.

Press, M.C. and Graves, J.D. (1991) Carbon relations of angiosperm parasites and their hosts. In: Wegmann, K. and Musselman, L.J. (eds) *Progress in* Orobanche *Research. Proceedings, International Workshop on* Orobanche *Research, Obermarchtal, 1989*. Eberhard-Karls-Universität, Tübingen, pp. 55–65.

Press, M.C., Shah, N. and Stewart, G.R. (1986) The parasitic habit: trends in metabolic reductionism. In: ter Borg, S.J. (ed.) *Proceedings, Workshop on Biology and Control of* Orobanche, *Wageningen, 1986*. LH/VPO, Wageningen, pp. 96–106.

Press, M.C., Shah, N., Tuohy, J.M. and Stewart, G.R. (1987) Carbon isotope ratios demonstrate carbon flux from C_4 host to C_3 parasite. *Plant Physiology* 85, 1143–1145.

Press, M.C., Graves, J.D. and Stewart, G.R. (1988) Transpiration and carbon acquisition in root hemiparasitic angiosperms. *Journal of Experimental Botany* 39, 1009–1014.

Press, M.C., Nour, J.J., Bebawi, F.F. and Stewart, G.R. (1989) Antitranspirant-induced heat stress in the parasitic plant *Striga hermonthica* – a novel method of control. *Journal of Experimental Botany* 40, 585–591.

Press, M.C., Graves, J.D. and Stewart, G.R. (1990) Physiology of the interaction of angiosperm parasites and their higher plant hosts. *Plant, Cell and Environment* 13, 91–104.

Preston, A.P. (1977) Effects of mistletoe (*Viscum album*) on young apple trees. *Horticultural Research* 17, 33–38.

Pundir, U.P.S. (1981) A note on the biological control of *Scurrula cordifolia* (Wall.) G.Don by another mistletoe in Sivalik Hills (India). *Weed Research* 21, 233–234.

Pustovoit, V.S. (1967) Selection for resistance to broomrape and moth. In: *Handbook of Selection and Seed Growing of Oil Plants*. Kolos, Moscow. (Translated by Israeli Program for Scientific Translations, Jerusalem, 1967), pp. 22–28.

Pustovoit, V.S. (1976) *Selection, Seed Culture and Some Agrotechnical Problems of Sunflower*. Translation from Russian. INSDOC, Delhi.

Puzzilli, M. (1983) Tobacco broomrapes and their control and some useful references to other parasite and host species. *Rivista di Agricultura Subtropicale e Tropicale* 77, 209–248.

Raaimakers, D., Raaimakers, J., ter Borg, S.J., Nassib, A.M. and Pieterse, A.H. (1988). Effect of sowing date on *Orobanche crenata* infestation in Egypt. *FABIS Newsletter* 22, 33–39.

Rabotnov, T.A. (1959) The effect of *Rhinanthus major* Ehrh. on the yield and composition of grasses of the floodland meadows (in Russian). *Moskov. Obsheh. Isp. Prirody B. Otd. Biol.* 64, 105–107.

Racovitza, A. (1973) Research on the susceptibility of various species of *Nicotiana* and of various cultivars of tobacco to *Orobanche ramosa* (in French). In: *Proceedings, Symposium on Parasitic Weeds, Malta, 1973*. European Weed Research Council, Wageningen, pp. 194–204.

Radwan, M.S., Abdalla, M.M.F., Fischbeck, G., Metwally, A.A. and Darwish, D.S. (1988a) Variation in reaction of faba bean lines to different accessions of *Orobanche crenata* Forsk. *Plant Breeding* 101, 208–216.

Radwan, M.S., Abdalla, M.M.F., Fischbeck, G., Metwally, A.A. and Darwish, D.S. (1988b) Selection in faba bean for tolerance to broomrape, *Orobanche crenata* Forsk. *Plant Breeding* 100, 289–298.

Raju, P.S., Osman, M.A., Soman, P. and Peacock, J.M. (1990) Effect of N, P and K on *Striga asiatica* (L.) Kuntze seed germination and infestation of sorghum. *Weed Research* 30, 139–144.

Rakhimov, U.K. (1967) Transpiration and diffusion pressure deficit of broomrape and the host plant. *Soviet Plant Physiology* 14, 631–632.

Ralston, D.M., Riches, C.R. and Musselman, L.J. (1987) Morphology and hosts of three *Striga* species (Scrophulariaceae) in Botswana. *Adansonia* 2, 195–215.

Ramaiah, K.V. (1981) *Striga* resistance breeding. In: *Annual Report 1981, ICRISAT/ Upper Volta Cooperative Programme, ICRISAT, Oagadougou*, pp. D 1–19.

Ramaiah, K.V. (1985a) Hand pulling of *Striga hermonthica* in pearl millet. *Tropical Pest Management* 31, 326–327.

Ramaiah, K.V. (1985b) ICSV 1002 HV – a promising sorghum variety from ICRISAT/Upper Volta. In: *Proceedings, OAU/FAO Workshop on* Striga, *Yaounde, 1985*. FAO, Rome, pp. 83–87.

Ramaiah, K.V. (1987a) Breeding cereal grains for resistance to witchweed. In: Musselman, L.J. (ed.) *Parasitic Weeds in Agriculture. Volume I.* Striga. CRC Press, Boca Raton, pp. 227–242.

Ramaiah, K.V. (1987b) Control of *Striga* and *Orobanche* species – a review. In: Weber, H.C. and Forstreuter, W. (eds) *Proceedings, 4th International Symposium on Parasitic Flowering Plants, Marburg, Germany, 1987*. Philipps-Universität, Marburg, pp. 637–664.

Ramaiah, K.V. (1991) Breeding for *Striga* resistance in sorghum and millet. In: Kim, S.K. (ed.) *Combating* Striga *in Africa. Proceedings, International Workshop, Ibadan, 1988*. IITA, Nigeria, pp. 75–80.

Ramaiah, K.V. and Vasudeva Rao, M.J. (eds) (1983) *Proceedings, 2nd International Workshop on* Striga, *Oagadougou, 1981*. ICRISAT, Patancheru, 129 pp.

Ramaiah, K.V., Parker, C., Vasudeva Rao, M.J. and Musselman, L.J. (1983) Striga *Identification and Control Handbook*. ICRISAT Information Bulletin 15. ICRISAT, Patancheru, 52 pp.

Ramaiah, K.V., Chidley, V.L. and House, L.R. (1990) Inheritance of *Striga* seed-germination stimulant in sorghum. *Euphytica* 45, 33–38.

Ramaiah, K.V., Chidley, V.L. and House, L.R. (1991) A time-course study of early establishment stages of parasitic angiosperm *Striga asiatica* on susceptible sorghum roots. *Annals of Applied Biology* 118 (2), 403–410.

Ransom, J.K. and Njoroge, J. (1991) Seasonal variation in ethylene-induced germination of *Striga hermonthica* in western Kenya. In: Weber, H.C. and Forstreuter, W. (eds) *Proceedings, 4th International Symposium on Parasitic Flowering Plants, Marburg, Germany, 1987*. Philipps-Universität, Marburg, pp. 391–396.

Ransom, J.K., Eplee, R.E. and Langston, M.A. (1990) Genetic variability for resistance to *Striga asiatica* in maize. *Cereal Research Communications* 18, 329–333.

Ransom, J.K., Musselman, L.J., Worsham, A.D. and Parker, C. (eds) (1991) *Proceedings, 5th International Symposium on Parasitic Weeds, Nairobi, 1991*. CIMMYT, Nairobi, 550 pp.

Rao, K.N. and Gupta, K.M. (1981) Chemical control of *Cuscuta* in pulses and other crops. In: *Proceedings, 8th Asian-Pacific Weed Science Society Conference, Bangalore, 1981*, pp. 215–217.

Rao, N.K. and Rao, R.S.N. (1991a) Studies on tolerance of black gram and green gram varieties to dodder (*Cuscuta chinensis* Lam.). In: Weber, H.C. and Forstreuter, W. (eds) *Proceedings, 4th International Symposium on Parasitic*

Flowering Plants, Marburg, Germany, 1987. Philipps-Universität, Marburg, pp. 170–175.

Rao, N.K. and Rao, R.S.N. (1991b) Evaluation of herbicides to control *Cuscuta* spp. in rice fallow black gram. In: Weber, H.C. and Forstreuter, W. (eds) *Proceedings, 4th International Symposium on Parasitic Flowering Plants, Marburg, Germany, 1987*. Philipps-Universität, Marburg, pp. 164–169.

Rao, P.N. and Reddy, A.R.S. (1987) Effect of china dodder on two pulses: green gram and cluster bean – the latter a possible trap crop to manage china dodder. In: Weber, H.C. and Forstreuter, W. (eds) *Parasitic Flowering Plants. Proceedings, 4th International Symposium on Parasitic Seed Plants, Marburg, 1987*. Philipps-Universität, Marburg, pp. 665–674.

Rao, P.N., Reddy, B.V.N. and Raghava Swamy, B.V. (1985a) Seed germination of *Striga angustifolia* in root eluates of 20 sorghum cultivars. In: *Proceedings, 10th Conference, Asian-Pacific Weed Science Society*, pp. 469–473.

Rao, K.N., Rao, A.S. and Reddy, R. (1985b) Effect of different herbicides on control of *Cuscuta* in legumes. *Pesticides* 19, 61–62.

Rao, P.N., Reddy, B.V.N. and Raghava Swamy, B.V. (1986) New hosts of *Striga angustifolia* (Scrophulariaceae). *Bulletin de la Musée Nationale d'Histoire Naturelle, Paris, 4ᵉ Séries* 8, 207–212.

Rapparini, G. (1989a) A new post-emergence weeding method for beet (in Italian). *Informatore Agrario* 45, 109–121.

Rapparini, G. (1989b) Chemical weed control in potatoes: how, when and why (in Italian). *Informatore Agrario* 45, 91–97.

Rath, G.C. and Mohanty, S.S. (1988) Observations of some grass hosts of *Cuscuta chinensis*. *Plant Pathology Newsletter* 6, 35–36.

Raynal-Roques, A. (1991) Diversification in the genus *Striga*. In: Ransom, J.K., Musselman, L.J., Worsham, A.D. and Parker, C. (eds) *Proceedings, 5th International Symposium on Parasitic Weeds, Nairobi, 1991*. CIMMYT, Nairobi, pp. 251–261.

Raynal-Roques, A. and Sallé, G. (1987) *Striga* Lour. (Scrophulariaceae), essai étymologique. *Journal d'Agriculture Traditionelle et de Botanique Appliquée* 34, 47–49.

Reddy, A.R.S. and Rao, P.N. (1987) Cluster bean – a possible herbicidal source for managing china dodder. In: *Proceedings, 11th Conference of the Asian Pacific Weed Science Society, Taipei, 1987*, pp. 265–270.

Reddy, B.V.N. and Rao, P.N. (1980) Colletors on the cotyledons of *in vitro* raised seedlings of witchweed – *Striga asiatica* (L.) Kuntze. *Current Science* 49, 595–597.

Reid, D.C. and Parker, C. (1979) Germination requirements of *Striga* species. In: Musselman, L.J., Worsham, A.D. and Eplee, R.E. (eds) *Proceedings, 2nd International Symposium on Parasitic Weeds, Raleigh, 1979*. North Carolina State University, Raleigh, pp. 202–210.

Rey, L., Sadik, A., Ferf, A. and Renaudin, S. (1991) Trophic relations of the dwarf mistletoe *Arceuthobium oxycedri* with its host *Juniperus oxycedrus*. *Journal of Plant Physiology* 138, 411–416.

Riches, C.R. (1987) The identification of resistance to *Alectra vogelii* Benth. (Scrophulariaceae) in cowpea. In: Weber, H.C. and Forstreuter, W. (eds) *Proceedings, 4th International Symposium on Parasitic Flowering Plants, Marburg,*

Germany, 1987. Philipps-Universität, Marburg, pp. 701–708.

Riches, C.R. (1989a) The biology and control of witchweeds of grain legume and cereal crops in Botswana. In: *Research Projects 1983–1986 – Summaries of the Final Reports. First Programme Science and Technology for Development; Subprogramme: Tropics and Sub-tropical Agriculture*. Commission of the European Communities, Brussels, pp. 318–322.

Riches, C.R. (1989b) *The biology and control of* A. vogelii *Benth. (Scrophulariaceae) in Botswana*. PhD Thesis, University of Reading, UK, 208 pp.

Riches, C.R., de Milliano, W.A.J., Obilana, A.T. and House, L.R. (1987) Witchweeds (*Striga* spp.) of sorghum and pearl millet in the SADCC regions: distribution and control. In: *Proceedings, 3rd Regional Workshop on Sorghums and Millets for Southern Africa, Lusaka, 1988*. SADCC/ICRISAT, Bulawayo, pp. 358–374.

Riches, C.R., Hamilton, K.A. and Parker, C. (1992) Parasitism of *Alectra* species by grain legumes. *Annals of Applied Biology* 121, 361–370.

Richter, A. and Popp, M. (1992) The physiological importance of cyclitols in *Viscum album*. *New Phytologist* 121, 431–438.

Riopel, J.L., Baird, W.V., Chang, M. and Lynn, D.G. (1990) Haustorial development in *Striga asiatica*. In: Sand, P.F., Eplee, R.E. and Westbrooks, R.G. (eds) *Witchweed Research and Control in the United States*. Weed Science Society of America, Champaign, pp. 27–36.

Rivera, J.R. (1953) Control of Bunja ng tubo (*Aeginetia indica* L.) with chemical weedkillers. *Philippine Agriculture* 37, 407–421.

Robbins, K., Johnson, D.W., Hawksworth, F.G. and Nicholls, T.H. (1989) Aerial application of ethephon is ineffective in controlling lodgepole pine dwarf mistletoe. *Western Journal of Applied Forestry* 4, 27–28.

Robinson, E.L. (1960) Growth of witchweed (*Striga asiatica*) as affected by soil types and soil and air temperature. *Weeds* 8, 576–581.

Robinson, E.L. and Dowler, C.C. (1966) Investigations on catch and trap crops to eradicate witchweed (*Striga asiatica*). *Weeds* 14, 275–276.

Robinson, E.L. and Stokes, I.E. (1960) Witchweed: A potential pest of sugar cane in the United States. *Sugar Journal* 23, 25–27.

Robson, T.O. and Broad, H.R. (eds) (1989) Striga – *Improved Management in Africa*. Proceedings of the FAO/OAU All-Africa Government Consultation on *Striga* control, Maroua, Cameroon, 1986. FAO Plant Production and Protection Paper 96, 205 pp.

Rogers, W.E. and Nelson, R.R. (1962) Penetration and nutrition of *Striga asiatica*. *Phytopathology* 52, 1064–1070.

Room, P.M. (1971) Some physiological aspects of the relationship between cocoa (*Theobroma cacao*) and the mistletoe *Tapinanthus bangwensis* (Engl. and Krause) Danser. *Annals of Botany* 35, 169–174.

Room, P.M. (1973) Ecology of the mistletoe *Tapinanthus bangwensis* growing on cocoa in Ghana. *Journal of Ecology* 61, 729–742.

Rubin, B. and Benjamin, A. (1984) Solar heating of soil: involvement of environmental factors in the weed control process. *Weed Science* 32, 138–142.

Rudakov, O.L. (1961) A bio-method for the destruction of dodder (in Russian). *Zashchita Rastenii* 6, 23–24.

Rudakov, O.L. (1963) The first results in the biological control of *Cuscuta* spp. (in

Russian). *Zashchita Rastenii* 8, 25–26.

Safa, S.B., Jones, B.M.G. and Musselman, L.J. (1984) Mechanisms favouring outbreeding in *Striga hermonthica* (Scrophulariaceae). *New Phytologist* 96, 299–305.

Saghir, A.R. (1986) Dormancy and germination of *Orobanche* seeds in relation to control methods. In: ter Borg, S.J. (ed.) *Proceedings, Workshop on Biology and Control of* Orobanche*, Wageningen, 1986*. LH/VPO, Wageningen, pp. 25–34.

Salako, E.A. (1984) Observations on the effect of *A. vogelii* infestation on the yield of groundnut. *Tropical Pest Management* 30, 209–211.

Salem, S.M., Harb, E.M.Z. and El-Shihy, O.M.A. (1991) Host–parasite relationship between *Vicia faba* and *Orobanche crenata* as affected by fast neutron presowing treatments. *Bulletin of Faculty of Agriculture, University of Cairo* 42, 533–536.

Sallé, G. (1991) *Striga* research for west Africa sponsored by the European Economic Community. In: Kim, S.K. (ed.) *Combating* Striga *in Africa. Proceedings, International Workshop, Ibadan, 1988*. IITA, Ibadan, pp. 117–121.

Sallé, G., Dembele, B., Raynal-Roques, A., Hallais, M.F. and Tuquet, C. (1987a) Biological aspects of *Striga* species, pest of food crops. In: Weber, H.C. and Forstreuter, W. (eds) *Proceedings, 4th International Symposium on Parasitic Flowering Plants, Marburg, Germany, 1987*. Philipps-Universität, Marburg, pp. 367–375.

Sallé, G., Raynal-Roques, A. and Tuquet, C. (1987b) Field identification of some Loranthaceae parasitizing *Butyrospermum paradoxum* (Gaertn. f.) Hepper (Sapotaceae). In: Weber, H.C. and Forstreuter, W. (eds) *Proceedings, 4th International Symposium on Parasitic Flowering Plants, Marburg, Germany, 1987*. Philipps-Universität, Marburg, pp. 719–731.

Sallé, G., Hariri, E.B. and Andary, C. (1991a) Some mechanisms involved in resistance of poplar (*Populus* spp.) to mistletoe (*Viscum album* L.). In: Ransom, J.K., Musselman, L.J., Worsham, A.D. and Parker, C. (eds) *Proceedings, 5th International Symposium on Parasitic Weeds, Nairobi, 1991*. CIMMYT, Nairobi, pp. 270–278.

Sallé, G., Hariri, E.B., Jeune, B. and Urech, K. (1991b) A new method to assess the level of resistance of oaks to mistletoe. In: Ransom, J.K., Musselman, L.J., Worsham, A.D. and Parker, C. (eds) *Proceedings, 5th International Symposium on Parasitic Weeds, Nairobi, 1991*. CIMMYT, Nairobi, pp. 525–526.

Samb, P.I. (1992) Contribution a l'étude de la biologie de phanérogames parasites du genre *Striga* et recherche d'une methode de lutte chimiquè utilisant deux herbicides phloeme-mobiles (dicamba et glyphosate). PhD Thesis, Université Cheikh Anta Diop de Dakar, Senegal, 111 pp.

Samb, P.I. and Chamel, A. (1991) Translocation and distribution of two labelled herbicides (C^{14}-glyphosate and C^{14}-dicamba) in plants parasitized by *Striga hermonthica* and *S. gesnerioides*. In: Ransom, J.K., Musselman, L.J., Worsham, A.D. and Parker, C. (eds) *Proceedings, 5th International Symposium on Parasitic Weeds, Nairobi, 1991*. CIMMYT, Nairobi, pp. 420–427.

Samb, P.I. and Chamel, A. (1992) Foliar absorption and translocation of ^{14}C-dicamba into host (pearl millet and cowpea) and parasite plants of the genus *Striga*. *Weed Research* 32, 129–136.

Sand, P.F. (1987) The American witchweed quarantine and eradication program.

In: Musselman, L.J. (ed.), *Parasitic Weeds in Agriculture*. Volume I. Striga. CRC Press, Boca Raton, pp. 207–223.

Sand, P.F. (1990) Growth, reproduction and economic significance of witchweed. In: Sand, P.F., Eplee, R.E. and Westbrooks, R.G. (eds) *Witchweed Research and Control in the United States*. Weed Science Society of America, Champaign, pp. 85–90.

Sand, P.F. and Manley, J.D. (1990) The witchweed eradication program survey, regulatory and control. In: Sand, P.F., Eplee, R.E. and Westbrooks, R.G. (eds) *Witchweed Research and Control in the United States*. Weed Science Society of America, Champaign, pp. 141–150.

Sand, P.F., Eplee, R.E. and Westbrooks, R.G. (eds) (1990) *Witchweed Research and Control in the United States of America*. Weed Science Society of America, Champaign, 154 pp.

Sankaran, T. and Rao, V.P. (1966) Insects attacking witchweed (*Striga*) in India. *Commonwealth Institute of Biological Control Technical Bulletin* 7, 63–73.

Sarpe, N., Haislau, D. and Guta, M. (1973) Research on the chemical control of dodder in lucerne and red clover. In: *Proceedings, Symposium on Parasitic Weeds, Malta, 1973*. European Weed Research Council, Wageningen, pp. 289–295.

Satour, M.M., El-Sherif, E.M., El-Ghareb, L. El-Hadad, S.A. and El-Wakili, H.R. (1991) Achievements in soil solarization in Egypt. In: *Proceedings of the First International Conference on Soil Solarization*. FAO Plant Production and Protection Paper 109, 200–212.

Sauerborn, J. (1991a) The economic importance of the phytoparasites *Orobanche* and *Striga*. In: Ransom, J.K., Musselman, L.J., Worsham, A.D. and Parker, C. (eds) *Proceedings, 5th International Symposium on Parasitic Weeds, Nairobi, 1991*. CIMMYT, Nairobi, pp. 137–143.

Sauerborn, J. (1991b) Possibilities for trap and catch crops in the control of parasitic seed plants (in German). *Gesunde Pflanzen* 43, 118–121.

Sauerborn, J. and Saxena, M.C. (1987) Effect of soil solarization on *Orobanche* spp. infestation and other pests in faba bean and lentil. In: Weber, H.C. and Forstreuter, W. (eds) *Proceedings, 4th International Symposium on Parasitic Flowering Plants, Marburg, Germany, 1987*. Philipps-Universität, Marburg, pp. 733–734.

Sauerborn, J., Wyrwal, G. and Linke, K.-H. (1991) Soil sampling for broomrape seeds in naturally infested fields. In: Wegmann, K. and Musselman, L.J. (eds) *Progress in* Orobanche *Research. Proceedings, International Workshop on* Orobanche *Research, Obermarchtal, Germany, 1989*. Eberhard-Karls-Universität, Tübingen, pp. 35–43.

Saunders, A.R. (1933) Studies in phanerogamic parasitism, with particular reference to *Striga lutea* Lour. *South Africa Department of Agriculture, Science Bulletin* 128, 56 pp.

Saunders, A.R. (1942) Field experiments at Potchefstroom. A summary of investigations conducted during the period 1903–1940. *South Africa Department of Agriculture and Forestry Science Bulletin* 14, 19–21, 49–50.

Sawyer, A.M. (1925) Result of investigations made by the Department of Agriculture, Burma, into the extent of damage caused by a parasitic plant known in Burmese as 'pwinbyu' (*Striga lutea*). *Burma Department of Agriculture Bulletin* 18, 12 pp.

Schaffer, A.A., Jacobsohn, R., Joel, D.M., Elliasi, E. and Fogelman, M. (1991) Effect of broomrape (*Orobanche* spp.) infection on sugar content of carrot roots. *HortScience* 26, 892–893.

Schaffer, B., Hawksworth, F.G., Wullschleger, S.D. and Reid, C.P.P. (1981) The role of cytokinins in host reactions to the dwarf mistletoes (*Arceuthobium*). *Phytopathology* 71, 902.

Scharpf, R.F. (1984) Host resistance to dwarf mistletoe. In: Hawksworth, F.G. and Scharpf, R.F. (eds) *Proceedings, Symposium Biology of Dwarf Mistletoes, Fort Collins, 1984*. USDA Forest Service General Technical Report RM-111, pp. 70–76.

Scharpf, R.F. (1987) Resistance of Jeffrey pine to dwarf mistletoe. In: Weber, H.C. and Forstreuter, W. (eds) *Proceedings, 4th International Symposium on Parasitic Flowering Plants, Marburg, Germany, 1987*. Philipps-Universität, Marburg, pp. 745–753.

Scharpf, R.F. and Hawksworth, F.G. (1974) *Mistletoes on Hardwoods in the United States*. USDA Forest Service Forest Pest Leaflet 147, 7 pp.

Scharpf, R.F. and Parmeter, J. (eds) (1978) *Proceedings of the Symposium on Dwarf Mistletoe Control through Forest Management, Berkeley, California 1978*. USDA Forest Service General Technical Report PSW-31, 190 pp.

Scharpf, R.F. and Roth, L.F. (1991) Dwarf mistletoe resistant ponderosa pines in Oregon. In: Ransom, J., Musselman, L.J., Worsham, A.D. and Parker, C. (eds) *Proceedings, 5th International Symposium on Parasitic Weeds, Nairobi, 1991*. CIMMYT, Nairobi, pp. 519–522.

Schauer, R.L. (1969) Control of dodder in potatoes with chlorthal dimethyl. In: *Proceedings, 23rd Northeastern Weed Control Conference*, pp. 95–98.

Schewe, A. and Staniforth, R. (1981) Effect of leachates from the shoots of red bartsia (*Odontites verna*) on four crops species. *Canadian Journal of Botany* 59, 421–423.

Schmitt, U., Schlüter, K. and Boorsma, P.A. (1979) Chemical control of *Orobanche crenata* in broad beans. *FAO Plant Protection Bulletin* 27, 88–91.

Selleck, G.W., Greider, R.S. and Creighton, J.F. (1979) Herbicides for dodder control in potatoes. In: *Proceedings, Northeastern Weed Science Society Conference* 33, pp. 191–195.

Seth, J.N. (1958) Comparative effect of certain herbicides on bandha and its hosts. *Science and Culture* 23, 424–426.

Severova, E. (1991) New taxonomic characters in the genus *Cuscuta* (Cuscutaceae). In: Weber, H.C. and Forstreuter, W. (eds) *Proceedings, 4th International Symposium on Parasitic Flowering Plants, Marburg, Germany, 1987*. Philipps-Universität, Marburg, pp. 514–518.

Shah, N., Smirnoff, N and Stewart, G.R. (1987) Photosynthesis and stomatal characteristics of *Striga hermonthica* in relation to its parasitic habit. *Physiologia Plantarum* 69, 699–703.

Shalom, N.G., Jacobsohn, R. and Cohen, Y. (1988) Effect of broomrape (Orobanchaceae) on sunflower yield. Abstract of paper presented at the 10th Conference of the Weed Science Society of Israel. *Phytoparasitica* 16, 375.

Shalom, N.G., Cohen, Y. and Jacobsohn, R. (1990) Ultrastructural differences in roots of sunflower cultivars resistant and susceptible to *Orobanche*. Abstract of paper presented at the 10th Conference of the Weed Science Society of Israel. *Phytoparasitica* 18, 249.

Sharma, S.L., Rao, D. and Trivedi, K.N. (1954) New hosts of three species of *Striga*. In: *Proceedings, Conference of Sugarcane Research and Development Workers Indian Union* 2, p. 75.

Shchory, Y. (1983) The yield potential of Indian *Brassica* varieties under Mediterranean conditions. Abstract in: *Proceedings, 6th International Rapeseed Conference*, p. 281.

Sherif, A.M. and Mohammed Gelma (1992) *Orobanche* control research in Nura Era Horticulture Enterprise. In: Rezene Fessehaie and Parker, C. (eds) *Problems and Control of Parasitic Weeds in Ethiopia, Proceedings, 2nd Ethiopian Weed Science Workshop, Addis Ababa, 1988*. Ethiopian Weed Science Committee, Addis Ababa, pp. 56–60.

Sherif, A.M. and Parker, C. (1988) Effect of nitrogen and organic matter on *Sorghum bicolor* Moench. and *Striga hermonthica* Benth. *Ethiopian Journal of Agricultural Science* 8, 19–35.

Sherif, A.M., Mohammed Gelma and Parker, C. (1989) Control of *Orobanche cernua* by soil fumigation at Era Horticulture Enterprise. In: Parker, C. and Rezene Fessehaie (eds) *Proceedings, Ethiopian Weed Science Committee 6th Annual Meeting, Addis Ababa, 1988*. Ethiopian Weed Science Committee, Addis Ababa, pp. 67–74.

Shinkarenko, V.A. (1982) A weevil and the control of dodder (in Russian). *Zashchita Rastenii* 7, 30–31.

Shlevin, E. and Golan, D. (1982) Selective control of dodder in carrots (in Hebrew). *Phytoparasitica* 10, 267.

Singh, B. (1958) Effect of temperature on different concentrations of diesel oil sprays suited to kill the bandha parasite (*Dendrophthoe falcata* (L.f.) Ettings). *Horticultural Advances* 2, 68–71.

Singh, B.B. and Emechebe, A.M. (1990) Inheritance of *Striga* resistance in cowpea genotype B301. *Crop Science* 30, 879–881.

Singh, B.B. and Emechebe, A.M. (1991) Breeding for resistance to *Striga* and *Alectra* in cowpea. In: Ransom, J.K., Musselman, L.J., Worsham, A.D. and Parker, C. (eds) *Proceedings, 5th International Symposium on Parasitic Weeds, Nairobi, 1991*. CIMMYT, Nairobi, pp. 303–305.

Singh, B.B. and Emechebe, A.M. (1992) Breeding for resistance to *Striga* and *Alectra* in cowpea. *IITA Research* 4, 5–8.

Singh, J.N., Rai, T.B. and Singh, J.N. (1972) Studies on the physiology of host–parasite relationship in *Orobanche*. III. Carbohydrate and nitrogen metabolism of host and parasite. *Physiologia Plantarum* 27, 347–353.

Singh, P. and Carew, G.C. (1989) Impact of eastern dwarf mistletoe on black spruce forests of Newfoundland. *European Journal of Forest Pathology* 19, 305–322.

Sitkin, R.S. (1976) Parasite–host interactions of field dodder (*Cuscuta campestris*). MSc Thesis, Cornell University. 64 pp.

Slater, C.H., Dawson, J.H., Furtick, W.R. and Appleby, A.P. (1969) Effects of chlorpropham vapors on dodder seedlings. *Weed Science* 17, 238–241.

Smith, C.E., Dudley, M.W. and Lynn, D.G. (1990) Vegetative/parasitic transition: control and plasticity in *Striga* development. *Plant Physiology* 93, 208–215.

Smith, P.L. and Stewart, G.R. (1987) *Striga gesnerioides* (Willd.) Vatke (*S. orobanchoides* Benth.) (Scrophulariaceae): haustorial ontogeny and a role for ergastic substances in the maintenance of physiological integrity. In: Weber, H.C. and Forstreuter, W. (eds) *Proceedings, 4th International Symposium on Parasitic Flowering Plants, Marburg, Germany, 1987*. Philipps-Universität, Marburg, pp. 763–774.

Smith, R.B. and Funk, A. (1980) Assessing *Nectria macrospora* as a biological control agent for hemlock dwarf mistletoe. *Bi-monthly Research Notes* 36, 3.

Smith, S. (1991) The stomatal behaviour of *Striga hermonthica*. In: Wegmann, K. and Musselman, L.J. (eds) *Progress in* Orobanche *Research. Proceedings, International Workshop on* Orobanche *Research, Obermarchtal, 1989*. Eberhard-Karls- Universität, Tübingen, pp. 329–338.

Sobrino-Vesperinas, E. (1985) Search for resistance to *Orobanche ramosa* L. in rapeseed. *Cruciferae Newsletter* 10, 120–121.

Southwood, O.R. (1971) The effect of superphosphate application, 2,4-DB and grazing on broomrape (*Orobanche minor* Sm.) in a subterranean clover pasture. *Weed Research* 11, 240–246.

Sprich, H., Sauerborn, J. and Koch, W. (1990) The solarizing effect of sprayable films (in German). *Zeitschrift für Planzenkrankheit und Pflanzenschutz* 12, 455–461.

Srivastava, H.S. and Chauhan, J.S. (1977) Seed germination, seedling growth and nitrogen and pigment concentration in dodder as affected by inorganic nitrogen. *Zeitschrift für Pflanzenphysiologie* 84, 391–398.

Stangle, C.M. and Musselman, L.J. (1981) *Some Growth Aspects of* Seymeria cassioides. Research Note SO 276, USDA Forest Service, Southern Forest Experiment Station, 3 pp.

Stevens, R.E. and Hawksworth, F.G. (1984) Insect– dwarf mistletoe associations: An update. In: Hawksworth, F.G. and Scharpf, R.F. (eds) *Biology of Dwarf Mistletoes: Proceedings of the Symposium. Fort Collins, Colorado, 1984*. USDA Forest Service General Technical Report RM-111, pp. 94–101.

Stewart, G.R. and Orebamjo, T.O. (1980) Nitrogen status and nitrate reductase activity of the parasitic angiosperm *Tapinanthus bangwensis* (Engl. and K. Krause) Danser growing on different hosts. *Annals of Botany* 45, 587–589.

Stewart, G.R. and Press, M.C. (1990) The physiology and biochemistry of parasitic angiosperms. *Annual Review of Plant Physiology and Plant Molecular Biology*, 41, 127–151.

Stewart, G.R., Nour, J., Macqueen, M. and Shah, N. (1984) Aspects of the biochemistry of *Striga*. In: Ayensu, E.S., Doggett, H., Keynes, R.D., Marton-Lefevre, J., Musselman, L.J., Parker, C. and Pickering, A. (eds), Striga *Biology and Control. Proceedings, International Workshop, Dakar, Senegal, 1983*. ICSU Press, Paris, pp. 161–178.

Stojanovic, D. and Mijàtovic, K. (1973) Distribution, biology and control of *Cuscuta* spp. in Yugoslavia. In: *Proceedings, Symposium on Parasitic Weeds, Malta, 1973*. European Weed Research Council, Wageningen, pp. 269–279.

Stojsin, V., Maric, A. and Jocic, B. (1991) Harmfulness of *Cuscuta campestris* Yunck. on sugar beet under varying mineral nutrition. *Plant Protection* (*Zastita Bilja*) 42, 357–363.

Strasser, E.G. (1988) Studies on the use of dry heat to decontaminate niger seed (*Guizotia abyssinica*) infested with dodder seeds (*Cuscuta* sp.). *Seed Science and Technology* 16, 501–505.

Sunderland, N. (1960) Production of the *Striga* and *Orobanche* germination stimulants by maize roots. I. The number and variety of stimulants. *Journal of Experimental Botany* 11, 236–245.

Takabayashi, M., Sato, H., Furuya, T. and Hashimoto, A. (1981) Distribution of dodder spp. and their parasitic damage caused by dodder. *Japan Pesticide Information* 38, 3–9.

Tashima, Y., Okamoto, K., Tanaka, Y. and Kuroki, S. (1972) Flower formation of *Aeginetia indica* Linn. in relation to the photperiodic treatment of the host plants (in Japanese). *Memoirs of the Faculty of Agriculture, Kagoshima University* 8, 121–125.

Taye Teferedegn (1973) Stimulation of *Striga* (Benth.) germination with sorghum root exudate. Master of Technology Thesis, Brunel University, Uxbridge, UK, 33 pp.

Tchalakov, C., Bozoukov, C. and Georgieva, I. (1991) Provocative method for broomrape (*Orobanche* sp.) control in tobacco – test results in field and production conditions. In: Wegmann, K. and Musselman, L.J. (eds) *Progress* in Orobanche *Research. Proceedings, International Workshop on* Orobanche *Research, Obermarchtal, August 1989*. Eberhard-Karls-Universität, Tübingen, pp. 234–238.

ter Borg, S.J. (ed.) (1986a) *Proceedings, Workshop on Biology and Control of* Orobanche, *Wageningen, 1986*. LH/PVO, Wageningen, vii + 206 pp.

ter Borg, S.J. (1986b) Effects of environmental factors on *Orobanche*-host relationships: a review and some recent results. In: ter Borg, S.J. (ed.) *Proceedings, Workshop on Biology and Control of* Orobanche, *Wageningen, 1986*. LH/VPO, Wageningen, pp. 57–69.

ter Borg, S.J. and van Ast, A. (1991) Soil moisture, root architecture and broomrape (*Orobanche crenata*) infestation in faba bean (*Vicia faba*). In: Wegmann, K. and Musselman, L.J. (eds), *Progress in* Orobanche *Research. Proceedings, International Workshop on* Orobanche *Research, Obermarchtal, 1989*. Eberhard-Karls-Universität, Tübingen, pp. 278–292.

ter Borg, S.J. and Bastiaans, J.C. (1973) Host–parasite relations in *Rhinathus serotinus* I. The effect of growth conditions and host: a preliminary report. In: *Proceedings, Symposium on Parasitic Weeds, Malta, 1973*. European Weed Research Council, Wageningen, pp. 236–246.

Thalouarn, P., Arnaud, M.C., Theodet, C. and Rey, L. (1991) Cytological, biochemical and genetic aspects of carbon fixation in *Striga hermonthica* and *Striga gesnerioides*. In: Ransom, J.K., Musselman, L.J., Worsham, A.D. and Parker, C. (eds) *Proceedings, 5th International Symposium on Parasitic Weeds, Nairobi, 1991*. CIMMYT, Nairobi, pp. 51–57.

Timson, S.D. (1945) Witchweed demonstration farm. Final progress report: December, 1944. *Rhodesia Agriculture Journal* 42, 404–409.

Torell, P.J. (1966) *Weed Control in Carrot Seed Fields*. Idaho University Current Information Series 23, 2 pp.

Tosh, G.C. and Patro, G.K. (1975) Control of dodder (*Cuscuta chinensis* Damk.) in niger (*Guizotia abyssinica* (L.f.) Cass.) with chlorpropham. *Weed Research* 15, 207–209.

Tosh, G.C., Patro, G.K. and Misra, A. (1978) Control of *Cuscuta* in niger seed. *Pest Articles and News Summaries* 24, 167–171.

Trenchev, G. (1981) The possibility of using *Phytomyza orobanchia* Kalt. for the control of broomrape (in Bulgarian). *Rasteniev"dni Nauki* 18, 112–119.

Trujillo, B. and Ordosgoitti, A. (1983) *Alectra fluminensis*, nuevo genero y especie para Venezuela. *Ernstia* 15, 4–12.

Truscott, F.H. (1958) On the regeneration of new shoots from isolated dodder haustoria. *American Journal of Botany* 45, 169–177.

Tsivion, Y. (1978a) Possible role of cytokinins in non-specific recognition of a host and in early growth of haustoria in the parasitic plant, *Cuscuta campestris*. *Botanical Gazette* 139, 27–31.

Tsivion, Y. (1978b) Loading of assimilates and some sugars into the translocation system of *Cuscuta*. *Australian Journal of Plant Physiology* 5, 851–857.

Tuohy, J.M. and Choinski, J.S. (1991) Water relations and photosynthesis in the mistletoe *Tieghemia bolusii* and its host tree *Diplorhynchus condylocarpon*. In: Ransom, J.K., Musselman, L.J., Worsham, A.D. and Parker, C. (eds) *Proceedings, 5th International Symposium on Parasitic Weeds, Nairobi, 1991*. CIMMYT, Nairobi, pp. 232–236.

Tuohy, J.M., Smith, E.A. and Stewart, G.R. (1986) The parasitic habit: trends in morphological and ultrastructural reductionism. In: ter Borg, S.J. (ed.) *Proceedings, Workshop on Biology and Control of* Orobanche, *Wageningen, 1986*. LH/VPO, Wageningen, pp. 86–95.

Tuquet, C. and Sallé, G. (1991) Biochemical features of chloroplasts isolated from hemiparasitic angiosperms. In: Weber, H.C. and Forstreuter, W. (eds) *Proceedings, 4th International Symposium on Parasitic Flowering Plants, Marburg, Germany, 1987*. Philipps-Universität, Marburg, pp. 262–269.

Tutin, T.G., Heywood, V.H., Burges, N.A., Moore D.M., Valentine, D.H., Walters, S.M. and Webb, D.A. (1972) *Flora Europaea, Volume 3: Diapensiaceae to Myoporaceae*. Cambridge University Press, Cambridge, 370 pp.

Tyurebaev, S.S. (1977) The use of the dodder gall beetle for the biological control of field dodder (in Russian). *Vestnik sel'skokhozyaistvennoi Nauki Kazakhastana* 20, 116–117.

Urech, K. (1987) Accumulation of arginine in *Viscum album* L. In: Weber, H.C. and Forstreuter, W. (eds) *Proceedings, 4th International Symposium on Parasitic Flowering Plants, Marburg, Germany, 1987*. Philipps-Universität, Marburg, pp. 781–788.

Uzunov, M., Benkov, B. and Bogdanovski, A. (1974) Methods for improving common bentgrass (*Agrostis vulgaris*) meadows in the western part of the Stara-Planina mountain range (in Bulgarian). *Rasteniev"dni Nauki* 11, 133–140.

Vail, S.L., Dailey, O.D., Connick, W.J. and Pepperman, A.B. (1985) Strigol synthesis and related structure-bioactivity studies. In: A.C.Thompson (ed.) *The Chemistry of Allelopathy. Biochemical Interactions Among Plants*. American Chemical Society Symposium Series 268, pp. 445–456.

Vail, S.L., Dailey, O.D., Blanchard, E.J., Pepperman, A.B. and Riopel, J.L. (1990a) Terpenoid precursors of strigol as seed germination stimulants of broomrape (*Orobanche ramosa*) and witchweed (*Striga asiatica*). *Journal of Plant Growth Regulation* 9, 77–83.

Vail, S.L., Eplee, R.E. and Norris, R.S. (1990b) Synthesis and testing of strigol and strigol analogues. In: Sand, P.F., Eplee, R.E. and Westbrooks, R.G. (eds) *Witchweed Research and Control in the United States*. Weed Science Society of America, Champaign, pp. 43–46.

Vallance, K.B. (1950) Studies on the germination of the seeds of *Striga hermonthica*. I. The influence of moisture-treatment, stimulant-dilution and after-ripening on germination. *Annals of Botany* 14, 348–363.

Vallance, K.B. (1951a) Studies on the germination of the seeds of *Striga hermonthica*. II. The effect of the stimulating solution on seed respiration. *Journal of Experimental Botany* 2, 31–40.

Vallance, K.B. (1951b) Studies on the germination of the seeds of *Striga hermonthica*. III. On the nature of pretreatment and after-ripening. *Annals of Botany* 15, 109–128.

Vallance, K.B. (1952) The germination of the seeds of *Rhinanthus crista-galli*. *Annals of Botany* 16, 409–420.

van der Goot, V.S. (1989) The broomrape root fly, *Chyliza extenuata*, found again in the Netherlands (in Dutch). *Entomologische Berichten* 49, 42–43.

van Hezewijk, M.J., Pieterse, A.H., Saxena, M.C. and ter Borg, S.J. (1987) Relationship between sowing date and *Orobanche* (broomrape) development on faba bean (*Vicia faba* L.) and lentil (*Lens culinaris* Medikus) in Syria. In: Weber, H.C. and Forstreuter, W. (eds) *Proceedings, 4th International Symposium on Parasitic Flowering Plants, Marburg, Germany, 1987*. Philipps-Universität, Marburg, pp. 377–390.

van Hezewijk, M.J., Verkleij, J.A.C. and Pieterse, A.H. (1991) Temperature dependence of germination in *Orobanche crenata*. In: Wegmann, K. and Musselman, L.J. (eds) *Proceedings, International Workshop on* Orobanche *Research, Obermarchtal, 1989*. Eberhard-Karls-Universität, Tübingen, pp. 125–133.

Vasilakakis, C.B., Vezirtjoglou, R.M. and Hadzistavros, C.S. (1988) The effect of polydynamic agrochemicals on the broomrape (*O. ramosa* L.) of tobacco in the field. *Bulletin d'Information CORESTA* 8, 146–147.

Vasudeva Rao, M.J. and Musselman, L.J. (1987) Host specificity in *Striga* spp. and physiological 'strains'. In: Musselman, L.J. (ed.) *Parasitic Weeds in Agriculture. Volume I*: Striga. CRC Press, Boca Raton, pp. 13–25.

Vasudeva Rao, M.J., Jayaswal, M.L. and Ranjit, J.D. (1988) Survey and assessment of *Orobanche* problem in brassica oilcrops in Nepal. *Oil Crops Newsletter* 5, 41–44.

Vasudeva Rao, M.J., Chidley, V.L. and House, L.R. (1989a) Estimates of grain yield loss caused in sorghum (*Sorghum bicolor* (L.) Moench) by *Striga asiatica* (L.) Kuntze obtained using the regression approach. *Agriculture, Ecosystems and Environment* 25, 139–149.

Vasudeva Rao, M.J., Vaidya, P.K., Chidley, V.L., Mukuru, S.Z. and House, L.R. (1989b) Registration of 'ICSV 145' *Striga asiatica* resistant sorghum cultivar. *Crop Science* 29, 488–489.

Verkleij, J.A.C., Egbers, W.S. and Pieterse, A.H. (1991a) Allozyme variations in populations of *Orobanche crenata* Forsk. from Syria. In: Wegmann, K. and Musselman, L.J. (eds) *Progress in* Orobanche *Research. Proceedings, International Workshop on* Orobanche *Research, Obermarchtal, 1989*. Eberhard-Karls-Universität, Tübingen, pp. 304–317.

Verkleij, J.A.C., Koevoets, P., López-Granados, F., Egbers, W.S., García-Torres, L. and Pieterse, A.H. (1991b) Genetic variation in populations of *Orobanche crenata* from Spain. In: Ransom, J.K., Musselman, L.J., Worsham, A.D. and Parker, C. (eds) *Proceedings, 5th International Symposium on Parasitic Weeds, Nairobi, Kenya 1991*. CIMMYT, Nairobi, pp. 462–469.

Vinogradov, V.A., Mironov, E.K., Strelyaeva, N.I. and Sarychev, Y.F. (1981) Breeding tobacco for resistance to *Orobanche* (in Russian). *Tabak* 4, 55–57.

Visser, J.H. (1975a) Germination stimulants of *Alectra vogelii* Benth. seed. *Zeitschrift für Planzenphysiologie* 75, 464–469.

Visser, J.H. (1975b) Production of *A. vogelii* Benth. germination stimulants by seed-borne fungi. *Zeitschrift für Planzenphysiologie* 75, 365–368.

Visser, J.H. (1978) The biology of *Alectra vogelii* Benth., an angiospermous root parasite. *Beiträge zür Chemischen Kommunikation in Bio- und Okosystemen, Witzenhausen* 1978, 279–294.

Visser, J.H. (1981) *South African Parasitic Flowering Plants*. Juta, Cape Town, 177 pp.

Visser, J.H. (1987) The susceptibility of some sugar cane cultivars to witchweed (*Striga asiatica* (L.) Kuntze). In: Weber, H.C. and Forstreuter, W. (eds) *Proceedings, 4th International Symposium on Parasitic Flowering Plants, Marburg, Germany, 1987*. Philipps-Universität, Marburg, pp. 789–795.

Visser, J.H. and Beck, B. (1989) The resistance of guar (*Cyamopsis tetragonoloba* (L.) Taub.) to attack by the root parasitic *Alectra vogelii* Benth. *South African Journal of Plant and Soil* 6, 124–128.

Visser, J.H. and Botha, P.J. (1974) Chromatographic investigation of the *Striga* seed germination stimulant. *Zeitschrift für Pflanzenphysiologie* 72, 352–358.

Visser, J.H. and Johnson, A.W. (1982) The effect of certain strigol analogues on the seed germination of *Alectra*. *South African Journal of Botany* 1, 75–76.

Visser, J.H. and Wentzel, L.F. (1980) Quantitative estimation of *Alectra* and *Striga* seed in soil. *Weed Research* 20, 77–81.

Visser, J.H., Dörr, I. and Kollmann, D. (1977) On the parasitism of *A. vogelii* Benth. (Scrophulariaceae) 1. Early development of the primary haustorium and initiation of the stem. *Zeitschrift für Pflanzenphysiologie* 84, 213–222.

Visser, J.H., Dörr, I. and Kollman, D. (1990) Compatibility of *Alectra vogelii* with different leguminous host species. *Journal of Plant Physiology*, 135, 737–745.

Vogt, W., Sauerborn, J. and Honisch, M. (1991) *Striga hermonthica* distribution and infestation in Ghana and Togo on grain crops. In: Ransom, J.K., Musselman, L.J., Worsham, A.D. and Parker, C. (eds) *Proceedings, 5th International Symposium on Parasitic Weeds, Nairobi, 1991*. CIMMYT, Nairobi, pp. 372–377.

Vranceanu, A.V., Tudor, V.A., Stoenescu, F.M. and Pirvu, N. (1981) The evolution of virulence in the parasite *Orobanche cumana* Wallr. and the corresponding resistance genes in sunflower (in Romanian). *Analele Institutului de Cercetari pentru Cereale si Plante Tehnice, Fundulea* 48, 37–43.

Wan, F.H. and Wang, R. (1991) Achievements of biological control in the world and its prospect in China (in Chinese). *Chinese Journal of Biological Control* 7, 81–87.

Wang, R. (1986) Current status and perspectives of biological weed control in China (in Chinese). *Chinese Journal of Biological Control* 2, 173–177.

Watt, J.M. and Breyer-Brandwijk, M.G. (1962) *The Medicinal and Poisonous Plants of Southern and Eastern Africa*, 2nd edition. Livingstone, Edinburgh, 1457 pp.

Watt, W.L. (1936) Control of *Striga* weed in Nyanza Province, Kenya. *East African Agricultural Journal* 1, 320–322.

Weber, H.C. and Forstreuter, W. (eds) (1987) *Parasitic Flowering Plants. Proceedings, 4th International Symposium on Parasitic Flowering Plants, Marburg, Germany, 1987*. Philipps-Universität, Marburg, 848 pp.

Weeraratna, W.G. (1959) The ecology and biology of parasitism of the Loranthaceae of Ceylon. In: Harper, J.L. (ed.) *The Biology of Weeds. Proceedings, British Ecological Society Symposium No. 1*. Blackwell, Oxford, pp. 189–202.

Wegmann, K. (1986) Biochemistry of osmoregulation and possible biochemical reasons of resistance against *Orobanche*. In: ter Borg, S.J. (ed.) *Proceedings, Workshop on Biology and Control of* Orobanche, *Wageningen, 1986*. LH/VPO, Wageningen, pp. 107–113.

Wegmann, K. and Musselman, L.J. (eds) (1991) *Progress in* Orobanche *Research. Proceedings, International Workshop on* Orobanche *Research, Obermarchtal, 1989*. Eberhard-Karls-Universität, Tübingen, 360 pp.

Wegmann, K., von Elert, E., Harloff, H.-J. and Stadler, M. (1991) Tolerance and resistance to *Orobanche*. In: Wegmann, K. and Musselman, L.J. (eds) *Progress in* Orobanche *Research. Proceedings, International Workshop on* Orobanche *Research, Obermarchtal, 1989*. Eberhard-Karls-Universität, Tübingen, pp. 318–321.

Weinert, H. and Barckhaus, R.H. (1975) Fortified synthesis of cutin at the contact zones between *Cuscuta odorata* and *Pelargonium zonale*. *Cytobios* 13, 17–22.

Werth, C.R., Pusateri, W.P., Eshbaugh, W.H. and Wilson, T.K. (1979) Field observations on the natural history of *Cassytha filiformis* L. In: Musselman, L.J., Worsham, A.D. and Eplee, R.E. (eds) *Proceedings, 2nd International Symposium on Parasitic Weeds, Raleigh, 1979*. North Carolina State University, Raleigh, pp. 94–110.

Werth, C.R., Riopel, J.L. and Gillespie, N.W. (1984) Genetic uniformity in an introduced population of witchweed (*Striga asiatica*) in the United States. *Weed Science* 32, 645–648.

Whitney, P.J. (1972) The carbohydrate and water balance of beans (*Vicia faba*) attacked by broomrape (*Orobanche crenata*). *Annals of Applied Biology* 70, 59–66.

Whitney, P.J. (1973) Transport across the region of fusion between bean (*Vicia faba*) and broomrape (*Orobanche crenata*). In: *Proceedings, Symposium on Parasitic Weeds, Malta, 1973*. European Weed Research Council, Wageningen, pp. 154–166.

Whitney, P.J. and Carsten, C. (1981). Chemotropic response of broomrape radicles to host root exudates. *Annals of Botany* 48, 919–921.

Wiens, D. and Calvin, C.L. (1987) Epiparasitism in mistletoes. *The Golden Bough* 9, 2–4.

Wild, H. (1948) A suggestion for the control of tobacco witchweed (*S. gesnerioides* (Willd.) Vatke) by leguminous trap crops. *Rhodesia Agricultural Journal* 51, 330–341.

Wild, H. (1951) Rhodesian witchweeds (*Striga* and *Alectra* species). *Rhodesian Agricultural Journal* 51, 330–341.

Wilhelm, S. (1962) History of broomrape (*Orobanche ramosa* and *O. ludoviciana*) and their control by preplant soil injection with methyl bromide solutions. In: *Proceedings, 16th International Horticultural Congress, Brussels*, pp. 392–399.

Williams, C.N. (1959) Resistance of *Sorghum* to witchweed. *Nature* 184, 1511–1512.

Williams, C.N. (1961) Tropism and morphogenesis of *Striga* seedlings in the host rhizosphere. *Annals of Botany* 25, 407–415.

Wilson, C.L. (1969) Use of plant pathogens in weed control. *Annual Review of Phytopathology* 7, 411–434.

Wolswinkel, P. (1973) The disturbance of the development of broad bean (*Vicia faba* L.) and the setting and growth of pods after infection by *Cuscuta*: experiments about translocation of assimilates. In: *Proceedings, Symposium on Parasitic Weeds, Malta, 1973.* European Weed Research Council, Wageningen, pp. 177–187.

Wolswinkel, P. (1974a) Complete inhibition of setting and growth of fruits of *Vicia faba* L. resulting from the draining of the phloem system by *Cuscuta* species. *Acta Botanica Neerlandica* 23, 48–60.

Wolswinkel. P. (1974b) Enhanced rate of ^{14}C-solute release to the free space by the phloem of *Vicia faba* stems parasitised by *Cuscuta. Acta Botanica Neerlandica* 23, 177–188.

Wolswinkel, P. (1975) The active role of the host (*Vicia faba* L.) in the transfer of nutrient elements from the phloem to the parasite (*Cuscuta* species): metabolically controlled K^+ and Mg^+ release to the free space. *Acta Botanica Neerlandica* 24, 211–224.

Wolswinkel, P. (1979) Transport of assimilates and mineral elements at the site of attachment of *Cuscuta*. The role of phloem unloading in the parasitic relationship. In: Musselman, L.J., Worsham, A.D. and Eplee, R.E. (eds) *Proceedings, 2nd International Symposium on Parasitic Weeds, Raleigh, 1979.* North Carolina State University, Raleigh, pp. 156–163.

Wolswinkel, P. (1989) *Cuscuta*. In: Helevy, H.A. (ed.) *Handbook of Flowering*, Volume 6. CRC Press, Florida, pp. 270–274.

Worsham, A.D. and Egley, G.H. (1990) Physiology of witchweed seed dormancy and germination. In: Sand, P.F., Eplee, R.E. and Westbrooks, R.G. (eds) *Witchweed Research and Control in the United States.* Weed Science Society of America, Champaign, pp. 11–26.

Wylde, A. (1991) Qualitative and quantitative studies on the effect of *Striga hermonthica* upon the quality and quantity of grain yield from millet (*Pennisetum typhoides*). In: Wegmann, K. and Musselman, L.J. (eds) *Progress in* Orobanche *Research. Proceedings, International Workshop on* Orobanche *Research, Obermarchtal, Germany, 1989.* Eberhard-Karls-Universität, Tübingen, pp. 322–328.

Yan, Z.G. (1990) Host specificity of *Lysiana exocarpi* subsp. *exocarpi* and other mistletoes in southern South Australia. *Australian Journal of Botany* 38, 475–486.

Yoshikawa, F., Worsham, A.D., Moreland, D.E. and Eplee, R.E. (1978) Biochemical requirements for seed germination and development of witchweed (*Striga asiatica*). *Weed Science* 26, 119–122.

Yuncker, T.G. (1932) The genus *Cuscuta. Memoirs of the Torrey Botanical Club* 18, 113–331.

Zahran, M.A. (1982) Control of parasitic plants (broomrape and dodder) in different crops in Egypt. *Final Technical Report, Agricultural Research Program, PL 480*, 53 pp.

Zaitoun, F.M.F., Al-Menoufi, O.A. and Weber, H.C. (1991a) Loss assessment and forecasting work on plant diseases: 1. A new method for assessment of loss in *Vicia faba* through infection with *Orobanche crenata*. In: Wegmann, K. and Musselman, L.J. (eds) *Progress in* Orobanche *Research. Proceedings, International Workshop on* Orobanche *Research, Obermarchtal, Germany, 1989*. Eberhard-Karls-Universität, Tübingen, pp. 167–184.

Zaitoun, F.M.F., Al-Menoufi, O.A. and Weber, H.C. (1991b) Mechanisms of tolerance and susceptibility of three *Vicia faba* varieties to the infection with *Orobanche crenata*. In: Ransom, J.K., Musselman, L.J., Worsham, A.D. and Parker, C. (eds) *Proceedings, 5th International Symposium on Parasitic Weeds, Nairobi, 1991*. CIMMYT, Nairobi, pp. 195–207.

Zaroug, M.S. and Ito, K. (1988) Host range and susceptibility of some cultivated plants to dodder (*Cuscuta japonica* Choisy) in Hokkaido. *Weed Research, Japan* 33, 129–135.

Zazzerini, A., della Torre, G., and Tosi, L. (1981) Broomrape of tobacco: epidemiology and control (in Italian). *Informatore Fitopatologico* 31, 15–23.

Zhang, S.S. (1988) Damage to *Pinus massoniana* caused by a parasitic seed-plant, *Cassytha filiformis* (in Chinese). *Forest Pest and Disease* 4, 3.

Zhang, T.Y. (1985) A forma specialis of *Colletotrichum gloeosporioides* on *Cuscuta* spp. (in Chinese). *Acta Mycologica Sinica* 4, 234–239.

Zhelev, N. (1987) The biological role of exogenous factors in broomrape seed germination (in Bulgarian). *Rastenievudni Nauki* 24, 36–43.

Zimmerman, C.E. (1962) Autotrophic development of dodder (*Cuscuta pentagona* Engelm.) *in vitro*. *Crop Science* 2, 449–450.

Zimmerman, G.T. and Laven, R.D. (1984) Ecological relationships of dwarf mistletoe and fire in lodgepole pine forests. In: Hawksworth, F.G. and Scharpf, R.F. (eds) *Biology of Dwarf Mistletoes: Proceedings of the Symposium. Fort Collins, Colorado, 1984*. USDA Forest Service General Technical Report RM-111, pp. 123–131.

Zummo, N. (1974) *Annual Report: Cereal Pathology*. Institute of Agricultural Research, Samaru, Zaria, Nigeria, 20 pp.

Zummo, N. (1977) Diseases of giant witchweed, *Striga hermonthica*, in West Africa. *Plant Disease Reporter* 61, 428–430.

Zwanenburg, B., Mhehe, G.L., t'Lam, G.K., Dommerholt, F.J. and Kishimba, M.A. (1986) The search for new germination stimulants of *Striga* spp. In: ter Borg, S.J. (ed.) *Proceedings, Workshop on Biology and Control of* Orobanche, *Wageningen, 1986*. LH/VPO, Wageningen, pp. 35–41.

Index

Page numbers in italics indicate positions of relevant figures.